QoS for Fixed and Mobile Ultra-Broadband

QoS for Fixed and Mobile Ultra-Broadband

Toni Janevski

Ss. Cyril and Methodius University
Skopje, Macedonia

Registered Offices
John Wiley & Sons, Inc., 111 River Street, Hoboken, NJ 07030, USA
John Wiley & Sons Ltd, The Atrium, Southern Gate, Chichester, West Sussex, PO19 8SQ, UK

Editorial Office
The Atrium, Southern Gate, Chichester, West Sussex, PO19 8SQ, UK

For details of our global editorial offices, customer services, and more information about Wiley products visit us at www.wiley.com.

Wiley also publishes its books in a variety of electronic formats and by print-on-demand. Some content that appears in standard print versions of this book may not be available in other formats.

Library of Congress Cataloging-in-Publication Data

Names: Janevski, Toni, author.
Title: QoS for fixed and mobile ultra-broadband / Toni Janevski, Ss. Cyril
    and Methodius University, Skopje, Macedonia.
Description: Hoboken, NJ, USA : Wiley IEEE Press, 2019. | Includes
    bibliographical references and index. |
Identifiers: LCCN 2018060368 (print) | LCCN 2019000192 (ebook) | ISBN
    9781119470496 (Adobe PDF) | ISBN 9781119470489 (ePub) | ISBN 9781119470502
    (hardcover)
Subjects: LCSH: Mobile communication systems–Quality control. | Wireless
    communication systems–Quality control. | Broadband communication
    systems–Quality control.
Classification: LCC TK5102.84 (ebook) | LCC TK5102.84 .J36 2019 (print) | DDC
    384.3068/5–dc23
LC record available at https://lccn.loc.gov/2018060368

Cover Design: Wiley
Cover Image: © monsitj/iStock.com

Set in 10/12pt WarnockPro by SPi Global, Chennai, India

Printed in Great Britain by TJ International Ltd, Padstow, Cornwall

10 9 8 7 6 5 4 3 2 1

To my great sons, Dario and Antonio, and to the most precious woman in my life, Jasmina.

Contents

1

Introduction

The telecommunications world has been developing at a rapid pace since the growth of the Internet in the 1990s and 2000s. Nowadays telecommunications is also referred to as information and communication technologies (ICT), as stated by the largest telecommunications agency in the world, the International Telecommunication Union (ITU) [1], a specialized agency of the United Nations. Telecommunications has been around for more than 150 years, starting with telegraphy in the nineteenth century. In fact, the telecommunications world and the ITU have been interrelated since 1865 when the ITU was formed as the International Telegraph Union. Nowadays, telegraphy belongs to history (it has become redundant since the appearance of email and other messaging services available worldwide today). But speaking about the history of telecommunications, after Alexander Graham Bell invented the telephone in 1876 the following century was marked by the development and deployment of telephony, with fixed telephony until the 1980s accompanied by mobile telephony worldwide from the 1990s. Of course, one should not forget to mention television and radio as important telecommunication services during the twentieth century, and they continue to be so in the twenty-first century.

ICT has created a globally connected world, not only giving people the ability to communicate with each other but also opening up access to information and facilitating the exchange of information. The foundation of such an ICT world (the terms ICT and telecommunications will be used interchangeably in this book) lies in the introduction of digital systems and networks in the 1970s and 1980s, which provided the possibility for all information from different sources and of different types (e.g. voice, video, data, multimedia) to be transferred over the global telecommunications infrastructure by using series of bits and zeros (i.e. in a digital form). When information is coded at the source as series of ones and zeros, all such information can be transferred over the same telecommunications networks and accessed via the same devices – if, of course, the networks are created in such a manner. The Internet has provided the required openness to transfer all different types of information over the same network, which is present and working well in the second decade of the twenty-first century and is expected to continue in a similar manner in the future. However, telecommunication networks are interconnected on a local, regional, and global scale to be able to transfer the information between any two or more communication endpoints on the Earth, so the telecommunication services are global. Therefore, the quality of telecommunication services which is applied in a single network or in a single country has influence on the end-to-end quality of that service. So, the quality cannot be considered only at national or regional level, it needs

QoS for Fixed and Mobile Ultra-Broadband, First Edition. Toni Janevski.
© 2019 John Wiley & Sons Ltd. Published 2019 by John Wiley & Sons Ltd.

to be considered globally. Today, citizens around the world rely on ICT to conduct their everyday activities in personal or business life, and that requires having certain quality of services (QoS). Therefore guaranteeing QoS in the socio-economic environment of users is becoming very important. For that purpose there is a need for technical mechanisms and functions for implementation of the QoS in networks and end-users' devices, and for its regulation in a harmonized and globally accepted approach. That will enable greater quality of services provided to users as customers or consumers or content generators, irrespective of their location or service provider (the entity that provides access to telecommunications services), where the provider can be a telecom operator (on local or national levels) or global over-the-top (OTT) provider (e.g. Google, Amazon, etc.).

1.1 The Telecommunications/ICT Sector in the Twenty-First Century

The telecommunication/ICT world in the twenty-first century is characterized by two important developments:

- It is becoming fully based on Internet technologies, including all networks (with fixed and mobile access) and all services (including all applications working over the Internet).
- It is becoming broadband, which means that there are enough high bitrates in access networks (and also end-to-end) which provide all available applications/services to run smoothly and with satisfactory quality experienced by end-users (we will define what the quality means later in the chapter).

So, the ICT world is becoming all-IP (Internet protocol) and broadband on a global scale (Figure 1.1) [1]. However, it is even more interesting to note that mobile communications are spreading at a faster pace than fixed communications regarding access

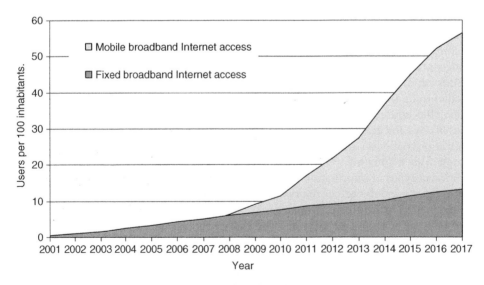

Figure 1.1 Global telecommunication/ICT broadband developments.

networks. One may note that in the 1990s only about 1% of the global population had a mobile cellular subscription and about 11% had a fixed telephone subscription. Nowadays, mobile subscribers' penetration is near saturation, with almost everyone on this planet having a mobile phone/device. Mobile broadband is the most dynamic market segment, surpassing fixed broadband access in the second decade of the twenty-first century. The number of Internet users is also constantly growing over the entire world population (Figure 1.1).

What is broadband? One may define broadband as access network bitrates and end-to-end network bitrates in both directions (downstream and upstream) which support all available types of services with satisfactory quality. For example, in the first decade of the twenty-first century broadband access meant offering hundreds of kbit/s, while a decade later it means access with Mbit/s or tens of Mbit/s (as measurement units for data speeds), enabling, for example, the provision of HD (high definition) video and ultra-HD video streaming (some of the most demanding services).

Broadband can be divided into two main categories (similar to type of access networks):

- *Fixed broadband.* Fixed broadband access technologies are provided by copper (twisted pairs) by reusing local loops deployed for fixed telephony for IP-based access via ADSL (asymmetric digital subscriber line), VDSL (very-high-bit-rate digital subscriber line), or generally x digital subscriber line (xDSL); or by cable access (by reusing coaxial cable networks, primarily developed for TV distribution, and FTTH (fiber to the home) or more general FTTx (fiber to the x), which is a long-term future for fixed broadband access in all regions. On the other side, almost all transport networks nowadays are fiber-based (accompanied by satellite networks, where optical transport is not present), so the differences remain mainly in the last mile.
- *Mobile broadband.* Mobile broadband access technologies appeared with the 3G (third-generation) mobile networks in the late 2000s, and they continue with 4G (fourth-generation) and 5G (fifth-generation) mobile networks in the second and third decades of the twenty-first century. The widespread mobile technologies which belong to the mobile broadband world include UMTS/HSPA (universal mobile telecommunication system/high speed packet access) as part of the 3G standards umbrella, Mobile WiMAX (3G and 4G umbrella), and LTE/LTE-Advanced (LTE stands for long-term evolution) as the most successful 4G technologies. One may note that WiFi is not presented under a separate bullet here, due to the fact that WiFi is more a local wireless extension of fixed or mobile broadband access networks.

According to the ITU [2], by early 2016 total international Internet bandwidth had reached 185,000 Gbit/s, a significant increase from the globally available bandwidth of about 30,000 Gbit/s in 2008. However, there is no equal connectivity of all regions on the global scale. For example, Africa has the lowest international connectivity of all regions, while there is twice as much bandwidth per inhabitant available in Asia and the Pacific, eight times as much in the Americas, and more than 20 times as much in Europe [3]. Also, one may note that always-on access, mobile broadband penetration, as well as the massive adoption of broadband-enabled devices have irreversibly changed the consumers of telecommunication/ICT services, including their social and economic behaviors as well as QoS expectations, which are constantly increasing over time for certain services (e.g. video).

The more diverse telecommunications/ICT world requires more effort in maintaining the QoS, including provision, measurements, and enforcement. QoS and quality of experience (QoE) are becoming more and more complex because the quality can be impacted by many different factors coming from the telecommunication networks as well as along the value chain, which includes the end-user's device, the available hardware, the network infrastructure, the offered services/applications, etc. As usual, in this process influenced by many factors (which may appear to be different to different users of the same telecommunication service), some differences may arise between perceived and assessed QoS.

QoS is important for all parties involved with the telecommunication services, including both customers (or end-users) and service providers. Therefore there is a need for QoS standards that can form a basis for establishing QoS policies in each country – this is the remit of the appropriate authority in the telecommunication/ICT area (e.g. the national regulatory authority (NRA), ICT ministry, or other ministry or government body). Then the QoS provisioning for the services offered to the customers should be monitored as well as encouraged and/or enforced when necessary.

1.2 Convergence of the Telecom and Internet Worlds and QoS

The telecom and Internet worlds developed in parallel in the 1970s and 1980s. On one side, the telecom world was focused on traditional telecommunication services, telephony (i.e. voice) and television (i.e. TV), which were primary telecommunication services during most of the twentieth century. Traditional telephony was initially fixed based, which refers to the fixed access network via so-called twisted pairs (a pair of two copper wires twisted with the aim of reducing the echo in the opposite direction). A pair of wires also is needed to create a circuit between the telephone (as end-user terminal) and the telecom operator's equipment (e.g. the exchange), therefore such communication was also referred to as circuit-switching (exchanges are network nodes placed on the operator's side, and they provide switching between number of channels on their input and their output).

The first condition for the appearance of the Internet on a global scale was the digitalization of the telecommunication networks built for telephony as the primary service. The process of digitalization of initially analogue telecommunication networks (which used to carry analogue voice signals end-to-end) happened in the 1970s and 1980s, supported by the appearance and development of computers. The telecommunication networks were among the first to accept computers in place of the older analogue technology (based on electrical equipment), with the new digital equipment being based on hardware (i.e. electronics) and software. Digitalization changed the design of telecommunication networks so they started to carry digits as signals (instead of analogue audio for telephony or analogue video and audio for TV), where typically the digits were bits (i.e. ones and zeros) as the most appropriate form for representing different types of media (only a single threshold was needed to distinguish between one and zero at the receiving end of the transmission link, and also computers' architectures are based on storing and processing the data in a digital form, consisting of ones and zeros). With digitalization, the path was traced to transmit all types of information (audio, video, data) over the same network because all signals were in fact converted into ones and zeros before the transmission (at the sender's end) and vice versa (to the original form of the information) at the receiver's end. So, after the digitalization of telecom networks, data services started to increase in importance; however, there was also a need for a new technology which would suit all types of services, including voice, television, and various data. The next technology was packet-switching. Unlike with circuit-switching where a given channel is occupied all the time during the connection between two ends (e.g. between two telephones), regardless of whether or not there is information to transmit (e.g. voice), the new technology was thought to be based on a different approach. That approach was transmitting a chunk of information in a unit called a packet where each packet has a header (which is heading the data) which includes the address information of the sending and/or receiving end and certain control information regarding the given type of information carried in the packet, called a payload. This technology was packet-switching and there were two main competitors for it in the 1990s: the European-based ATM (asynchronous transfer mode) and the US-based Internet. The ATM was mainly based on the philosophy of traditional telecommunications where all intelligence was placed in the network nodes on the side of the telecom operators while the users had simpler equipment (such as telephone devices). The virtual circuits (introduced by the ATM) are established on signaling messages between the end nodes (which is similar to establishing a telephone call in a telephone network). Meanwhile, the Internet was created on several principles that made it a global success, from which the following ones can be considered as the most important:

- There is separation of applications and services from the underlying transport technologies (e.g. mobile or fixed access networks, transport networks).
- All network nodes and user terminals have the main IP stack based on transport layer protocols, which are primarily UDP (user datagram protocol) [4] and TCP (transmission control protocol) [5], over the IP in its two existing versions, IP version 4 [6] and IP version 6 [7].

The traditional telecommunications layering protocol and the IP model are compared in Figure 1.2. Initially, in the early days (the 1970s) the IP model was based on three layers: the Interface layer at the bottom, network control program (NCP) in the middle, and the application layer on the top. In 1981 the NCP split into TCP (or UDP) over IP, so they became four protocol layers as the native Internet model from the 1980s. However, the network interface layer is typically split into the physical layer and data-link layer by all standard development organizations (SDOs), so with such classification the basic IP layering model has five layers.

The network layer of the Internet is the IP – version 4 (IPv4) or version 6 (IPv6) is currently present in every host, router, and gateway in every network. So, the Internet had won the packet-switched networking technologies battle by the end of the 1990s, which further resulted in the telecommunication/ICT world moving toward Internet-based networking and Internet technologies, a process that continued in the 2000s and 2010s.

But was the Internet a separate network from the traditional telecommunication networks? Well, corporations connected to the Internet via the Ethernet-based networks (where Ethernet is the IEEE 802.3 family of standards) or WiFi (which is the IEEE 802.11

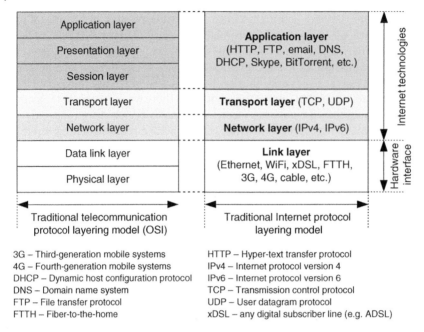

3G – Third-generation mobile systems
4G – Fourth-generation mobile systems
DHCP – Dynamic host configuration protocol
DNS – Domain name system
FTP – File transfer protocol
FTTH – Fiber-to-the-home

HTTP – Hyper-text transfer protocol
IPv4 – Internet protocol version 4
IPv6 – Internet protocol version 6
TCP – Transmission control protocol
UDP – User datagram protocol
xDSL – any digital subscriber line (e.g. ADSL)

Figure 1.2 Comparison of traditional protocol layering model (Open Systems Interconnection (OSI) model) and Internet protocol layering model.

family of standards) were initially denoted as computer networks (in the 1990s) because they were connecting computers (called hosts, either clients or servers). Meanwhile, the telecommunication networks typically connected "dumb" devices to the "smart" network, which included end-user devices such as telephones and TV sets. But residential users started to connect to the Internet in the 1990s using dial-up modems over their telephone lines, that is, over the established global telecommunication infrastructure for telephony. Also, the Ethernet access networks were interconnected via the global telecommunication infrastructure created for transmission primarily of digital telephony, based on 64 kbit/s dedicated bitrate in each direction of the voice communication (i.e. ITU-T G.711 voice codec standard, based on pulse code modulation (PCM)). The "chunk" bitrate of 64 kbit/s was the basis for all digital telecommunication systems in the 1990s, and even in the 2000s. For example, the main transport technology in digital PSTNs (public switched telephone networks), the SDH (synchronous digital hierarchy) was based on the 64 kbit/s chunks made for voice, so the STM-1 (synchronous transport module, level 1) has a bitrate of 155.52 Mbit/s, which equals exactly 2430 × 64 kbit/s, and higher transport modules, STM-N, have bitrates of N × 155.52 Mbit/s, which again is a multiple of 64 kbit/s as basic voice throughput in one direction (from user A to user B) in digital telephony. However, the appearance of the Internet on the global ICT scene in the 1990s (sped up by the invention of the World Wide Web and its growth in that period) resulted finally in transport of Internet traffic (various data carried with IP packets and transported by using the IP stack end-to-end) over such SDH transport networks created for transport of digital voice signals. With the convergence of ICTs to the Internet networking principle, Internet networking principles started to penetrate from access networks (e.g. Ethernet) to metro and transport networks, thus making SDH and other technologies from the digital PSTN era obsolete.

What did this convergence of the telecom and Internet worlds mean? Instead of having separate networks for transmission of different services (e.g. a telephone network for telephony, broadcast network for television, and separate network for data transmission) as the main characteristic of the traditional telecom/ICT world, the transition to Internet networking principles (based on separation of underlying transport technologies from the networking protocol above them, as well as separation of networking protocols from the application on the top) provided the possibility for the realization of one network for all telecommunication/ICT services. So, Internet networking and Internet technologies (standardized primarily by the Internet Engineering Task Force (IETF)) have become the main approach in the telecommunication world, and have become the telecommunications from the network layer up to the application layer. Thus the Internet is not something separate from the telecommunication world, as it was considered to be at the beginning (in the 1980s and even the 1990s), nowadays it is the main part of that world.

With this convergence, certain issues have also arisen. One of the main such issues is the QoS. Why? Because the telecom world based on telephony and TV had strict specifications for QoS requirements for those services, and the telephone and TV networks were designed to provide end-to-end QoS (we will define QoS later). For example, in circuit-switching, PSTNs typically are established two channels end-to-end (each 64 kbit/s), one per direction between the two users (talking to each other over the telephone), and such bitrate is dedicated to the given call for its duration regardless of whether there are voice signals (i.e. talk) to transport over the line or not (i.e. silence). The same approach is present in TV broadcast networks; however, the bandwidth (in bit/s) for TV (i.e. for video) is many times higher than the one needed for voice. On the other side, the native Internet was built on best-effort principles, which means that the network will make the best effort to carry each IP packet from its source to its destination address; however, there are no strict guarantees that the packet will be delivered. So, the Internet in its native design does not contain mandatory QoS mechanisms. Overall, the traditional Internet world has no mandatory QoS, while the traditional telecom world has mandatory QoS mechanisms and functions implemented in the network.

Regarding the Internet, one should note that there were efforts to provide certain QoS options in IP standards from the start. IPv4 (IP and IPv4 are used interchangeably in the following text) has a type of service (ToS) field in which it can specify QoS requirements on precedence, delay, throughput, and reliability. In a similar manner, IPv6 has a DSCP (differentiated service code point) field which can provide support for QoS per flow on the network layer. Neither IPv4 nor IPv6 guarantees the actual end-to-end QoS as there is no reservation of network resources, which is something that should be provided by other mechanisms in IP-based networks.

What about the main Internet architecture? The Internet is built on the basis of autonomous systems (ASs). Each AS is in fact an autonomous administrative domain and it is identified by a 16-bit or 32-bit AS number, which is allocated by the IANA (Internet Assigned Numbers Authority), a department of the ICANN (Internet Corporation for Assigned Names and Numbers), which governs the Internet in terms of domain naming and IP addressing, as well as other well-known numbers from various standardized protocols for IP-based networks (e.g. port numbers for different protocols on the application layer). The autonomous system is called "autonomous" for a reason, that is, it can apply within its administrative domain the traffic management schemes and routing protocols independently from other ASs, which directly impact the QoS. However, one company or operator can have several ASs and can apply similar traffic management and QoS functions in them. In general, the Internet and the global telecommunication networks are based on IP networks consisting of about 50,000–100,000 active ASs (this number is constantly increasing) [8], which are interconnected. Every AS is connected with either one other AS (e.g. so-called stub AS), or with several adjacent ASs, thus creating the global telecommunication network of today (which is completely based on Internet networking technologies). The global Internet infrastructure consisting of interconnected ASs is crucial in understanding why the traditional way of QoS implementation and enforcement (that is, the same approach in all countries, e.g. for digital telephony, i.e. PSTN/Integrated Services Digital Network (ISDN)) is no longer possible in an IP-based environment. Why? Because in IP environments there is heterogeneity of various IP networks, variety of applied network and traffic management techniques, and the plethora of services and applications which are constantly being offered (e.g. huge OTT applications/services ecosystems, with millions of applications). So, one may note that the QoS as an end-to-end characteristic is becoming more complex in the telecommunications/ICT completely based on Internet networking and Internet technologies.

How is QoS transitioning from the traditional telecom world (made for telephony and TV) to the all-IP world? It is based on strict standardization of certain functions and certain protocols. For example, the traditional telecommunication approach by default includes end-to-end QoS support in the network. Also, signalization in a standardized manner is required for establishing calls/sessions between any two ends on any two devices connected to any two networks regardless of the types of application, device, or network. That led to implementation of certain functions and approaches from the traditional telecom world in the Internet networks, which were initially best-effort based. The main standardization for such convergence was carried by the ITU in its umbrella of specifications called next generation networks (NGNs), which initially started as an idea around 2003. One of the main purposes of the NGN standardization framework was (and still is) to standardize the end-to-end QoS support in all-IP networks (including all needed functions in transport and service stratums) that is essential for real-time services, such as VoIP (voice over Internet protocol) and IPTV (Internet protocol television). Such services have strict requirements regarding QoS (guaranteed bitrates, losses, delay, delay variation, jitter, etc.). So, NGN provides a standardized implementation of QoS (instead of proprietary case-by-case implementations) [9], which is mandatory for the transition from PSTN and public land mobile network (PLMN) to all-IP networks.

So, the telecommunication world is transiting from circuit-switched networks (e.g. PSTN/ISDN) to all-IP networks, including fixed and mobile access networks, as well as core and transit networks. First, the transition was completed in transit and core networks, then in fixed access networks, and lastly in mobile access networks (due to mobility of users, which makes the continuous QoS provisioning more complex). When the transition to all-IP-based networks/services is completed in the telecommunications/ICT world, the Internet technologies are also used for traffic management and for QoS standardization, monitoring, and enforcement. The transition from separate

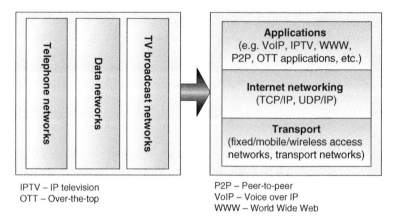

IPTV – IP television
OTT – Over-the-top

P2P – Peer-to-peer
VoIP – Voice over IP
WWW – World Wide Web

Figure 1.3 Transition from separate networks for different services (the traditional telecom approach) to horizontal separation of services/applications provided via broadband IP networks (the new way).

networks for different services (traditional telecom approach) to a horizontal layered approach in an all-IP environment with broadband access is illustrated in Figure 1.3.

Once the path to packet-switching in the telecom world had been well established by the start of the 2000s, the ITU defined an architectural framework for the support of QoS in packet-switched networks. Nowadays, although there are unified packet-switched networks, the IP-based networks, other different packet-switched networks exist which are also standardized, including the SS7 signaling (as usual, every new technology enters first into the signaling segment of the telecom networks) as well as the already mentioned ATM.

1.3 Introduction to QoS, QoE, and Network Performance

With the convergence of the telecom and Internet worlds, the QoS functions and requirements apply not only to traditional telecommunication and broadcast services but also to broadband Internet-based services.

Overall, telecommunications/ICT services in the twenty-first century are increasingly being delivered using IP based networks, including:

- IP-based networks (access networks, core/backbone networks, and transit networks), and
- IP-based services, which include two main types from the QoS viewpoint:
  - QoS-enabled, i.e. managed services, such as voice, TV, and any other service provided by telecom operators with QoS guarantees end-to-end based on a signed agreement between the telecom operator and the customer (in such case the end-user becomes a customer for the operator);
  - OTT services, which are provided in a best-effort manner over Internet access (either fixed or mobile), without end-to-end QoS, and based on the network neutrality principle being implemented in the Internet (we will refer to network neutrality in more detail in Chapter 8).

So, QoS is clearly moving from its initial definitions targeted at traditional telecommunication networks (e.g. PSTN/ISDN, broadcast networks) to QoS in IP networks and services.

Networks and systems are gradually being designed in consideration of the end-to-end performance required by user applications. In the following subsection we define the QoS, QoE, and network performance (NP).

1.3.1 Quality of Service (QoS) Definition

Traditionally, QoS was mainly addressed from the perspective of the end-user being a person (e.g. telephony), with the ability to hear and see and be tolerant of some degradations of services (e.g. low packet loss ratio is acceptable for voice, while end-to-end delay for voice should be less than 400 ms). But with the advent of new types of communications where services may not require real-time delivery and where the sender or the end-user may not be a person but could be a machine (e.g. Internet of Things), it is important to keep in mind that not all services are the same, and even similar services can be treated in different ways depending on whether they are used by machines or by humans on one or both ends of a given communication session or connection.

> Quality of service (QoS), as defined by the ITU [10], is totality of characteristics of a telecommunications service that bear on its ability to satisfy stated and implied needs of the user of the service.

Similar definitions of QoS are used by other SDOs in their standards. However, from the telecommunications/ICT point of view the QoS is always an end-to-end characteristic, which however can be split into different network segments between the ends (e.g. between two hosts on the Internet, or two telephones). However, the end-user's perception of a given telecommunication/ICT service is also influenced by different factors which may include (but are not limited to) social trends, advertising, tariffs, and costs, which are interrelated with the customer expectation of QoS. For example, social trends may be in terms of popular devices (e.g. smartphones), services (e.g. some services are more popular than other similar services over the Internet), applications (e.g. there is different popularity of different applications in their ecosystems), etc. Further, end-user perception of the QoS is not limited only to objective characteristics at the man-machine interface. For end-users, the QoS refers to the quality that they personally experience during their use of a given telecommunication service. The end-user may be satisfied with the QoS of a given service at a certain time and the same user may not be satisfied with the same service 10 years in the future. For example, when a user accommodates to higher resolution of TV or video streaming due to higher bitrates in the access networks, then the same user will increase the expectations of such a service (i.e. TV or video streaming).

Figure 1.4 illustrates how the QoS depends on the end-to-end technical aspects, which include network performance and terminal performance, and non-technical aspects (those that are not directly related to the equipment), which include customer care and point of sale.

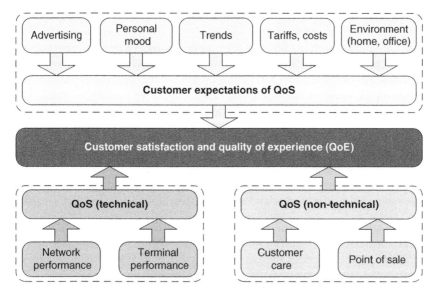

Figure 1.4 Technical and non-technical points of view for quality of service.

1.3.2 Quality of Experience (QoE)

Initially QoE was defined as the overall acceptability of an application or service, as perceived subjectively by the end-user (according to ITU-T Recommendation P.10/G.100, [11]).

ITU-T has replaced the QoE definition developed in 2007 with a new definition adopted in 2016, which is currently the actual QoE definition given as follows [11]:

> Quality of experience (QoE) is the degree of delight or annoyance of the user of an application or service.

However, one should note that there is continuous research on the QoE topic, so the definition is expected to evolve further.

Regarding the QoE there are two topics that need to be addressed together with the definition:

- *QoE influencing factors.* They include the type and characteristics of the application or service, context of use, the user's expectations with respect to the application or service and their fulfillment, the user's cultural background, socio-economic issues, psychological profiles, emotional state of the user, and other similar factors.
- *QoE assessment.* This is the process of measuring or estimating the QoE for a given number of end-users of a given application or a service. The assessment is typically based on an established procedure, and taking into account all important influencing factors. The output of the QoE assessment may result in a scalar value, various multi-dimensional representations of the results, as well as verbal descriptors (e.g. good, bad). In theory, all assessments of QoE should be accompanied by the description of the influencing factors that are included.

Overall, the QoE includes complete end-to-end system effects (end-user equipment and application, various influencing factors on the user for the given services, as well as network and service infrastructure). So, the QoE may be influenced by user expectations and the context (e.g. for the same obtained bitrate for a given service, a user who had lower expectations will enjoy higher QoE than a user who had higher expectations for the same service on the same equipment offered via the same network). QoE takes into consideration certain additional parameters, such as:

- user expectations;
- user context (what is in trend, user's personal mood, environment where the service is being used such as work/home/outside environments, etc.);
- the potential difference between the service being offered to the user and the individual user awareness about the service and additional features (if any) for that service.

Regarding the QoE assessment, the most used measure is the mean opinion score (MOS). Initially, the MOS scale referred to voice service only (ITU-T P.800), but nowadays it is also used for other services such as video (e.g. for IPTV). MOS is expressed as a single number in the range between 1 and 5, where MOS with a value of 1 denotes the worst and a value of 5 denotes the best quality experienced by the user (Table 1.1).

1.3.3 Network Performance (NP)

Network performance differs from QoS because it relates only to the network part of the service provisioning, without taking into account different user influencing factors. On one side, the QoS is the outcome of the user's experience of using a given service and the user's perception of it, while on the other side the NP is determined by the performances of network elements one by one, or by the performance of the network as a whole. So, the NP has an influence on the QoS, and it represents a part of it. However, the QoS is not influenced only by the NP but also by non-network performance parameters. Simply said, QoS consists of network performance and non-network performance, as shown in Figure 1.5.

The NP concept is applied for purely technical purposes, i.e. assessment and analysis of technical functions. NP is the ability of a network portion or the whole network to provide the QoS functions related to communications between the users. Network performance is determined by the performance of network elements one-by-one. The

Table 1.1 Mean opinion score for the quality of experience.

Mean opinion score (MOS)	Quality classification
5	Excellent
4	Good
3	Fair
2	Poor
1	Bad

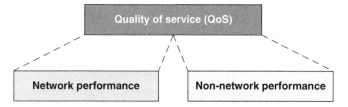

Figure 1.5 Network performance (NP) and quality of service (QoS).

performance of the network as a whole (end-to-end) is determined by the combination of the performance of all single elements along with their interconnections on the communication path between the end-user devices. NP is specified in terms of objective performance parameters, which can be measured, and then the performance value is assigned quantitatively [10].

1.3.4 QoS, QoE, and NP Relations

All three terms – QoS, QoE, and NP – are related one to another. QoE is different from QoS and NP as it has a subjective nature by definition, and because it depends on the end-user's perception. Clearly QoE is impacted by QoS and NP. For example, if NP is lower, it will result in lower quality experienced by the end-user.

Further, NP applies to various aspects of the network provider's functioning, including planning, development, operations, and maintenance of network elements and the whole network. Also, there can be several interconnected networks (along the path between the endpoints of the established communication session), which may be operated by different network providers (i.e. telecom operators). As shown in Figure 1.6,

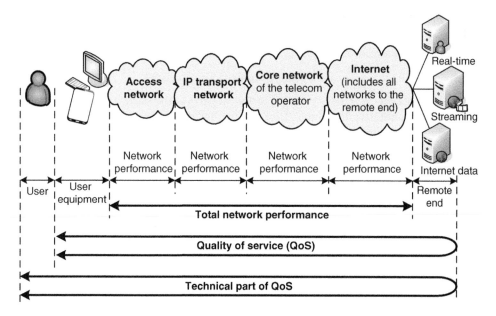

Figure 1.6 Network performance, QoS, and QoE.

NP is the detailed technical part of the offered QoS. Also, NP contributes to QoS as experienced by the user [12]. The functions of a service depend on the performance of the network elements and the performance of the user's terminal equipment. QoS is always end-to-end, which can be user-to-user or user-to-content (one may also add here QoS for machine-to-machine communications, which may be directly not related to a human end-user). Hence, QoS is always an end-to-end characteristic, where it depends on the contributions from different components (Figure 1.6), including the end-user, its equipment (smartphone, computer, etc.), access network (fixed or mobile), IP transport network, core network, and the rest of the path end-to-end (e.g. through the Internet). QoE has a broader scope because it is impacted by QoS as well as by user expectations and the context. Meanwhile, QoS has broader scope than NP.

1.4 ITU's QoS Framework

ITU has developed a QoS framework to suit different networks and services, initially defined in ITU-T G.1000 [12] and then in ITU-T 802 [13]. The framework is continuously evolving. To provide QoS support for a given service, QoS criteria and then QoS parameters based on the criteria need to be defined. Such definitions of QoS criteria were initially given in ITU-T recommendation G.1000, which provides the general QoS framework (by the ITU). Before defining QoS parameters, QoS criteria relevant to the user are required. The main goal is to establish a list of all aspects that could influence the QoS. There are three models for this purpose [13]:

- *Universal model.* This is a generic and a conceptual model.
- *Performance model.* This model is more suited for determining the performance criteria of a telecommunication service.
- *Four-market model.* This model is particularly suited for multimedia services.

1.4.1 Universal Model

In this model all QoS criteria are grouped under four categories:

- *Performance criteria.* These cover the technical parts of the service, and can be qualitative or quantitative (or both). They are further defined within the performance model (next section).
- *Esthetic criteria.* These refer to ease of interaction between the user and the telecommunication service (ergonomic aspects, ease of use of the service, style and look, design of functionalities of the service, etc.).
- *Presentational criteria.* These reflect the presentational aspects of the service to the customer, including the packaging of the service, tariff models for that service, billing options for the service, etc.).
- *Ethical criteria.* These are related to the ethical aspects, such as conditions specified for cutting off the given service, services for disabled users, etc.

The service is split into functional elements, where each of the is cross-checked against the given four quality components and criteria. However, the definitions and measurement methods of the quality parameters are not a part of the universal model, but they are defined within the performance model.

1.4.2 Performance Model

The main aim in the performance model is to determine performance criteria which are further used for defining QoS parameters important to both users and providers [13]. In total, there are seven specified QoS criteria [13], as shown in Table 1.2. They are identified to provide easy translation into QoS parameters.

The given quality criteria (Table 1.2) are mapped on a set of service functions, which include service management (sales and pre-contract activities, service provision, alteration, service support, repair, and cessation), connection quality (connection establishment, information transfer, connection release), billing, and network or service management by the customer. The mapping between the service functions and service quality criteria is referred to as a performance model (in ITU-T E.802).

The QoE is influenced by all seven QoS criteria given in the performance model. For example, speed impacts the available bitrates (in downlink and uplink) and latencies (i.e. delays), which is crucial for the end-user's experience of the service. Upgrading to higher access bitrates (including fixed and mobile networks) improves the overall QoE. Further, availability and reliability are also very important and are directly related to planning and dimensioning of the network (for a given number of users and for a given service) as well as to its operation and maintenance functions and procedures. For example, one typical quality metric for network availability which is used in the traditional telecom world (at the end of the twentieth century) is so-called "five nines," that is, 99.999% of the time service to be available to the end-users, and that poses certain requirements regarding the survivability mechanisms which need to be implemented in the network (e.g. automatic traffic redirections over alternative paths in the network in a case of link or path failures). Also, one may note that security aspects – accuracy (e.g. billing accuracy), simplicity of use of the service (the user should not be required to read a manual in order to use an offered telecommunication service), and flexibility regarding the service use (e.g. ease of change of tariff model or billing method, or even changing the operator in a case of QoS degradations) – influence the QoE.

Table 1.2 QoS criteria.

No.	QoS criteria	Applying QoS criteria to different service functions
1	Speed	Service supply time, call setup time, one-way delay, release time, billing frequency, etc.
2	Accuracy	Unsuccessful call ratio, speech quality, bill correctness, etc.
3	Availability	Coverage, availability of call center, service availability, etc.
4	Reliability	Dropped calls ratio, number of billing complaints within a specific time period, etc.
5	Security	Fraud protection and prevention, etc.
6	Simplicity	Professionalism of help line (i.e. customer care), ease of software updates, ease of contract cessation procedure, etc.
7	Flexibility	Ease of change in contract, availability of different billing methods including online and offline billing, etc.

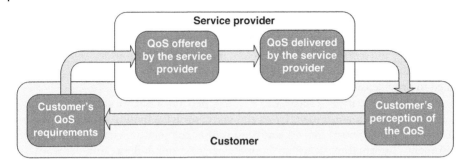

Figure 1.7 QoS viewpoints.

The ITU's performance model is based on four viewpoints of QoS initially defined in ITU-T G.1000 [12]. These fours viewpoints cover QoS from both customer and service aspects, as given in Figure 1.7:

- *Customer's QoS requirements.* This is the QoS level required by the subscriber, which may be specified in non-technical language also (e.g. good service is required), because the customer is interested not in how a given service is provided (e.g. by the telecom operator's network) but primarily in the obtained end-to-end QoS (expressed in terms they can understand, such as bitrates, data volume, etc.).
- *QoS offered by the provider (or planned/targeted QoS).* This is a statement made by the service provider (e.g. telecom operator) to customers about the QoS offered for a given service. This viewpoint is primarily used for a service level agreement (SLA), which serves as a bilateral agreement signed between the customer and the service provider. This QoS can be specified in terms understandable to the customer on one side and with technical terms for the purposes of implementation of such QoS on the side of the operator. This can also serve as a merit for subscribers to make the best choice from the given service provider's offerings.
- *QoS achieved (i.e. delivered) by the provider.* This viewpoint refers to the actual level of QoS achieved or delivered by the service provider, and for purposes of comparison it should be expressed through the same QoS parameters as the QoS offered to the customer (e.g. specified in the SLA). This QoS viewpoint can be used by the regulator for the purposes of QoS regulation, including publication of the results from QoS audit in the telecom operators' networks and then QoS encouragement or enforcement (when and where needed).
- *Customer QoS perception.* This is the QoS level experienced by the customer, typically obtained from user ratings of the QoS provided by the service operator. This is also a customer viewpoint, so it is not expressed in technical terms but in terms of degrees of satisfaction (e.g. from "not satisfied" up to "very satisfied"). For example, a customer may rate the service on a 5-point scale, with grade 1 being the worst and grade 5 being the best experience with the service. The perceived QoS can be used by service providers or regulators to determine the customer's satisfaction, which may further lead to corrective actions by providers or regulators (e.g. in situations when there is a significant mismatch between the perceived QoS and the QoS offered by the provider).

1.4.3 Four-Market Model

The four-market model is created for multimedia services [13], such as web browsing or video streaming. It defines the chain of actions going from content creation on one side, to service provision (e.g. by media servers of the service provider), service transport (via interconnected telecommunication networks, including access, core, and transit networks) and finally customers' equipment (personal computer, smartphone, etc.). These four elements can be provided independently of each other, and in such case different parties may be in charge for installation, operation, and maintenance of each of them. For example, a movie studio may produce the content, a video streaming service (e.g. YouTube, Netflix) may provide the service, the networks of telecommunication operators will carry the video traffic (e.g. as Internet traffic), and finally the customer's equipment (e.g. PC, smartphone) will be used for watching the video content. However, two or more elements in this chain may be offered by the same party. For example, the network operator can also be a service provider (e.g. for IP-based TV), and also it can provide the setup box at the customer's premises. Another example is when the content is created by the service provider (a typical example at the present time is Netflix, a service provider offering in-house content).

The four-market model is shown in Figure 1.8. However, different multimedia services have different QoS requirements, and the four-market model provides the possibility to identify QoS criteria for one or more components in the model. For example, for multimedia streaming (i.e. video streaming with accompanied audio and data), the following criteria can be applied considering the four-market model:

- *Content creation.* Suitability and popularity of the content, codec quality, piracy aspects, etc.
- *Service provision.* Ease of navigation to requested multimedia content, security, pricing, customer care, etc.
- *Service transport.* Bandwidth (in bit/s), latency (i.e. one-way end-to-end delay) and round-trip delay, jitter (i.e. delay variation), error rate, traffic congestion, distortion, etc.
- *Customer equipment.* Network interface (e.g. supported bitrates in downlink and uplink), ease of selection and playback of the video content, ease of navigation through the content, ergonomic aspects of the user device, etc.

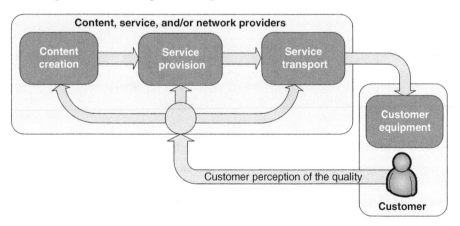

Figure 1.8 Four-market model.

1.5 QoE Concepts and Standards

QoE covers entire end-to-end system effects, including customers' equipment as well as network and service infrastructure. Additionally, end-user perception of telecommunication services is influenced by trends, advertising, tariffs, and costs, which are interrelated with the customer expectation of the QoS.

So, on one side, QoE is influenced by technical QoS (that is, all network equipment and all customer equipment on the path end-to-end) and non-technical QoS (including points of sale, customer care, etc.). On the other side, QoE is also influenced by customers' expectations of QoS, which include current trends (at a given time), advertising (it has also an influence on customers), tariffs, and costs for enjoying the given service. The influences on QoS (technical and non-technical) and customer expectation of QoS are shown in Figure 1.9.

QoE can take into account additional parameters, such as customer experience (more experienced users may have better experience of service use), user context (mood, environment, etc.), and potential discrepancy between the offered service and individual user feelings about the quality of the service delivery and its features.

1.5.1 QoE and QoS Comparison

If one compares QoE and QoS, it appears that QoE is wider and bigger. But let's discuss QoE versus QoS further. On one side, QoS can be seen as an objective measure at predefined measurements points, such as network interfaces in hosts and network nodes. On the other side, there is a need for subjective merit for the quality of the given service perceived by the end-users, and that is where QoE comes onto the scene.

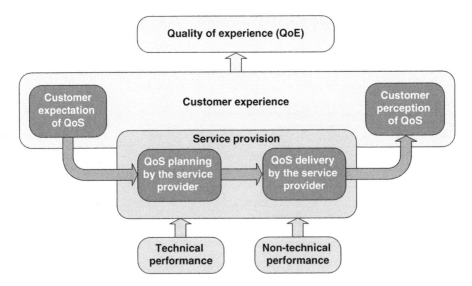

Figure 1.9 Customer experience dependence on QoS and customer expectation of QoS.

If one considers the four viewpoints in the ITU-T QoS framework, QoE can be considered as a viewpoint of QoS from the customer's perspective. However, QoE also includes the customer's decision on retaining the service or withdrawing from that service.

So, QoE can be considered as being on the edge between the perceived QoS and the assessed QoS. On one side, it includes the QoS perceived by the customer and QoS required by the customer, which belongs to the customer's side of ITU's QoS framework (the one with four viewpoints). On the other side, QoE also includes qualitative terms that refer to customer satisfaction from the provided service and additionally customer attraction to the service.

Considering the performance model for QoS criteria, QoE is influenced by all seven defined criteria. For example, speed (e.g. bitrate in downlink and uplink) influences the latency as well as the resolution of the video (e.g. is it HD, full HD, or higher picture resolution), which directly impacts the customer's experience of that service (the video streaming service in this example).

As with any other defined quality, QoE should also be assessed. The QoE assessment is performed by using subjective tests with metrics such as MOS, and that is in contrast to QoS which is assessed with objective metrics.

Overall, it is clear that QoE must be assessed using subjective tests with metrics such as MOS, as standardized in ITU-T Rec. G.1011 [14]. However, it is also possible, and sometimes more convenient, to estimate QoE based on objective testing and then to apply associated quality estimation models. Using different quality estimation models, one can measure or calculate the objective parameters affecting QoE, with the aim of evaluating QoE.

Why is such an approach – assessment of QoE as a subjective merit with measurement of objective parameters – needed? Because subjective tests require more resources and effort (e.g. they require human subjects), while objective measurement and automatic calculations using appropriate quality estimation models are generally faster and cheaper.

1.5.2 QoS and QoE Standards

With the aim of defining and then assessing QoS and QoE, there are required standards which can provide globally acceptable parameters that can be assessed and that are important to customers, service providers, and the public (e.g. NRAs in the telecommunications/ICT area).

SDOs, such as the ITU, the European Telecommunications Standards Institute (ETSI), or the IETF, have collective knowledge and expertise with respect to QoS and QoE. Because QoS and QoE are always considered end-to-end (between the endpoints of the communication call or session), there are required global standards on QoS matters. However, SDOs are seeing an increase in the number of cases where some parties decide to rely on industry standards relating to QoS/QoE instead of globally recognized standards. Regarding global standards, for example, ITU-T develops recommendations on performance, QoS, and QoE for all terminals, networks, and services, ranging from circuit-switched traditional telecom networks and services to multimedia applications over fixed or mobile IP-based networks. Table 1.3 gives examples of certain important ITU-T recommendations on QoS and QoE matters from different recommendation

Table 1.3 Some important ITU-T recommendations on QoS and QoE.

ITU recommendation series	Example ITU-T recommendations on QoS/QoE
E-Series: Overall network operation, telephone service, telephone operation, and human factors	• E.800: Definition of terms related to quality of service • E.802: Framework and methodologies for the determination and application of QoS parameters • E.803: Quality of service parameters for supporting service aspects • E.804: QoS aspects for popular services in mobile networks • E.807: Definitions and associated measurement methods of user-centric parameters for call handling in cellular mobile voice service • Supplement 9 to E-series: Guidelines on regulatory aspects of QoS
G-Series: Transmission systems and media, digital systems and networks	• G.1000: Communications quality of service: A framework and definitions • G.1010: End-user multimedia QoS categories • G.1011: Reference guide to quality of experience assessment methodologies
P-Series: Terminals, subjective and objective test methods	• P.10/G.100: Vocabulary for performance and quality of service • P.800: Methods for subjective determination of transmission quality • P.863: Perceptual objective listening quality assessment • P.1200-P.1299: Models and tools for quality assessment of streamed media
Y-series: Global information infrastructure, Internet protocol aspects, and next generation networks (NGN)	• Y.1540: Internet protocol data communication service – IP packet transfer and availability performance parameters • Y.1541: Network performance objectives for IP-based services • Y.1542: Framework for achieving end-to-end IP performance objectives • Y.1543: Measurements in IP networks for inter-domain performance assessment • Y.1545.1: Framework for monitoring the quality of service of IP network services

series, which shows that QoS/QoE topics are considered in all types of telecommunication networks and for all types of telecommunication services.

1.6 General QoS Terminology

Terminology standardization in general is necessary for two main reasons: first, to avoid confusion to readers by introducing conflicting terms and definitions; and second, to

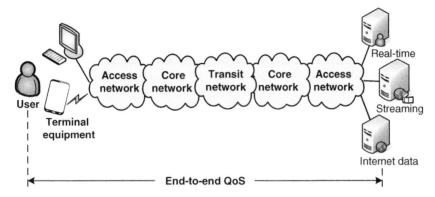

Note: on both sides of the end-to-end path there can be humans with terminal equipment (PC, smartphone, etc.) or machines (various servers, things, etc.).

Figure 1.10 End-to-end QoS.

assist alignment between the various groups involved in telecommunication standards development. The terminology for QoS is standardized in ITU-T Rec. E.800 [10].

As discussed in this chapter, the end-to-end QoS depends on the contributions made by all components on the end-to-end communication path, as shown in Figure 1.10.

General QoS terminology, which is also used in this book, is given in Table 1.4 [10]. Besides the general QoS terms, there are service-related terms (call setup time, conversational quality, cybersecurity, etc.), network-related terms (customer premises equipment, network/user interface, interconnection, etc.), and management-related terms (QoS resource management, class of service, fault, etc.). So, QoS terms have been classified into general terms and three broad areas which include service, network, and management.

1.7 Discussion

This introductory chapter has provided information regarding the convergence of the telecom and Internet worlds which is being realized in the first two decades of the twenty-first century and it is expected to continue toward a full, all-IP world. However, the traditional telecom world is based on strict end-to-end QoS provisioning, such as QoS in telephone networks for voice services (including fixed PSTN as well as PLMN such as the Global System for Mobile communication (GSM)), TV broadcast networks for TV, and leased lines for business users. Meanwhile, the Internet has been developed from the beginning for best-effort services (i.e. each IP network gives its best effort to transfer the IP packets through it) and it is network neutral by nature (e.g. no differentiation of IP packets by Internet service providers (ISPs) regarding the source or destination address, application, or type of content carried by the packets).

However, the approaches from both sides (telecoms and the Internet) have merged, including QoS support as the main contribution and requirement from the telecom side (e.g. for real-time services, such as voice and TV) to influence the initially best-effort and network-neutral Internet. The convergence of all telecommunication services onto the

Table 1.4 General QoS terms [12].

QoS term	Definition
QoS requirements of user/customer (QoSR)	A statement of QoS requirements by customers/users with unique performance requirements or needs.
QoS offered/planned by service provider (QoSO)	A statement of the level of quality planned and offered to the customer by the service provider.
QoS delivered/achieved by service provider (QoSD)	A statement of the level of QoS achieved or delivered to the customer.
QoS experienced/perceived by customer/user (QoSE)	A statement showing the level of quality that customers/users believe they have experienced with the service usage.
Criterion	Collections of appropriate characteristics used to describe benefit to user of the service.
Parameter	A quantifiable characteristic of a service with given scope and specified boundaries (minimum or maximum).
Objective (quantitative) parameters	Parameters measurable with instruments or observations.
Subjective (qualitative) parameters	Parameters expressed by using human judgment.
Measure	A unit by which a parameter can be expressed at measurement.
Metric (i.e. "indicator")	Value which is calculated from observed attributes of a measure.
Service	A set of functions offered to a user by a service provider.
User	A person or entity external to the telecommunication network, which uses connections through the network for communication.
Customer	A user who is responsible for payment for the services based on agreement with the service provider (e.g. telecom operator).
Network performance	The ability of a network or its portion to provide the technical functions related to communications between users.
Network provider	An organization/company which owns a telecommunications network for the purpose of transporting bearers of telecommunication services.
Service provider	An organization/company which provides telecommunication services to users and customers.

Internet (including video as the most demanding service regarding bitrates, i.e. bandwidth) was starting to happen with the development of broadband access to the IP-based networks and provision of IP-based services (via such broadband access).

The convergence onto the Internet as a single networking platform for all services is possible only with development of broadband access to the Internet, including fixed access and mobile access networks. Although broadband is a relative terminology with respect to the bitrates which provide seamless access to all existing applications and services at the present time, the term ultra-broadband is used (e.g. in many national broadband strategies for 2020 and beyond) to point to higher bitrates than the existing broadband worldwide. While existing broadband access (on average) is in the range of Mbit/s per user to tens of Mbit/s per user, ultra-broadband access is targeted at bitrates in the range of several hundreds of Mbit/s per user to Gbit/s per user. However, the defined QoS and QoE refer to current and future broadband and ultra-broadband speeds. The principles of QoS and QoE as well as the QoS framework will not change significantly in the near future; however, they will evolve together with the appearance of new innovative services/applications in the global telecommunications/ICT world.

References

1 International Telecommunication Union (ITU), www.itu.int, accessed October 2017.
2 ITU, ICT Facts and Figures 2016, 2016.
3 ITU ICT-Eye, http://www.itu.int/icteye, accessed October 2017.
4 RFC 768, User Datagram Protocol. IETF, August 1980.
5 RFC 793, Transmission Control Protocol, IETF, September 1981.
6 RFC 791, Internet Protocol, IETF, September 1981.
7 IETF 2460, Internet Protocol, Version 6 (IPv6), December 1998.
8 http://bgp.potaroo.net, accessed October 2017.
9 Janevski, T. (2014). *NGN Architectures, Protocols and Services*. Chichester: Wiley.
10 ITU-T Recommendation E.800, Definitions of Terms Related to Quality of Service, September 2008.
11 ITU-T Recommendation P.10/G.100, Amendment 5, "New definitions for inclusion in Rec. ITU-T P10/G100," 2016.
12 ITU-T Rec. G.1000, Communications Quality of Service: A Framework and Definitions, November 2001.
13 ITU-T Rec. E.802, Framework and Methodologies for the Determination and Application of QoS Parameters, February 2007.
14 ITU-T Rec. G.1011, Reference Guide to Quality of Experience Assessment Methodologies, July 2016.

2

Internet QoS

The Internet was created to provide only best-effort services, which means that every connection is admitted into the network and every packet is transferred from source to destination with the best effort by the network, and without any guarantees regarding quality of service.

However, from the beginning the Internet designers allocated for QoS support even in the fundamental Internet Protocol in both versions, IPv4 (IP version 4, which is also simply referred to as IP) and IPv6 (IP version 6). That was done because it was known that different type of media such as audio, video, and data, have different requirements on certain quality metrics (bitrate in downlink and uplink, delay, losses, etc.). Those protocols are standardized by the IETF, the main standardization body for Internet technologies (i.e. it includes protocols and architectures from the network protocol layer up to the application layer, which are used in hosts and nodes attached to the Internet network).

Since the beginning of the twenty-first century the whole telecommunication world has been transiting to the Internet as a single networking platform for all telecommunication services, including native Internet services (e.g. Web, email) and traditional telecommunication services (e.g. telephony, television). That makes for different QoS techniques which are standardized for legacy telecommunication networks (e.g. PSTNs) to be implemented in a certain manner in IP networks (here, an IP network is based on the Internet networking principles and has IP protocol implemented in all hosts and nodes in the given network). Also, there were certain QoS solutions specifically standardized for the Internet by the IETF, and such solutions have further influenced QoS mechanisms in the telecommunication/Information and Communication Technologies (ICTs) world based on Internet technology protocols.

It has been said that the long-term goal of networking is to design a network that has the flexibility and low cost of the Internet and at the same time can provide the end-to-end QoS guarantees of the telephone network. However, that is not a simple process when there are many services with different quality requirements over a single network.

2.1 Overview of Internet Technology Protocols

Internet technology protocols are placed in the upper layers of the protocol layering model, covering the range from the network layer (layer 3) up to the application layer

QoS for Fixed and Mobile Ultra-Broadband, First Edition. Toni Janevski.
© 2019 John Wiley & Sons Ltd. Published 2019 by John Wiley & Sons Ltd.

(layer 7). However, the main protocols are those implemented at the network layer and the transport layer (layer 4). The main protocol is the IP, which currently is standardized in two versions, IPv4 [1] and IPv6 [2]. The main transport layer protocols are the user datagram protocol [3] and the transmission control protocol [4]. However, there are others that may exist in layer 4, such as SCTP (stream control transmission protocol) [5] and DCCP (datagram congestion control protocol) [6].

2.1.1   Internet Network Layer Protocols: IPv4 and IPv6

IP is the main communication protocol in the IP stack. It is positioned on the network protocol layer (layer 3) and provides functions for transmission of datagrams (i.e. blocks of data) over interconnected systems of packet-switched networks from any source to any destination.

IP is a connectionless protocol because it does not require a connection to be established between the source host and the destination host prior to the datagrams' transmission. When the IP first appeared (at the beginning of the 1980s) it was completely different from the traditional telecommunication approach in PSTNs where connections were established (prior to the transfer, such as voice transfer) by using signaling between the two end devices and the network.

Internet datagrams are also called IP packets, so the terms are used interchangeably throughout this book. In general, a datagram is a variable length packet consisting of two parts: IP header and data (in the payload of the packet).

The IP has exactly two versions (not more, not less):

- IPv4, standardized in 1981 by the IETF [1];
- IPv6, standardized in 1998 by the IETF [2].

The IPv4 and IPv6 packet headers are shown in Figure 2.1. The main difference is that IPv6 addresses are four times longer (128 bits) than IPv4 addresses (which have a length of 32 bits). The main driver for introduction of IPv6 was in fact IPv4 address space exhaustion, which started to become evident with the exponential growth of the Internet in the 1990s. However, IPv6 does offer several other improvements to the IPv4 header, such as reducing some fields which refer to packet IP packet fragmentation (that is fragmentation of a single IP packet into multiple fragment packets at a certain point in the network, such as a router, and then doing the reverse process at some point in the downstream direction) which has proven to be inefficient for practical use (e.g. if some of the routers on the way do not handle the fragmentation of the IP packets well, the IP communication will suffer interruptions), hence these fields are omitted in the newer IPv6 header. However, due to longer IP addresses, IPv6 has a header that is at least two times longer (a minimum of 40 bytes) than its predecessor IPv4 (a minimum of 20 bytes), which influences certain traffic types such as voice which have low-length payload (due to lower traffic intensity of the voice). Another difference is the Flow Label field introduced in IPv6 with the intention of labeling individual IP flows on layer 3 (network layer), which may be important for QoS provisioning per flow. Nevertheless, the Internet was created for IPv4 networks and QoS solutions targeting IPv4 networks, so one may expect such mechanisms (e.g. for traffic management) to continue to be used for IPv6 networks, which limits the possibilities of end-to-end use of the Flow Label.

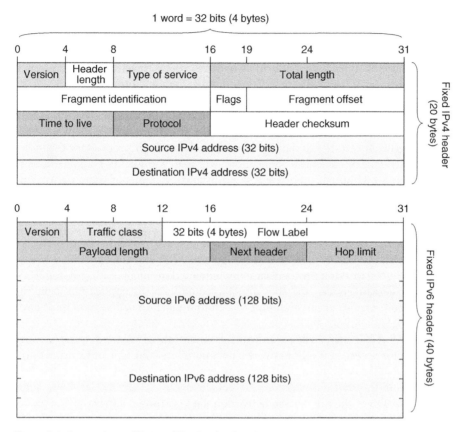

Figure 2.1 Comparison of IPv4 and IPv 6packet headers.

Regarding the QoS on the network layer, the most important field in IP headers is Type of Service (ToS) in IPv4 as well as DSCP (Differentiated Services Code Point) in IPv6. Both fields have the same length (1 byte, i.e. 1 octet of bits) and can be used for traffic differentiation on the network layer, for marking packets (with this field in the IP headers) which belong to the same aggregate traffic type (also called a traffic class or QoS class), such as voice, TV, various data, VPN (virtual private network), etc.

IP packets travel across different IP-based networks which are using different under-lying transport technologies (this refers to network interfaces defined with layers 1 and 2 of the protocol stack). For example, the most used local area network (LAN) for Inter-net access in home and business environments is Ethernet (IEEE 802.3 standard) and WiFi (IEEE 802.11 family of standards) as unified wireless LAN (WLAN). Both main Internet access standards, Ethernet and WiFi, are standardized by the Institute of Elec-trical and Electronics Engineers (IEEE) only on layer 1 and layer 2 (i.e. physical layer and data-link layer, respectively), while on layer 3 it is expected to carry primarily IP pack-ets, that is, to have IP on the protocol layer 3. So, from inception (Ethernet appeared in the 1980s and WiFi at the end of the 1990s) they were built to carry Internet traffic in the local area networks. Of course, one should note that these access networks are typically connected via other broadband access technologies (fixed or mobile) to the core networks of the network providers. However, due to their influence globally on the Internet network infrastructures, the size limit in their layer 2 payloads (that is Medium access control (MAC) frame payload), which is approximately 1500 bytes, in fact defines the maximum size of IP packets (either IPv4 or IPv6) which travel across the Internet, regardless of the type of traffic they carry (voice, video, or some other data).

IPs as well as all other Internet technologies are being standardized by the IETF (www .ietf.org) in a form of RFCs (request for comments). Not all RFCs are Internet standards, some are informational only. Also, there are Internet technologies which are proprietary and not standardized (e.g. developed by certain vendors). Such proprietary technologies are typically developed on the application layer (e.g. various OTT applications in the public Internet), although there are also proprietary versions of some networking protocols, such as proprietary TCP versions and proprietary routing protocols.

2.1.2 Main Internet Transport Layer Protocols: TCP and UDP

Internet networking is based on network layer protocol (IP) and transport layer protocols (TCP, UDP). Networking protocols are implemented in the operating system (OS) of the hosts or network nodes and they are connected with the application using the abstraction called "socket," as an endpoint of a communication call/session in IP environments.

What is a "socket"? It is an abstraction which connects the IP address of the network interface (used for access to the IP network), port numbers, and type of transport protocol being used in the communication. From an application point of view, it hides from the application all Internet networking protocols below (e.g. TCP/IP, UDP/IP), which are built into the OS of the host or the network node (e.g. router, gateway). The socket interface for the application is defined with an application programming interface (API), which is specific to the OS of the given host or network node. Therefore there is required development of a given application for a given OS (different API exists in Windows, Linux, Android, iOS, etc.).

From the network point of view, the socket is the endpoint of a given Internet connection. So, if a typical connection includes two end parties (e.g. two Internet hosts), then the connection (a call, a session) is established between two open sockets at the two end hosts. If one of the sockets closes, it means that the given connection is also considered as closed. So, to maintain a given connection means to keep open the sockets between the two ends. That refers to both fixed and mobile Internet access. However, it is more complex to keep the socket open during a connection for a moving host which typically changes a point of attachment to the network (while it is moving), and hence it may change the IP network to which it is attached, which means that the IP address must change in such a case. But if any of the parameters (IP address, port number, transport protocol type) which define a given socket change, it means that a new socket should be opened and the previous one closed. That results in closing (i.e. dropping) the connection and opening a new one (that is, a new connection with a new socket). However, in such a case there is no mobility because mobility means maintaining the same connection of mobile hosts regardless of their movement. Due to such mobility problems in IP network environments, there are certain solutions on the network layer for Internet hosts, which include standardized protocols such as Mobile IP (for IPv4 mobile networks) [7] and Mobile IPv6 (for IPv6 mobile networks) [8]. But in mobile networks (until and including 4G) the mobility is typically handled on the data-link layer (below the IP layer), which hides the mobility from the network layer up and hence eliminates potential problems that may arise from the IP addressing principles (which were initially created for fixed hosts only).

After defining the sockets that bind the IP (on the network layer) and the transport layer protocols, let's see which are the main transport protocols and their main principles influencing QoS. The main transport protocols are:

- UDP, standardized with RFC 798 in 1980 [3];
- TCP, initially standardized in 1981 [4] but later updated with congestion control mechanisms in the 1990s [9]. TCP mechanisms are continuously being updated [10] simply because there is no single best congestion control mechanism for all times.

UDP/IP is typically used for real-time data (e.g. VoIP, IPTV) and certain control traffic (e.g. the domain name system (DNS) uses UDP/IP). In practice, UDP adds only source and destination port numbers (Figure 2.2), which are being used by sockets opened on both ends of the communication connection. The port numbers identify the application using the given transport protocol layer (UDP in this case). All well-known protocol numbers are defined by the IANA, the non-governmental body governing the Internet on a global scale. So, UDP has no particular direct function regarding QoS. However, thanks to its simplicity (no guarantees for datagram delivery end-to-end), UDP provides the lowest end-to-end delay (because there are no retransmissions of lost packets, which requires additional time and hence adds delay), which is useful for some services that are delay sensitive (voice is the most typical example of a service that uses UDP/IP). Yet since there are no guarantees or feedback control (from the receiving host to the sending host), typically the control is implemented into the application (e.g. DNS) or another protocol used over UDP, such as RTP (real-time transport protocol) [11]. RTP provides synchronization of multiple media streams (e.g. video and accompanied audio streams) as well as control messages exchange in both directions between sender and recipient by using its counterpart protocol called RTCP (real-time transport control protocol). For example, voice communication over IP uses voice application over RTP/UDP/IP

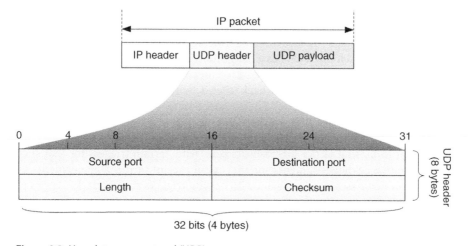

Figure 2.2 User datagram protocol (UDP).

protocol stack, although it is possible for RTP-like functionalities to be implemented in a certain manner in the VoIP application, something that is typical for OTT voice services.

Unlike UDP, which is very light (regarding implementation and resource consumption at the host) and has no particular function except providing port numbers (which is good for some real-time applications), TCP can be considered as one of the most important Internet technology protocols. Why? Because the initial concept of Internet networking for end-to-end QoS support is implemented with TCP at the end hosts. While in traditional telecommunication networks, the PSTNs, the intelligence was located inside the network and hosts were very simple devices (traditional telephones), in the Internet the intelligence for dealing with end-to-end QoS (e.g. congestion control) is implemented at end hosts (which are typically computers such as personal computers, smartphones, or other type of devices with hardware and software on them) via TCP. But TCP has also evolved with the Internet. During the ARPANET era in the 1970s, TCP functionalities were in fact implemented in one protocol together with network functionalities (which are now part of IP), called a network control program (NCP), as a single protocol implemented between the applications on the top and the network interface at the bottom of the hosts. However, with the standardization of IP and TCP in 1981, the NCP split into two main protocols: IP as a network protocol and transport protocols (TCP and UDP) over it. This was first implemented in the BSD (Berkeley Software Distribution) Unix OS in the 1980s and implemented in all Internet hosts (several hundred of them) on January 1, 1983, known as Internet flag day. Because of the low number of hosts connected to the Internet at the time, it was possible to change the OS in all of them in a short period of time. However, is another flag day possible? Well, when the Internet has billions and tens of billions of hosts (including things attached to the Internet, besides humans with their devices), it becomes almost impossible to implement such change in all hosts at the same time (nowadays the changes must be gradual and backward compatible).

The TCP standardized in 1981 had no congestion control mechanisms. When the Internet began to grow and expand outside the research community with the appearance of NSFNET (National Science Foundation Network) in 1985, which made it possible for companies to connect to the network, congestion started to appear more frequently. The initial solution was standardization of congestion control mechanisms for TCP [9, 10]. Such mechanisms include Slow Start, Congestion Avoidance, Fast Retransmission, and Fast Recovery. In fact, TCP uses acknowledgements sent by the TCP receiver to the TCP sender to acknowledge the last received TCP segment in sequence (with no lost or reordered segments before). TCP has a congestion window (CW), which defines the number of TCP segments that can be transmitted (of course, the TCP segment is placed into a payload of an IP packet) back-to-back, that is, without waiting for acknowledgments to arrive. Each non-duplicate acknowledgment increases the number of segments received and hence pushes the CW, so the CW is in fact a sliding window. It is also a dynamic window, which means that its size in almost all TCP versions is never a constant value over time, so it either increases its size (with certain triggers for that such as received positive acknowledgments for successful receipt of the TCP segments by the other end) or decreases its size (again, with certain triggers such as packet losses or timeouts). Bigger CW gives higher bitrates, that is, achievable bitrates for a given TCP connection are proportional to the CW size. Let us explain this through an example.

If a TCP segment is carried in 1500 bytes long IP packets (typical size for Web or video traffic), then (as an example) 10 Gbit/s for a given TCP connection for a given

RTT (round trip time) between the TCP sender and the TCP receiver can be reached with a CW of size:

$$CW = \frac{10Gbit/s}{8bits/byte * 1500bytes} * RTT = 833333[\text{sec}^{-1}] * RTT \qquad (2.1)$$

For RTT = 100 ms = 0.1 s, the required CW size of TCP to reach the maximum bitrate of 10 Gbit/s will be CW = 83 333, while the CW size required in the case of RTT = 10 ms will be CW = 8333. So, the CW size depends upon the capacity of the bottleneck link between the two ends (which communicate by using TCP/IP) and the RTT. One may note that bitrate of the application that is using TCP/IP is directly related to the CW size of the TCP in use. Hence, bigger CW is needed to reach higher speeds and vice versa.

However, the main target of TCP is not only to reach maximum capacity (i.e. bitrate) available end-to-end but also to provide fairness based on its congestion avoidance mechanism.

The following are the main TCP congestion control mechanisms (Figure 2.3):

- Slow start provides an exponential increase in the CW size with the aim of reaching the capacity of the bottleneck for the given TCP connection sooner (in the default version, in slow start CW is doubled for each RTT interval). Usually, slow start begins with a CW size equal to one segment, although there are TCP versions which provide slow starts with higher initial values (two TCP segments or more).
- *Congestion avoidance*. It provides linear increase of the CW size with the aim of postponing congestion (which results in packet losses).
- *Fast retransmission*. After a lost TCP segment, TCP reacts after receiving a certain number of acknowledgments for TCP segments successfully transmitted after the lost one (e.g. three duplicated acknowledgments) and reacts by reducing the size of CW (e.g. to half).
- *Fast recovery*. This mechanism provides the possibility for TCP to stay in the congestion avoidance phase after packet loss, thus avoiding going into slow start phase and avoiding significant disruption of the bitrate of the given connection. However, multiple TCP losses within the CW may lead to expiration of TCP timeouts (e.g. maximum time for waiting for acknowledgements), which may result in transition to slow start.

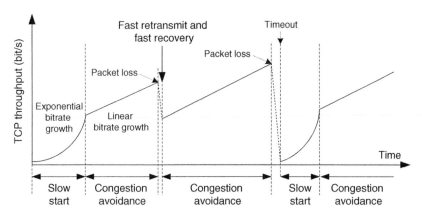

Figure 2.3 TCP congestion control.

Different TCP versions may implement the above basic congestion control mechanisms partially, modify them in certain aspects (e.g. regarding the phases and their trigger), or combine TCP and UDP characteristics to achieve reliable and low-delay transfer (e.g. for services such as video streaming or IPTV).

TCP payload contains data from application protocols that use TCP (e.g. the Web, i.e. Hypertext Transfer Protocol (HTTP)), email protocols, File Transfer Protocol (FTP)). TCP is a client-server protocol which establishes a connection between two end hosts prior to the data transfer. For connection establishment it uses a three-way handshake. "Fake" TCP connection establishment requests are the basis for denial of service (DoS) attacks, which affect the availability of the application or service provided via TCP.

2.1.3 Dynamic Host Configuration Protocol – DHCP

The Dynamic Host Configuration Protocol (DHCP) is a standardized protocol for dynamic provision of network configuration parameters to Internet hosts in a given IP network. DHCP exists in every network, including telecom internal networks, and enterprise or home networks. DHCP has prolonged the life of IP version 4 (the initial IP) [12], providing the possibility to serve more users with a limited pool of public IP addresses. However, DHCP has fundamental importance for the global Internet and for the telecommunication world based on the Internet technologies for providing a plug-and-play approach for access to the Internet. How? Well, besides the dynamically allocated IP address, the DHCP sets other networking information, of which the most important are the IP addresses of the DNS servers because without access to a DNS server a given host cannot resolve domain names into IP addresses, which is necessary for Internet-based communication. Therefore DHCP is also standardized for dynamic assignment of IPv6 addresses, called DHCPv6 [13].

From the QoS point of view, proper functioning of the DHCP server in the IP network to which a given host is attached is crucial in order for a host to obtain an IP address on its network interface in a plug-and-play manner and to have Internet connectivity.

2.1.4 Domain Name System – DNS

While in traditional telephone networks the main naming and addressing space was (and it still is) the telephone numbering space, standardized and governed on a global scale by the ITU (defined with ITU-T E.164 telecommunication numbering plan [14]), on the Internet there are exactly two name spaces, which are governed on a global scale by ICANN and its department IANA. These two name spaces are:

- IP addresses, including:
  - IPv4 addresses, e.g. 194.149.144.1 (typically given in decimal-dot notation, with a length of 32 bits)
  - IPv6 addresses, e.g. 2001:0000:0000:0000:0008:0800:200C:417A (typically given in a hexadecimal notation, with a length of 128 bits);
- domain names [15], e.g. "www.itu.int," "www.ietf.org."

DNS was created to provide a means for using names (domains) instead of IP addresses to access certain machines on the network as people more easily use names than long numbers. However, machines communicate over the Internet only via IP addresses, so the domain name must be resolved to an IP address (which corresponds to the given domain name according to the DNS system records). For that purpose, DNS is a distributed hierarchically-based system which translates a domain name to an IP address associated with that domain.

DNS is a fundamental Internet technology which provides translation between the two name spaces on the Internet. However, it is characterized by two conceptually independent aspects:

- The first is the definition of names and rules for delegation of authority for the names.
- The second is specification of a distributed computer system which will provide efficient mapping of domain names to IP addresses and vice versa.

The domain name space is defined as a tree structure, with the root on the top. Each domain name consists of a sequence of labels, which are separated with dots, and each label corresponds to a separate node in the domain tree. For example, in the domain name "www.itu.int" one will observe that it consists of three labels divided by dots. The label on the right side is always higher in the name hierarchy. So, in the given example the top-level domain is the domain "int."

Regarding QoS, DNS availability is very important for Internet services' accessibility. In fact, besides a physical connection to the network, to have availability of Internet access service (as a whole) requires an IP address (which is allocated dynamically with DHCP) and at least one available DNS server (and its IP address). Typically, IP addresses of two or three DNS servers are set up in each host connected to the Internet, due to the importance of the DNS functionalities. Each host has a DNS client built into it, which contacts the DNS server (from a defined list). In order to have lower RTT for DNS resolution (e.g. resolution of a domain name into an IP address), it is important to have DNS servers closer to the clients (e.g. in the same IP network, where possible), although a DNS client can send a request to any available DNS server on the Internet. However, not all DNS servers have records for all possible domain names. If a certain record is not available, the contacted DNS server sends a request to another DNS server either iteratively or recursively (to different DNS servers, according to obtained responses). If there is no possible DNS communication, the user can still connect to a server or a peer host by specifying directly the IP address of the destination host. However, that is not feasible in most cases (e.g. the requesting user usually does not know the destination IP address, and it is not convenient to include IPv4 and especially longer IPv6 addresses).

DNS is implemented on the application layer (regarding the protocol stack) and initially it used UDP (as a transport protocol) due to its lower delay. However, with the deployment of high-speed Internet links and access around the globe, as well as with the higher processing power in all hosts (which doubles every 1.5–2 years according to Moore's law), there is also a standardized approach for DNS communication by using TCP as transport layer protocol [16].

Overall, proper functioning of DNS as well as lowest possible DNS resolution times are key performance parameters for Internet QoS because they directly refer to Internet service availability and accessibility.

2.1.5   Internet Fundamental Applications

2.1.5.1   Web Technology

The World Wide Web (WWW) is a global system started in 1989 by Tim Berners Lee and completed with the development of its main standards during the 1990s. The Web (i.e. WWW) is made up of a large number of documents called web pages, each of which is a hypermedia document. "Hyper" means that it contains links to other websites and "media" means that it can contain other objects besides text (pictures, videos, various files, etc.).

The Web is based on the client-server principle. The main protocol for Web end-to-end communication, for access to web pages, is called HTTP. It is based on client-server TCP/IP protocol model, where:

- Web servers (i.e. HTTP servers) listen on the well-known TCP port 80;
- Web application on the users' side is called a browser (e.g. Chrome, Firefox, Opera), which has an HTTP client built into it.

So, HTTP is only a communication protocol for Web traffic. The presentation of Web content is done with other tools [17], and the most known markup language for creating Web pages is Hypertext Markup Language (HTML).

HTTP was initially standardized with HTTP/1.0 version, which uses a new HTTP connection for each object transfer (non-persistent mode), where an object can be the text, a picture, etc. However, the widely spread standard over the years was HTTP/1.1, in which one HTTP connection is used for transfer of multiple objects between a client and a server (called persistent mode, or pipelining). It was standardized with RFC 2616 in 1999 [18]. After more than 15 years of HTTP 1.1 the IETF published HTTP version 2 (HTTP/2) in 2015, standardized with RFC 7540 [19].

HTTP/2 addresses several issues that were considered to be disadvantages in HTTP/1.0 and HTTP/1.1 regarding performance of the HTTP connection and Web traffic in general. HTTP/2 defines an optimized mapping of HTTP semantics to an underlying TCP connection. Also, it allows interleaving of request and response messages on the same connection and uses an efficient coding for HTTP header fields. Further, HTTP2 provides the possibility for prioritization of HTTP requests (from HTTP clients, i.e. Web browsers) by allowing more important requests to complete more quickly, which should result in improved performance of Web-based applications.

2.1.5.2   File Transfer Protocol (FTP)

FTP is an application for copying files from one machine (i.e. host) to another. It was one of the first important application-layer protocols on the Internet (the first file transfer mechanisms were proposed in 1971 with RFC 114) [17].

It is also a client-server protocol that uses TCP/IP protocol stack. It consumes more resources on the computer than HTTP because FTP requires two processes to be established for file transfer (e.g. file upload or download), a control process (e.g. for user authentication, file administration), and data process (for file transfer). Therefore, it was rarely used after the global spread of HTTP and WWW, which are also used for file upload and download (among other possibilities) and hence make FTP redundant.

Nevertheless, even in the second decade of the twenty-first century, FTP as a protocol is still used for Internet connection measurements (e.g. bitrate measurements in downlink or uplink).

2.1.5.3   Email Protocols

Email communication is done with standardized IETF protocols [17]. The main email protocol is SMTP (Simple Mail Transfer Protocol), which is used for sending email from a standalone mail client (Thunderbird, Outlook, etc.) to a mail server and for communication between email servers (e.g. for sending an email from the sending email server to the destination email server, of course after resolving the domain name of that email server via DNS).

For access to the email messages received in the mailbox of the email server, there are standardized access email protocols, which include POP3 (Post Office Protocol version 3) and IMAP4 (Internet Message Access Protocol version 4). Popularization of free email accounts since the 2000s also increased the portion of Web-based access to email. In such cases the protocol used for access to email is HTTP.

2.2   Fundamental Internet Network Architectures

2.2.1   Client-Server Internet Networking

Most of the Internet applications use client-server communication, which means that there are two types of entities (e.g. hosts, network nodes) in each such communication:

- *Client (e.g. a Web browser).* It sends requests to the server.
- *Server (e.g. Web server).* It sends responses to the client's requests.

The principle of the client-server communication is such that clients are the entities which request certain information or action from the server. The server is not intended to initiate any action by the client. This is native Internet network architecture, where clients are simpler machines which in average have lower processing power and lower memory capacity than servers because servers must handle multiple requests from multiple clients in parallel. When there are many expected requests from many clients, a single server cannot serve them all, so in such cases multiple servers are serving the clients for a given (same) application or service. The servers can be centralized into a single data center or distributed among several geographically dispersed data centers (e.g. content distribution network (CDN) [17]). For example, global OTT service providers have many data centers, which are situated as close as possible to the end-users in a given region because besides balancing of the servers' load, another important parameter for client-server communication is the delay in receiving a response from a server for a given client request (also referred to as RTT).

Overall, client-server architectures are Internet native and fundamental; however, they are not symmetric regarding client and server functionalities and their roles. Both main types of transport protocols (TCP, UDP) as well as others (SCTP, DCCP) are also based on the client-server principles. The transport protocol on the side of the client application acts as a client (e.g. TCP client, UDP client) and the transport protocol on the side of the server acts as a server (e.g. TCP server, UDP server). Each protocol communication is peer-to-peer (P2P), which means that a given protocol (e.g. application, transport protocol) must communicate with exactly the same protocol on the other side of the connection (i.e. the same application protocol and the same transport protocol, respectively).

2.2.2 Peer-to-Peer Internet Networking

P2P networking is based on network architectures in which each host (computer, smartphone, etc.) or network node (e.g. router) has similar capabilities and functionalities as other peers.

P2P applications for residential users appeared by the end of 1990s. One of the first globally known P2P architectures (besides those used in research) was Napster, which was created and used for sharing of music files (mp3) around the year 2000. It was followed by other file-sharing systems (e.g. Gnutella, Kazaa, eDonkey, BitTorrent).

All conversational communications over IP networks, such as VoIP or video telephony over IP, are based on P2P communication because in such a case both end-hosts have similar capabilities and functionalities. So, QoS-enabled VoIP as a replacement of PSTN (implemented by telecom operators) is in fact implemented with P2P architectures for transfer of voice data.

2.2.3 Basic Internet Network Architectures

Basic Internet architecture consists of interconnected IP networks. In each such IP network, every Internet host and router must have implemented the IP on the network layer (in the OSs).

According to the network size (number of hosts as well as distance between the hosts), terrestrial IP networks are typically classified into the following types [20]:

- *Local area network* (*LAN*). It connects several hosts to several hundreds of hosts via network switches which work on layers 1 and 2. It is connected to the public Internet via a router, typically called a gateway router. LANs are used everywhere for access to IP-based infrastructure, including home networks, corporate networks, or networks in public places (hotels, cafeterias, etc.). LANs are typically represented with Ethernet and WiFi as unified wired and wireless local access to the Internet network, respectively.
- *Metropolitan area network* (*MAN*). This refers to a network which provides access to hosts on a territory the size of a metropolitan area (e.g. a city) and may be used for direct connection of subscribers (e.g. WiMAX [20]) or for connections of different LANs between each other or with the Internet network.
- *Wide area network* (*WAN*). This refers to nationwide transport networks which are used to connect various core networks and/or access networks (e.g. LANs, MANs). Typically WANs are deployed through fiber links (due to highest capacity of the fiber and its longest reach). They are used for building Internet transport and transit networks on national, regional, continental, or intercontinental levels.
- *Radio access network* (*RAN*). This type of access network refers to access of the mobile networks (2G, 3G, 4G, 5G, etc.). For example, the RAN in 3G mobile networks, standardized by 3GPP (3G Partnership Project), is called UTRAN (Universal Mobile Telecommunication Systems Terrestrial Radio Access Network), while the RAN in 4G mobile networks from 3GPP is based on LTE (long-term evolution) technology and is called E-UTRAN (Evolved Universal Mobile Telecommunication Systems Terrestrial Radio Access Network), as an evolution of its predecessor.

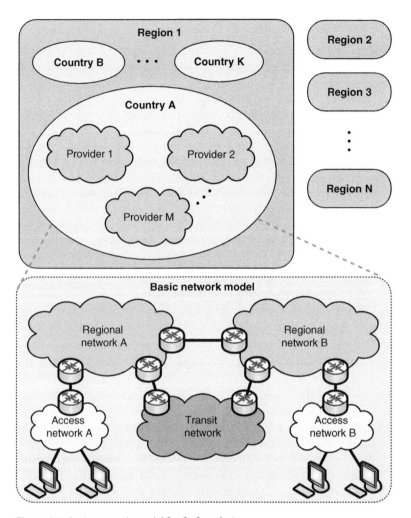

Figure 2.4 Basic network model for QoS analysis.

However, from the QoS point of view, end-to-end IP networks can be divided into three main groups (Figure 2.4):

- *Access networks.* They include all types of fixed and mobile access networks, either LAN or MAN, through which the end-hosts (either users with devices or various "things") are attached.
- *Regional networks.* They connect different access networks which are under administrative control of a single network provider such as a telecom operator, and provide their connection to the Internet and to a set of common functionalities (e.g. control functions, management functions). They consist of core networks (where the main gateways are located) and transport networks (which are used for interconnection of different nodes in the core and in the access parts).
- *Transit networks.* These types of networks are used for interconnection of different regional networks on a national or regional basis.

2.2.4 Autonomous Systems on the Internet

The Internet as a global network is built on the principle of a network of many networks that are interconnected through routers. Certainly, there must be routers between different IP networks. However, several smaller or larger networks can be administered by a company, such as an Internet service provider. In this case we actually have independently operated networks. Each of these networks of networks (where a network is identified by a unique routing prefix or network ID, which is the part of the IP address that is the same for all network interfaces attached to the given IP network) actually forms an autonomous system. So, an AS is a collection of routers under the control of one authority (a company, telecom operator, etc.). These ASs are connected with each other through gateway routers to the Internet (i.e. to other ASs). Each AS is identified via a 16-bit or 32-bit AS number, where AS numbers are globally allocated and managed by IANA.

With ASs as the basic architectural approach at present, we actually have two-tiers routing on the Internet [17]:

- *Intra-AS routing*. This is between routers belonging to a single AS (e.g. Routing Information Protocol (RIP), Open Shortest Path First (OSPF)).
- *Inter-AS routing*. This is routing between routers belonging to different ASs. Currently, the globally accepted protocol for inter-AS routing is BGP-4 (Border Gateway Protocol version 4).

So, while different routing protocols for routing inside a given AS can be used, the one used for routing between different ASs (inter-AS routing) is the same everywhere: BGP-4. BGP is in fact the "glue" that connects all autonomous parts of the global Internet network.

How does BGP work? It is not a typical routing protocol which uses the shortest path routing principle, it is also based on policies that are influenced by business decisions about peering of the given network with other networks. BGP routers are typically implemented at the edge of the AS (toward the outside Internet network, from the AS's point of view), and they advertise (to other BGP routers with which they have an established BGP peering connection) reachability of other ASs or IP network prefixes through them.

The Internet has become more global and interlinked over time, while at the same time dependence on the largest ISPs has decreased. The number of ASs continues to grow [21]. However, the average path length, which is expressed as a number of AS hops (not router hops), is stable [21]. This implies that ASs are more richly connected than in the past. In 2017, the entire Internet consisted of approximately 60 000 active ASs, with the average path length (end-to-end) span across 3–5 ASs. Regarding end-to-end QoS, more ASs on the path of the IP packets (i.e. the traffic) add more delay and a higher probability of bottlenecks somewhere on the path between the endpoints of a communication call/session. This shows that the Internet has good scalability, that is, with its increase in size, the number of ASs that the traffic passes end-to-end remains almost the same, which directly influences the end-to-end QoS (fewer ASs on the path is better).

The different types of ASs lead to different types of business relationships between them, which in turn translate to different policies for exchanging and selecting routes. Overall, there are two main types of exchange of traffic between ASs at inter-As interconnections [22]:

- *Transit*. In this case one ISP provides access to most or all destinations in its routing tables. Transit is typically based on financial settlement for an inter-AS relationship where the provider ISP charges its customers (other ISPs or companies which have their own ASs) for forwarding/routing their IP packets on behalf of customers to destinations (and vice versa).
- *Peering*. In this case two ASs (which are typically ISPs) provide mutual access to a subset of each other's routing tables. Similar to the transit, the peering is also a business deal; however, it may be set up without particular financial settlement. There are also cases for paid peering in some parts of the world, though in most of those cases it is based on reciprocal agreements. That is acceptable from a business point of view in cases where the traffic ratio between the concerned ASs is not highly asymmetric (e.g. a ratio of 1:4 may be considered as some boundary [22]).

Large network operators (such as national telecom operators) typically exchange traffic with other comparably large network operators through private peering arrangements, which are based on direct connections between the operators. Such peering agreements are usually subject to contracts and to certain nondisclosure commitments.

In terms of the amount of Internet traffic carried through peering interconnections, they are very important because most of the Internet traffic travels across them. In the 2010s it is becoming common for interconnection between IP networks of large network operators to be realized via Internet Protocol eXchange (IPX) points on a national or international level, which are typically used by operators in a given country or region.

2.3 Internet Traffic Characterization

There are different applications/services present on the public Internet. Also, there are different requirements for telecom-provided services over managed IP networks, with certain guaranteed QoS and based on signed agreements with customers (e.g. fixed or mobile telephony and TV, business services). All Internet packets on a given network interface (e.g. on a host, on a router), or in a given network or a network segment, are referred to as IP traffic. The traffic notion comes from legacy telecommunication networks, where it was used to denote mainly voice traffic (in PSTNs).

There are different traffic classifications. Although there are many different applications for traffic nowadays (e.g. millions of applications are available in different ecosystems of global OTT service providers), they all can be grouped into three main types: audio, video, and data. The first two, audio and videos, are based upon the human senses and their "design" for hearing (audio) and viewing (video), while the data traffic type includes everything else. Each of these traffic types has specific traffic characteristics which influence network performances and QoS solutions. So, typically we have the following types of traffic on the Internet:

- *Voice traffic*. This is conversational type of traffic (with similar requirements in both directions, from calling party A to called party B, and vice versa) and has constant bitrate when sending or receiving. Also, voice requires relatively small IP packets (e.g. 50–200 bytes).
- *Video traffic*. This generally has high variable bitrate, which is dependent upon the codec efficiency of the moving pictures (e.g. Moving Picture Experts Group

(MPEG-2), MPEG-4 various codec types). Video is typically unidirectional traffic (in downstream, i.e. downlink, direction regarding the end-user), so it can use the largest possible IP packets since it can eliminate IP packet delay variations with delayed playback on the receiving side. Since local access to the Internet in homes, offices, or public places is unified to Ethernet or WiFi (except in mobile networks), the maximum IP packet size for video traffic is limited by the MTU (message transfer unit) of Ethernet (and similarly WiFi), which is 1500 bytes.

- *Data traffic.* This is typically the non-real-time (NRT) traffic on the Internet, the most prominent examples being the WWW and email. This type of traffic is typically TCP-based (regarding transport protocol, which goes over IP, thus giving TCP/IP).

2.3.1 Audio Traffic Characterization

Audio traffic usually refers to voice. The conversational voice (i.e. telephony) is most sensitive to delay and jitter. Regarding telephony, the recommended delay in PSTN is below 150 ms, while above 400 ms is not acceptable because the people talking will start interrupting each other [23]. However, in IP environments the delay is always higher than in PSTN/ISDN due to packetization (creation of IP packets with payload many times bigger than the IP header due to efficiency reasons, that is, lower redundancy), buffering in network nodes (every router first buffers the packets and then sends them over a given network interface on the next hop, which is different than PSTN where bits and bytes are being transferred from one end to the other end without any notable buffering), propagation delay (due to limited speed of the signals, which is always around the speed of light on any transmission media), and so on.

In all-IP environments, different packets may enter buffers with different queue length, which may introduce different delays to IP packets from the same end-to-end connection. The average delay variation is also referred to as jitter. To compensate for the jitter, the voice has a playback point, similar to video.

However, audio is more tolerant to errors than video due to the human ear, which is more error resistant (up to several percent of the error ratio). Audio must synchronize with video in multimedia communication, such as real-time video streaming (with audio typically synchronized with the video content), video telephony (voice as audio traffic, with synchronized video, in both directions), etc.

Because voice is a two-way continuous audio streaming with strict delay requirements, it is typically carried by using RTP/UDP/IP protocol stack. The voice is a most typical example of real-time traffic. However, there are differences in how the voice is treated when provided by telecom operators with guaranteed QoS and when it is being provided as any other data by OTT voice service providers (e.g. Skype, Viber, WhatsApp) over the public Internet.

2.3.2 Video Traffic Characterization

Most of the video traffic is based on video information which is statistically compressed. The statistical compression is performed by source coding in the video codec (includes video coder on the sending side and video decoder on the receiving side). The most used video codecs nowadays are based on the MPEG-4 standard. Before this came its predecessors, MPEG-2 targeted to DVD-quality video, which was also used for transmission over telecommunication networks, and MPEG-1, which was used for recording video on storage media only (not suitable for transmission over telecommunication networks).

The MPEG standards, from MPEG-1 and further in MPEG-2 and MPEG-4, follow the same pattern regarding compression of the video content. Typically the video is organized in series of pictures (called frames) which appear 25 or 30 times per second (at equidistant time intervals), according to the legacy TV standards from the twentieth century. Each such picture is compressed by removing its redundancy with respect to other parts of the same picture or in regard to other neighboring pictures. In MPEG standards there are reference pictures, which are not coded in respect to the other pictures (i.e. frames), and such frames are highest in size and are referred to as "I" frames (according to the MPEG terminology). Besides the reference frames, there are "P" pictures, which are compressed by removing temporal redundancy in respect to neighboring "I" frames. Finally, MPEG also has "B" pictures in which temporal redundancy in respect to neighboring "P" and "I" frames is removed. Hence, "B" frames are on average smallest, while "P" frames have sizes between the size of "B" and "I" frames, and "I" frames have the largest average size. Of course, all pictures are also spatially compressed by removing as much as possible of their own redundancy.

Why is the approach used in video codecs such as MPEG important for video traffic? From the 1990s onwards almost all video codecs were based on various versions of the MPEG standards. For example, MPEG-4 is currently the most used video coding standard (it also includes audio coding, where audio typically accompanies the video traffic) and it absorbs all features of the previous MPEG standards (note that there is no MPEG-3 standard). In fact, MPEG-4 is an umbrella standard, which was standardized in 1998 by ISO (International Organization for Standardization); since then many new parts of the standards have been standardized (over 30 such parts of MPEG-4 were standardized by ISO in the first two decades after the initial standard and their number continues to grow). Then, bearing in mind the type of pictures in MPEG, i.e. "I", "P," and "B" frames, one may conclude that there will be variations in sizes of frames of different types as well as of the same type due to different video content and different compression possibilities. So, all digital video content is compressed with source (statistical) coding techniques before transmission (or such video files are stored on video content servers), which results in highly varying bitrates for video.

Because video traffic has high peak to mean ratio, in telecommunication networks it may be shaped (up to predefined maximum bitrates), which is typical in cases when video content over managed IP-based networks (not part of the public Internet) is provided with QoS guarantees end-to-end (e.g. IPTV services provided by telecom operators). Such video traffic shaping is shown in Figure 2.5.

At the receiving side, the application (e.g. video player) buffers the video data with the aim of ensuring consistency in the reproduction of the video content. So, playback point is introduced with a certain delay in respect of the average delay of the packets to compensate for the delay variation of most of the packets. Figure 2.6 shows the typical positioning of the playback point and typical probability distribution function of IP packet delays for video traffic. It is up to the video player to calculate the playback point to provide smooth watching of the content by the end-user (however, this is dependent upon the availability of consistent bitrate end-to-end from the video sender downstream to the video player of the end-user).

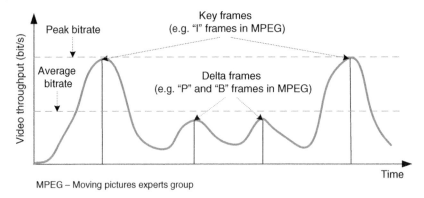

Figure 2.5 Video traffic shaping and effect of the packet delay.

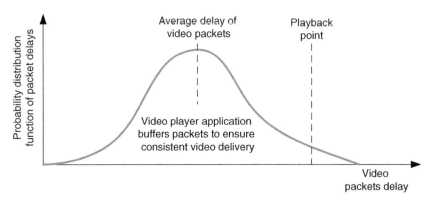

Figure 2.6 Video playback point.

Most of the live video traffic is based on the RTP/UDP/IP protocol stack. However, with the availability of broadband access to the Internet, many video sharing sites provide content by using the HTTP/TCP/IP protocol stack, with longer delays for playback point due to retransmission of all lost segments by TCP. Also, proprietary transport protocol implementations, based on certain characteristic of UDP (e.g. datagram transfer) and TCP (e.g. acknowledgments and retransmissions of lost IP packets), are being used for video traffic (e.g. for IPTV). However, video streaming (as unidirectional video, from video servers toward the video player of the end-user) is also considered as a real-time service, although it has less strict end-to-end delay requirements when compared with voice (e.g. it can tolerate delays up to several seconds).

2.3.3 Non-Real-Time Traffic Characterization

NRT traffic has no strict requirements on delay. Such traffic may include applications which are interactive (e.g. Web, FTP) or message based (e.g. email). Most of this NRT traffic is carried by using TCP/IP protocol stack, so it has lossless transmission end-to-end, which is provided by TCP on the transport layer at the end hosts (e.g. Web browser and Web server as two end sides for Web communication). NRT is considered to be one of the main reasons for Internet congestion at bottlenecks in the networks for

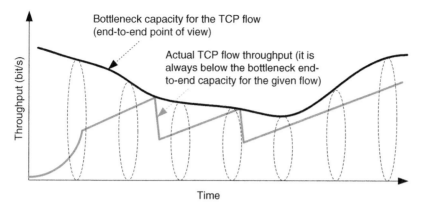

Figure 2.7 TCP behavior at congestion.

a given connection end-to-end (a bottleneck may appear anywhere in access, core, or transit networks). Congestion typically appears on network interfaces when all packets that should be sent through them cannot be sent in the required time interval, which results in long buffer queues and finally in packet losses (packets are dropped when the buffers are filled over a certain threshold).

What are the reasons for Internet congestion? Well, most Internet traffic is TCP-based (and the rest is UDP) because the dominant type of traffic is from the WWW and Web-based services, including here also video and picture sharing sites, and other multimedia sharing websites. TCP congestion control mechanisms assure maximal utilization of bottleneck (Figure 2.7), so TCP traffic flow consumes all available bitrate (i.e. bandwidth) which is limited by the bottleneck anywhere on the end-to-end path of the packets. If there are 10 TCP file downloads (e.g. Web based) over a 10 Mbit/s link (as a bottleneck link), then after enough time each of the TCP flows will have up to approximately 1/10th of the total available bandwidth (i.e. bitrate). TCP mechanisms do not allow TCP to use 100% of any link bandwidth because CWs in most TCP implementations are never constant – either they rise until a packet loss occurs, or they decrease after losses (e.g. due to fast retransmission and fast recovery, or due to TCP timeout and then slow start phase).

TCP mechanisms at the end-hosts are completely independent of the functions of the network nodes (e.g. switches and routers) and hence accidentally may cause congestion. That results in long queues without a control mechanism (under assumed heavy congestion at the given network node). The congestion creates back-to-back packet losses, which may trigger two or more losses in a given TCP CW (for each TCP connection which goes through that congestion bottleneck). This arises from multiple loss behavior of TCP, and may result in TCP timeouts and entering the slow start phase, which results in higher oscillations in the bitrate of a given connection. In an attempt to avoid multiple packet losses from a single TCP connection as much as possible, network nodes (such as routers) implement random discard of packets (from various positions in the buffer queue) when the buffer is filled over a certain upper congestion threshold (e.g. 80% filled buffers).

Summarizing Internet congestion, it is mainly due to TCP, which is implemented and controlled by end-user devices (i.e. hosts, including clients and servers). Then the question is how to provide QoS guarantees to this NRT traffic, which is TCP-based. That can be accomplished either by adding certain QoS mechanisms in an initially best-effort Internet network or by ensuring that sufficient capacity is available.

So, something that was the main Internet networking paradigm (simple network which uses the best-effort principle for transfer of packets, and complex hosts which deal with congestion that may occur anywhere along the path between the endpoints of communication by using TCP at the end-hosts) is becoming one of the main obstacles for provision of QoS in the network end-to-end for NRT traffic.

2.4 QoS on Different Protocols Layers

The networking protocols in all-IP telecommunication networks are based on IP on the network layer and on TCP or UDP on the transport protocol layer. The lowest protocol layer in the protocol stack which functions end to end is the network layer, where the IP is positioned in all hosts and network nodes (e.g. routers). End-to-end means that the sending hosts creates an IP packet which may be received as a whole (IP packet) through the Internet by the receiving host. Of course, certain types of network nodes may change the contents of the IP packet, such as NAT (network address translation) gateways or proxies along the path. So, network layer QoS is related to the IP in both IPv4 and IPv6. However, protocols over and below the IP layer influence the QoS being provided.

The performance of IP-based services depends upon the performance of other protocol layers below or above the IP. Regarding the lower protocol layers (below IP), these belong to certain links (e.g. between two network elements such as routers and switches, or between network elements and end-hosts) which provide connection-oriented or connectionless transport for the IP layer (i.e. IP packets). In general, links between hosts and/or nodes in telecommunication networks may be based on different types of technologies, such as Ethernet, WiFi, mobile RANs, and optical links.

Higher protocol layers (above IP) include the transport protocol layer and above (e.g. presentation, session, and application layers, usually treated on the best-effort Internet as a single application layer), which include protocols (on those layers) such as TCP, UDP, RTP on the transport layer, HTTP, FTP, SMTP, POP3, and proprietary (not standardized) applications on the application layer. Higher layers may also influence the end-to-end performance (e.g. TCP provides congestion control end-to-end by the end-hosts, while RTP provides two-way control information needed for real-time services). Transport and network layer protocols are implemented in the OSs of the end-user equipment, so their implementation (e.g. TCP implementation) may influence the behavior of applications and services on the top. For example, different OSs may provide different user experiences at the same network conditions (throughput, congestion, etc.) due to different versions of TCP.

There are interrelationships of the QoS parameters on different layers, such as the link layer, network layer, and application layer (Figure 2.8). For example, link layer performance influences QoS, including the technology used for the given link (e.g. whether it is Ethernet or mobile access, and so on), as well as its QoS mechanisms and congestion occurrence at a given time period. Network factors, such as packet loss, jitter, or delay, in different networks as packets travel between the end-hosts influence the application factors, which may include overall packet loss, overall delay, end-to-end average jitter,

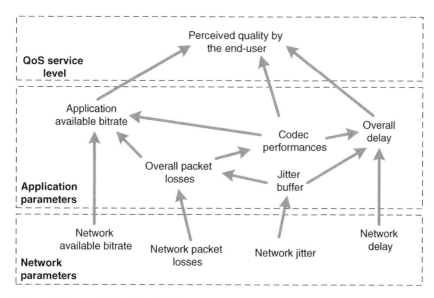

Figure 2.8 Interrelationship of QoS parameters.

and codec performance (for the given media type, such as audio or video). The over-all end-to-end application factors directly influence the perceived quality, which may be measured and discussed only end-to-end, at the point where the user is located. So, one may note that there is some interrelationship among QoS parameters on different protocol layers.

2.5 Traffic Management Techniques

When there is heterogeneous traffic over the same IP network, the issue is what to do with different traffic types on public or/and non-public networks. On one side, the Internet was initially designed to provide best-effort service, i.e. all IP packets are treated in the same manner. On the other side, not all packets are the same. For example, the Web as an interactive application is delay sensitive, but voice is more sensitive to delay and jitter. Video requirements are highest regarding bitrate because video is one of the most demanding traffic types on the Internet. Also, online games are delay and jitter sensitive, even more than voice. Some P2P applications for file sharing, where BitTorrent was the most used at the beginning of the twenty-first century [17], are generally insensitive to delay and jitter, while bandwidth does not matter a lot for them, although it is still impor-tant (e.g. file transfer will finish eventually, but having higher average download bitrate will help the user to be more satisfied with it). There are other examples, but the main points are given. This "big picture" refers to the public Internet, which is also referred to as Internet access service by many regulators and telecom operators.

When all IP traffic is multiplexed onto the same IP-based links and IP networks, including traditional telephony (as VoIP) and television (as IPTV), then the network needs to have techniques to give better quality to some packets for certain types of applications/services.

2.5.1 Classification of IP Packets

How do we classify IP packets? Well, it depends on the approach used, and there are different options on the table, which may include (but are not limited to) the following:

- *Classification based on ports.* Ports are used by transport layer protocols such as TCP and UDP to identity applications on the top. For example, port 80 (HTTP, i.e. the Web) may take precedence over port 21 (FTP).
- *Classification based on application type.* Different applications may be treated differently. For example, carrier grade VoIP takes precedence over HTTP, BitTorrent, and other OTT services used over Internet access service.
- *Classification based on user type.* There are typically different types of users of telecommunication services. For example, home and business users may get normal service, but hospitals, police, or fire departments get highest priority service and no mandatory requirement for users' authentication (e.g. emergency services number 112 in Europe, or 911 in the U.S.).
- *Classification based on subscription.* Different users may pay different prices for the same services with different QoS. For example, a user may need to pay $50 for high-speed Internet access (with guaranteed maximum access bitrates in downstream and upstream) and $10 for fair-usage policy-based Internet (e.g. bitrates are proportionally downgraded with higher usage).

Besides the above classification of IP packets and sessions or calls to which they belong, there are other options and possibilities which may be created for classification of user traffic and hence IP packets. Regardless of the approach used, it is always implemented via certain traffic management techniques. Thus there is technical classification of IP packets which is similar to port-based classification.

2.5.2 Packet Classification From the Technical Side

Packet classification is needed to sort packets into flows and then to group the flows into classes. This is further needed for scheduling purposes, such as FIFO (first in first out), PQ (priority queuing), or some of many other techniques. Considering the Internet networking principles, based on identification of a connection with the IP address and the port number (regarding both sides of the connection, the sending side and the receiving side), usually the following fields are used for classification (called 5-tuple):

- source address (from the IP header), which identifies the network interface of the sender;
- destination address (from the IP header), which identifies the network interface of the recipient;
- protocol type, such as TCP or UDP (read from the field "protocol number" in the IPv4 header or the "next header" field in the IPv6 header);
- source port (from the transport protocol header, e.g. the TCP or UDP port number);
- destination port (from the transport protocol header, e.g. the TCP or UDP port number).

In IPv6 networks, classification can be performed also with 3-tuple in cases where "Flow Label" in the IPv6 header is used. In such cases the 3-tuple is the:

- source IPv6 address;
- destination IPv6 address;
- "Flow Label" field from the IPv6 header, which identifies the flow, and in which case there is no need for use of port numbers from the transport layer protocol header.

After classification of IP packets, the next step is packet marking. The packets are typically classified and marked at the edge routers. After an IP packet enters the core network, core routers check the mark applied by the edge routers.

2.5.3 Packet Scheduling

Packet scheduling typically refers to mechanisms in network nodes (e.g. routers) regarding how to schedule the IP packets on the same outgoing interface (toward the next hop, which is the next router on their path). There are many scheduling mechanisms defined, and subsets of them are implemented by different vendors (in network equipment, such as routers, switches, mobile base stations). The typical default scheduling scheme is FIFO (sometimes called FCFS (first come first served)), which gives priority to the IP packets which first enter the buffer queue for the given outgoing network interface. FIFO suits the best-effort nature of Internet traffic, with the initial low cost of the networking equipment because no additional traffic management mechanisms are deployed.

However, when broadband IP access is used for both managed IP traffic (with QoS guarantees) and best-effort traffic over the Internet access service (for access to the public Internet), then there is a need for differentiation of traffic in the scheduling mechanism, so IP packets belonging to one traffic type (or class) will be served before packets of other traffic types (or classes). For example, delay-sensitive applications and services, which are referred to as real-time traffic, should be given priority over NRT traffic, which can tolerate longer delays (although not too long for most of the NRT services, for example Web browsing). Priority queuing is used for this purpose.

Priority queuing is the simplest and most commonly technique used to differentiate between traffic types (e.g. VoIP, IPTV, and best-effort Internet traffic). It places traffic into different queues and serves them in priority order (e.g. high priority for VoIP, medium for IPTV, and low for best-effort packets, as shown in Figure 2.9).

Overall, there is no single best scheduling mechanism. Which is best depends upon the traffic being scheduled, offered service packages to end-users, available bandwidth, traffic management techniques being used (e.g. traffic shaping for limiting the bitrate allocated to a given traffic type of a given IP flow), etc. Therefore, typically, scheduling mechanisms are not usually standardized; rather, it is left to the vendors to provide different scheduling options in the network equipment, and then to network designers on the side of telecom operators (i.e. network providers) to create traffic management solutions that best fit the given network and the telecom business regarding the offered services to their customers.

2.5.4 Admission Control

No matter how good the scheduler is, one may still have no QoS in practice. Why? Because when traffic demand exceeds available resources (i.e. bandwidth – bitrates

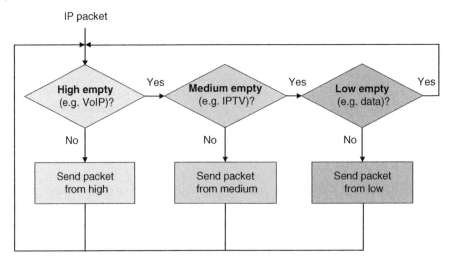

Figure 2.9 Priority queuing for VoIP, IPTV, and Internet access service.

in an uplink or downlink direction), there will be congestion, which means that QoS requirements cannot be maintained. With the aim of providing QoS, it is always necessary to provide more resources than those that can be spent (also referred to as overprovisioning) or to provide admission control (which means that the number of simultaneous connections over a given network link is limited, subject to the network design).

What is admission control? It is a traffic management technique which is used to reject a flow (a call or a session) if there are insufficient resources (e.g. in the access, core, or transit network). It is different from traffic policing (which is used to reject/drop a packet). For example, in the traditional telephone networks (e.g. PSTN, Public Land Mobile Network (PLMN), it is easy to manage admission control. Why? Due to circuit-switching (allocation of a pair of channels per call, i.e. fixed resources per call). Then, teletraffic engineering provides an efficient means for planning the telephone networks (e.g. the well-known Erlang-B formula for calculation of the number of channels required to serve certain traffic intensity under the given maximum acceptable new call blocking rate).

Internet flows from different applications/services have different QoS requirements. Therefore, admission control is much harder for implementation in IP-based environments where the same network is used for transfer of packets from many different services, each with potentially different QoS requirements and hence different admission control approaches.

Let us analyze a typical example at a telecom operator when multiple services are offered though broadband access, such as VoIP service, IPTV service, and Internet access service. If an average requirement per VoIP is 100 Kbit/s in each direction, IPTV service with standard definition (SD) quality requires up to 3 Mbit/s reserved bitrate (with traffic shaping of the video flow) per piece of end-user equipment (e.g. IPTV user equipment, such as a setup box), and there is 10 Mbit/s per second overall in the downlink direction, then there can be a maximum of three pieces of IPTV user equipment connected, with the aim of guaranteeing QoS for the IPTV. Also, more bandwidth dedicated to IPTV will result in less bandwidth (bitrate) in the downlink direction for the Internet access service. On the other side, services and applications offered through the Internet access services are not subject to admission control due to the network neutrality principle for the public Internet, and hence there is no admission control for the number of flows that can be established over it (e.g. a user can open different videos even in different browser tabs, and when the available bandwidth is consumed it will result in congestion, which will cause videos to freeze at a certain time period or constantly).

Overall, admission control is directly related to QoS provisioning per flow or per traffic class (where a class denotes traffic aggregated from many flows from the same traffic type carried over the same links or networks).

2.5.5 Traffic Management Versus Network Capacity

The extent and complexity of traffic management are associated with how much congestion is being experienced or how close network traffic is to the limit of the network's capacity, as given in Figure 2.10.

When traffic demand is much lower than the installed network capacity, there is no need for intervention. When there is moderate "space" between demand and available network capacity, simple traffic management is needed, which includes bandwidth allocation and packet prioritization. When traffic demand approaches network capacity, the probability of traffic congestion increases significantly. In that case, traffic management is strongly required to improve QoS.

In the long term, from a technical point of view, one may expect traffic management to remain as the main toolset for a response to congestion, as well as to maintain the highest possible level of QoS. One of the most important changes will be that packet

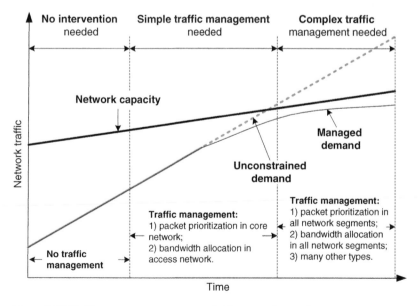

Figure 2.10 Traffic management versus available network capacity.

Table 2.1 Traffic management: positive and negative effects on QoS.

	Positive effects on QoS	Negative effects on QoS
Traffic management applied to user's traffic	Can guarantee or prioritize data for delay-sensitive or bandwidth-demanding applications	Can restrict traffic from certain applications
Traffic management applied to other users' traffic	Can reduce congestion, allowing fair use for all users	Other users' traffic may take priority over user's own traffic

inspection capabilities will migrate out of the core network toward the access networks, enabling a more finely graded and user-specific form of traffic management [24].

Overall, there are positive and negative effects of traffic management applied in the network, which are summarized in Table 2.1 [24].

2.6 Internet QoS Frameworks: the IETF and the ITU

Internet technologies are standardized by the IETF (e.g. IP, TCP, UDP, HTTP, DHCP, DNS, i.e. all important protocols on the Internet are Internet standards). The standards are required with the aim of making different vendors of hardware (e.g. network equipment, user equipment) and software (e.g. OSs, some applications such as Web browsers) interoperable.

The IETF standardized several QoS solutions in the 1990s, which provided an early Internet QoS standardization landscape. According to those standards and default traffic management principles used on the public Internet, one may distinguish an IETF Internet QoS framework consisting of the following approaches (including accepted scenarios and standardized mechanisms):

- *Best effort.* This is the traditional Internet model, without any QoS guarantees. The IP networks simply route packets until they reach their destination.
- *Integrated services (IntServ)* [25]. This was the first mechanism standardized by the IETF, based on resource reservation in routers on the path by using signaling. So, routers have to store traffic and QoS information per connection (i.e. per flow). It is an end-to-end QoS mechanism.
- *Differentiated services (DiffServ)* [26]. This is the most commonly used method for traffic differentiation, in which all packets are classified into a limited number of classes, so routers have to store only information per class (not per connection/flow). It is a hop-by-hop QoS mechanism.
- *Multi-protocol label switching (MPLS)* [27]. This is the "default" approach for QoS provisioning in transport IP networks, which may be combined with DiffServ as well as other protocols such as BGP, or VPN protocols.
- *Policy-based QoS.* This is typically used for QoS provisioning between network providers (e.g. BGP policies are used for routing of packets between interconnected ASs).

The ITU has also defined a QoS framework for telecommunication packet-switched networks (where the Internet and Internet networking technologies belong). As usual,

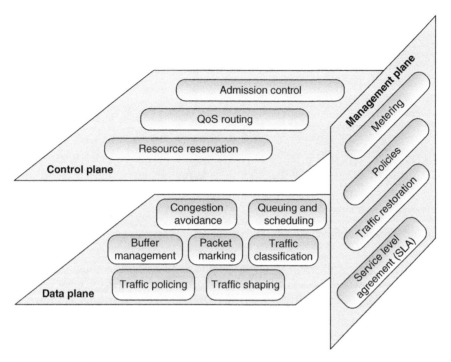

Figure 2.11 QoS architectural framework by the ITU.

since the PSTN/ISDN era, the ITU-T QoS architectural framework is organized into three planes [28] (Figure 2.11):

- *Control plane.* This includes the admission control, QoS routing, and resource reservation.
- *Data plane.* This includes several traffic management aspects such as buffer management, congestion avoidance, packet marking, queuing and scheduling, traffic classification, traffic policing, and traffic shaping.
- *Management plane.* This includes service level agreement (SLA), traffic restoration, metering and recording, and policies.

So, overall the ITU QoS framework includes all IETF standards on Internet QoS and puts them into the well-established telecommunication concept, which is based on the three planes. This way there is synergy between the standardization efforts on QoS by both standards development organizations (SDOs), the IETF and the ITU, which is beneficial to telecommunication/ICT networks and services which are converging on the Internet networking principles and Internet technologies for service provisioning.

2.7 Integrated Services (IntServ) and Differentiated Services (DiffServ)

Integrated Services (IntServ) is an architecture for providing QoS guarantees in IP networks for individual application sessions. It relies on use of Resource Reservation Protocol (RSVP) [29] and routers need to maintain records of allocated resources

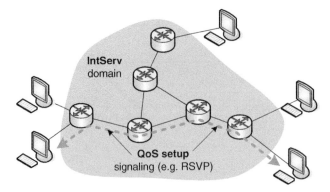

RSVP – Resource recervation protocol

Figure 2.12 QoS call setup signaling for integrated service (IntServ).

and to respond to new call setup signaling requests. In IntServ architecture every router between the two end-hosts which needs to establish a session must reserve required resources (e.g. bandwidth) by using the QoS call setup signaling, as shown in Figure 2.12.

Another IETF QoS standard on the network layer is the Differentiated Services (Diff-Serv) architecture [30], also defined in the 1990s (in 1998, after IntServ). DiffServ divides the Internet traffic into different classes, called behavior aggregates (BAs), and gives them differentiated treatment regarding QoS provisioning. It is based on per-hop behavior (PHB), which is in fact the externally observable forwarding behavior applied at a DiffServ-compliant node to a DiffServ behavior aggregate. The IETF's development of DiffServ in the 1990s was intended to address the following difficulties with IntServ and RSVP:

- *Scalability*. Maintaining states by routers in high-speed networks is difficult due to the very large number of flows.
- *Flexible service models*. IntServ includes only two classes, and there was a need to provide more qualitative service classes and "relative" service distinction.
- *Simpler signaling (compared with RSVP)*. Many applications and users may only want to specify a more qualitative notion of service.

Which are DiffServ functions? Well, it provides only simple functions in the core network (forwarding according to the PHB) and relatively complex functions at edge routers (or hosts). A DiffServ-capable host or first DiffServ-capable router requires the following actions (Figure 2.13):

- *Classification*. Network edge node marks packets according to classification rules to be specified (manually set by network administrators, or by some protocol);
- *Traffic conditioning*. Network edge node may delay and then forward, or they may discard packets.

In DiffServ, IP packets are marked in the IP headers, so this is truly a network-layer QoS mechanism (there is no need to read or change upper layer headers such as protocols on the transport or application layer). The marking in IP headers is completed by

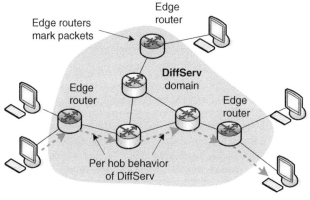

DiffServ – Differentiated services

Figure 2.13 Differentiated services (DiffServ) functions.

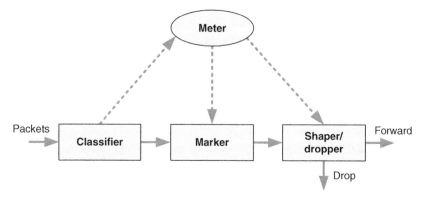

Figure 2.14 Classification and conditioning of IP traffic.

the edge routers in the DiffServ network domain (that is the network that implements DiffServ architecture) by marking specified fields in the IP headers:

- ToS field in the IPv4 header, and
- DSCP in the IPv6 header.

It may be desirable to limit the traffic injection bitrate of some classes, in which case the user declares the traffic profile (e.g. rate and burst size). Then IP traffic is metered and shaped if non-conforming (Figure 2.14).

Table 2.2 compares best-effort, DiffServ, and IntServ as QoS architectures. Best-effort is simplest, but it does not provide any QoS differentiation between calls/sessions or aggregates of IP traffic. However, if there is enough bandwidth available, best-effort can also provide services with satisfactory quality to end-users. When comparing IntServ and DiffServ, the main difference is in the scalability of the two approaches. DiffServ is much more scalable because all routers keep information per traffic class (the number of traffic classes is limited and it is not related to traffic volume), while in the IntServ architecture the network nodes (routers) make first reservation of resources regarding required QoS and then maintain this information in network nodes on a per flow basis,

Table 2.2 Comparison of best-effort, DiffServ, and IntServ.

	Best-effort	DiffServ	IntServ
Service	Connectivity No isolation No guarantees	Per aggregate isolation Per aggregate guarantee	Per flow isolation Per flow guarantee
Service scope	End-to-end	Domain	End-to-end
Complexity	No setup	Long-term setup	Per flow setup
Scalability	Highly scalable (nodes maintain only routing state)	Scalable (edge routers maintain per aggregate state; core routers per class state)	Not scalable (each router maintains per flow state)

which may lead to scalability problems (e.g. when there are many flows in parallel, which is typical in core and transport or transit networks).

2.8 QoS with Multi-Protocol Label Switching (MPLS)

The main IETF approach standardized for QoS traffic management in transport networks is MPLS, with its standard defined in 2001 [27].

In IP transport networks with implemented MPLS (a typical scenario in transport IP-based networks in the first two decades of the twenty-first century), label switching is used instead of routing based on IP addresses. So, QoS is not implemented with DiffServ fields in the IPv4 or IPv6 headers, but by adding information via label headers (between the MAC header on layer 2 and the IP header on layer 3). MPLS works between OSI layers 2 and 3 (below the IP and above the MAC) and adds labels over the IP packet.

In IP/MPLS networks, label switching path (LSP) technology is used to pre-provision a virtual multi-protocol label switching transport network (VMTN) for each service type (VoIP, IPTV, best-effort Internet traffic, corporate VPN, etc.). This can be performed manually (by setting the forwarding tables for MPLS) or automatically (by using RSVP for MPLS traffic engineering (RSVP-TE) [31], or Constraint-based Label Distribution Protocol (CR-LDP), [32], DiffServ-enabled MPLS traffic engineering (DS-TE) for optimization purposes [33], as well as BGP/MPLS with IP VPNs [34] for IP traffic exchange (e.g. peering) between network providers (e.g. telecom operators).

Regarding the MPLS network architecture, in short, fixed-length labels are associated to streams of data in the following manner (Figure 2.15):

- In the ingress edge MPLS router, called LER (label edge router), labels are added to IP streams (based on different constraints for traffic differentiation at the network edges).
- Packet belonging to those streams are forwarded along the LSPs based on their labels (not on IP addresses), so there is no IP header examination in LSRs (label switching routers), which are used for packet switching in the MPLS network domain (that is, network segment or a whole network which uses and understands the labels added to the packets by LERs).

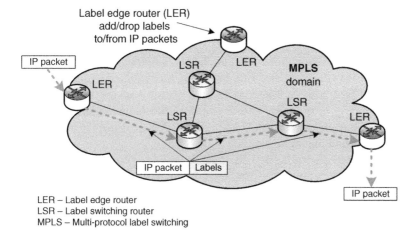

Figure 2.15 MPLS network architecture.

QoS technologies based on the differentiated services architecture by using MPLS with VPN are widely deployed within the transport networks of many providers.

2.9 Deep Packet Inspection (DPI)

Deep packet inspection is a type of packet filtering targeted at all headers and all payloads from network layer (layer 3) up to application layer (layer 7), i.e. all layers that make up the end-to-end communication between the end-hosts. However, in some cases layer 2 can be considered also in this group of layers for inspection. An illustration of DPI is given in Figure 2.16.

The reasons for packet inspections may vary from case to case, so the targets may include finding viruses, spam, intrusions, or certain protocol non-compliance, and so on. However, there can be other criteria associated with DPI filtering rules, which may include security as well as QoS topics.

In fact, DPI is directly related to the firewall concept, where firewalls have both types of implementation (stateful and stateless). Stateless refers to packet filtering where each packet is independently filtered in respect to other IP packets. Stateful inspection deals with the state of a given connection (e.g. between two end-hosts, a client and a server).

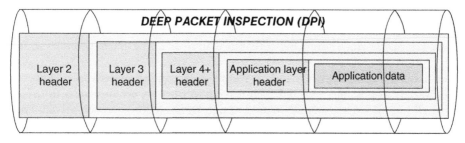

Figure 2.16 Deep packet inspection (DPI).

The state of the connections is a kind of vague definition because it very much depends upon the protocols being used (e.g. application layer protocols). In this regard, the DPI is in fact a stateful and dynamic packet inspection which identifies many applications and protocols (many times doing it with certain heuristic methods, i.e. not standardized ones, and methods being proprietary and developed by equipment vendors). DPI typically uses TCP and UDP port numbers as well as IP addresses in IP headers for matching packets belonging to the same connection (e.g. 5-tuple can identify any IP connection), but also it may inspect non-TCP and non-UDP IP protocols by using data packet inspections for certain matching values.

What is the potential QoS role of DPI? Well, it can be used to provide classification of individual sessions and their application/traffic type, which may be needed for traffic management. Also, it can be used to provide differentiated charging (e.g. user traffic from/to certain OTT application not to be charged). It can be used to locate and then throttle some traffic (e.g. traffic from P2P file sharing), thus providing better QoS for the other OTT services offered via public Internet access.

Overall, one may note that DPI is implemented in different ways by different vendors of Internet network equipment (e.g. routers, gateways). However, the importance of DPI is also considered by the ITU-T in its next generation network (NGN) series of recommendations. The requirements regarding DPI in NGN [35] are targeted (but are not limited to) to perform application identification by inspecting application payload, to support various kinds of policy rules, to be in line with IETF standards for IP flow information exports [36], and so on. The potential use of DPI for QoS and performance measurements according to the ITU's framework for DPI [37] is given in Table 2.3.

As one may note from Table 2.3, the DPI framework includes different depths of packet inspection, considering all headers from layer 2 up to layer 7 (according to the OSI protocol layering model with seven layers), and application payload (which is added by the sending application on the sender's side, and used by the receiving application at the recipient's side). So, one may distinguish deep header inspection (DHI), based on inspection of all headers from layer 3 to layer 7, and deep application identification (DAI), based on layer 7 payload inspection. Performance measurement (locally or remotely) uses DHI and DAI (only layer 2 is excluded here because it refers to the link-layer header, not to a header or payload which is transferred end-to-end). For QoS support, such as traffic management techniques, DHI can be used for traffic-shaping purposes. Further, for network analysis regarding user behavior or usage patterns, headers from layer 3 up and layer 7 as well as application payload can be used. Similar use of DPI can be considered for SLAs between the end-user as a customer and the service provider. Inspection of all packet headers including layer 2 (the link layer which is limited to a network link or segment) is usually performed for layer 2 related policy conditions. In general, one may distinguish [37] between link-layer DPI (limited to the L2 network domain) and network-layer DPI (which covers network information on L3, i.e. the network layer, and higher protocol layers).

Overall, the use of DPI is not mandatory in telecommunication networks. However, it has a variety of possible uses and then it is up to the network designers to decide whether DPI is needed and should be used or not. DPI also draws certain privacy concerns regarding end-users, so legislation may be needed to regulate the purposes for which DPI can be used and the purposes for which it is not allowed (e.g. for sniffing user behavior and the type of content a given user consumes over the public Internet).

Table 2.3 Framework for DPI on QoS.

Network application	Deep packet inspection (DPI)			
		Deep header inspection (DHI)		Deep application identification (DAI)
		Shallow packet inspection (SPI)	Medium-depth packet inspection (MPI)	
	Layer 2 header inspection (L_2HI)	Layer 3, 4 header inspection ($L_{3,4}$HI)	Layer 4+ header inspection (L_{4+}HI)	Layer 7 payload inspection
Performance measurements (key performance indicators [KPIs])				
Collection of remote measurements	—	X	X	X
Generation of local measurements	—	X	X	X
Usage parameter monitoring				
Service level agreements	—	X	X	X
Traffic parameter control	—	X	X	X
L3 policing (peak rate, sustainable rate, burst size)	—	X	—	—
L7 policing ("application payload" size, byte rate)	—	X	X	—
QoS support				
Traffic shaping (L3 byte shaping)	—	X	—	—
Network analysis				
User behavior	—	X	X	X
Usage patterns	—	X	X	X
Charging/billing support				
Time-based information	—	X	—	—
Traffic volume-based information	—	X	—	X
Event-based information (associated with content)	—	X	X	X
Link-oriented DPI				
Layer 2-related policy conditions	X	X	X	X

2.10 Basic Inter-Provider QoS Model

The QoS model which is used inside the given network provider's domain (e.g. given telecom operator, either fixed or mobile) is targeted to a limited number of services offered to end-users (e.g. VoIP as a replacement for PSTN/PLMN telephony, and data service which is in fact Internet access service). One may add IPTV as a service; however, this is typically packetized (this refers to transformation of digital TV into an IPTV stream based on Internet technologies) inside the telecom operator's own network. Also, there are services offered to business customers via VPNs as a replacement for leased lines from the PSTN era. However, the number of QoS classes is not a constant value for every network (QoS classes will be discussed in more detail in the following chapters). In different networks the traffic can be grouped in different numbers of QoS classes, depending upon operators' service offerings. That creates a slightly different picture: consider QoS for a network of a single network provider and for the interconnection between providers.

2.10.1 Basic DiffServ Model for a Single Provider

In the case of a single network provider for the whole path of a connection (session/call), customer sites connect to the provider network via CE (customer edge) devices [38]. The provider's routers which connect to customer sites are referred to as PE (provider edge) devices (or nodes).

The CE router can therefore prioritize and police the traffic based on the DiffServ concept (classifying and marking the traffic at the CE toward the PE, and vice versa from PE to CE), so different classes will be treated differently on the link (or network segment) between CE and PE on both ends of the connection. This basic DiffServ model for a single network provider is shown in Figure 2.17.

2.10.2 Basic DiffServ Inter-Provider Model

In this section the DiffServ model is extended to a simple inter-provider scenario, as shown in Figure 2.18. In general, there can be multiple network providers on the path of the user information transfer between two endpoints (two CEs). However, the simple inter-provider scenario is to have just two providers in the path between the two sites of each customer, which are interconnected via (as a typical example) an IPX.

With the aim of providing end-to-end QoS for the given connection, packets from the end-users should be correctly marked on the inter-provider link and avoid modifications of packets. The two providers, A and B, should have a peering agreement

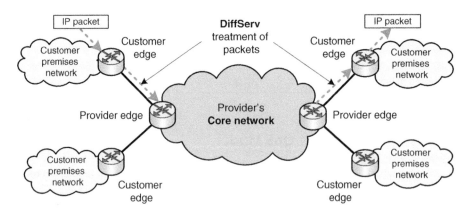

Figure 2.17 Basic model for a single provider.

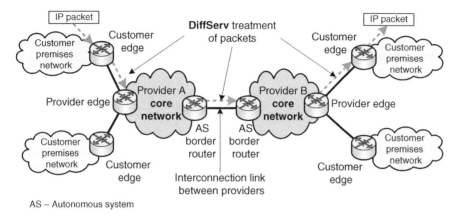

Figure 2.18 Basic inter-provider model.

for an end-to-end service with the appropriate QoS mechanism applied on their interconnection. Also, when there is a need for monitoring of the customer (of either of the two providers in this scenario), such action is likely to involve both providers.

Which solution can be used for the interconnection between two network providers? Well, the interconnection should be possible with MPLS VPN [34], with a simple IP interconnection (with differentiation on the network layer, e.g. by using DiffServ fields in the IP headers), as well as with a layer 2 interconnection.

In general, in Internet networking philosophy, each of the networks which is a separate administrative domain (consisting of one or more ASs) is completely autonomous to choose traffic management mechanisms and QoS techniques, including the number of classes used inside the network provider's administrative domain. So, it is possible for one provider to group (aggregate) the traffic in three classes only (as an example), while the other operator may distinguish among five or six (or more) classes. However, at the interconnection point between the network providers there is a need to map classes which are being used by one provider to classes which are being used by the other provider and vice versa. This is usually implemented at the edge routers of each of the network providers.

2.11 IP Network Architectures for End-to-End QoS

There is a fundamental difficulty in the IP-based platforms and networks due to heterogeneity in all aspects, including network infrastructures, type of applications and services carried through the network, management infrastructure, and end-user devices. Simple use of IP as a transport technology does not mean networks and platforms are the same or even compatible. So, QoS is not assured by itself in IP networks, especially for traffic being transported through multiple network providers. There is an excessive number of different configurations among different networks, although all of them are using IP (and Internet networking principles) as a transport means. Thus it is hard to provide certain QoS for different services from one end to the other, simply because each network has different mechanisms and the levels of control and provision are also different. For example, as shown in Figure 2.19, one network

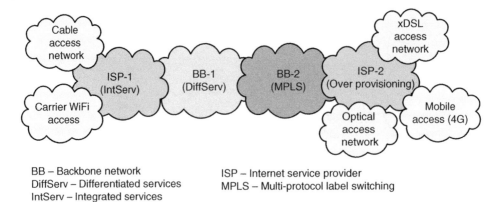

BB – Backbone network
DiffServ – Differentiated services
IntServ – Integrated services

ISP – Internet service provider
MPLS – Multi-protocol label switching

Figure 2.19 IP networks with heterogeneous QoS mechanisms.

provider may use an IntServ-based approach, another one on the packets' route may use the DiffServ approach for QoS provisioning, a third one may use MPLS traffic engineering, while a fourth may use over-provisioning (without explicit QoS support for given traffic types).

Possible mapping between DiffServ layer 3 marking, MPLS class of services, and Ethernet priority code points (for Ethernet versions which support it) is shown in Table 2.4 [39].

The basic network model (as given in the previous subsection) should be applied commonly to the providers, but the detailed QoS technologies to compose such a network model would be different for different providers and countries. Additionally, there are differences in the regulation environment, which is a fundamental framework for provision of QoS and QoE.

Table 2.4 Mapping between DiffServ, MPLS, and Ethernet.

Packet network QoS class	Description	Layer 3 packet marking: DSCP (DiffServ code point)	Layer 2 packet marking		Applications
			MPLS (class of service)	Ethernet (priority code point)	
Classes 0, 1	Jitter sensitive	EF (expedited forward)	5	5 (default) or 6	Telephony
Classes 2, 3, 4	Low latency	AF (assured forward)	4, 3 or 2	4, 3 or 2	Signaling, interactive data
Class 5	Best efforts	DF (default forward)	0	0	Web browsing, email

2.12   Discussion

Internet QoS was a very important topic in the 1990s when the two main IETF standards on QoS were developed, IntServ and DiffServ. They are not widely deployed today exactly as standardized, yet their concepts are still alive.

Internet capacity exploded in the 2000s with the spread of broadband access networks to residential users, via either fixed access (e.g. x digital subscriber line [xDSL] technologies over twisted pair copper lines, cable access, or fiber access) or mobile access (e.g. 3G, 4G, 5G mobile access networks). When there is a lot of capacity (although one should note that "a lot" is a relative term which refers only to a given point in time), packets are dropped only when capacity is reached. Then, QoS solutions can be very useful when capacity is saturated.

With the expansion of the Internet in the 2000s, a huge number of fiber links have been deployed, which give a lot of network capacity. Then, when there is lots of spare capacity, there is less requirement for Internet QoS end-to-end. However, carrier-grade VoIP and IPTV, as well as VPNs as a replacement for leased lines for business users, require QoS support end-to-end because Internet traffic is unpredictable (bursty by nature, with a high degree of self-similarity [40]). In short, IP-based networks cannot be planned in the same manner as PSTNs, which were developed for a single ToS, the voice (i.e. telephony).

Additionally, emerging cloud computing and the Internet of Things may need certain QoS guarantees for specific types of users (business users, industry, etc.). Also, there is a need to guarantee QoS for access to the Internet as a whole. That is not access to particular applications over open Internet access (a particular application can be the Web, Skype, Viber, a social networking application, P2P applications such as BitTorrent, etc.) but to Internet access as a service provided by telecom operators (as ISPs) and used by end-users to access all applications and services available through the public Internet network.

IP does not provide QoS mechanisms, but it has defined the ToS field in IPv4 and the DSCP field in IPv6 to specify QoS requirements on precedence, delay, throughput, and reliability, or to provide traffic differentiation. Overall, Internet architecture influences the QoS provisioning end-to-end. ASs are autonomous in every aspect regarding applied routing protocols and traffic management; only inter-AS routing is standardized to go with a single routing protocol (that is BGP-4).

Which is the default (and native) QoS in Internet? Well, the default QoS approach is the best-effort approach (i.e. no QoS). That means every IP network gives its best-effort to transfer every IP packet that travels through it, however that is done, without any guarantees about whether the packet will reach its destination and what will be the quality parameters associated with it (such as delay, losses). To introduce QoS in what was initially a no-QoS Internet, the framework included IntServ (as the oldest), DiffServ (most used as a concept), and MPLS-TE (the most used approach for QoS provisioning in transport IP networks) [41]. Nevertheless, other proprietary vendor solutions (e.g. DPI) for Internet QoS are also being deployed in some IP networks.

References

1 RFC 791, Internet Protocol, IETF, September 1981.

2 RFC 2460, Internet Protocol, Version 6 (IPv6), December 1998.

3 RFC 768, User Datagram Protocol, August 1980.

4 RFC 793, Transmission Control Protocol, September 1981.

5 RFC 4960, Stream Control Transmission Protocol, September 2007.

6 RFC 4340, Datagram Congestion Control Protocol (DCCP), March 2006.

7 RFC 5944, IP Mobility Support for IPv4, Revised, November 2010.

8 RFC 6275, Mobility Support in IPv6, July 2011.

9 RFC 2581, TCP Slow Start, Congestion Avoidance, Fast Retransmit, and Fast Recovery Algorithms, January 1997.

10 RFC 5681, TCP Congestion Control, September 2009.

11 RFC 3550, RTP: A Transport Protocol for Real-Time Applications, IETF, July 2003.

12 RFC 2131, Dynamic Host Configuration Protocol, IETF, March 1997.

13 RFC 3315, Dynamic Host Configuration Protocol for IPv6 (DHCPv6), IETF, July 2003.

14 ITU-T Recommendation E.164, The International Public Telecommunication Numbering Plan, November 2010.

15 RFC 1035, Domain Names – Implementation and Specification, November 1987.

16 RFC 7766, DNS Transport over TCP – Implementation Requirements, IETF, March 2016.

17 Janevski, T. (2015). *Internet Technologies for Fixed and Mobile Networks*. Norwood, MA: Artech House.

18 RFC 2616, Hypertext Transfer Protocol – HTTP/1.1, IETF, June 1999.

19 RFC 7540, Hypertext Transfer Protocol Version 2 (HTTP/2), IETF, May 2015.

20 Janevski, T. (2014). *NGN Architectures, Protocols and Services*. Chichester: Wiley.

21 http://bgp.potaroo.net, BGP Routing Table Analysis Reports, accessed in November 2017.

22 Massachusetts Institute of Technology, Department of Electrical Engineering and Computer Science, Wide-Area Internet Routing, January 2009.

23 ITU-T Recommendation G.114, One-Way Transmission Time, May 2003.

24 Ofcom UK, Traffic Management and Quality of Experience, 2011.

25 RFC 1633, Integrated Services in the Internet Architecture: An Overview, June 1994.

26 RFC 2474, Definition of the Differentiated Services Field (DS Field) in the IPv4 and IPv6 Headers, IETF, December 1998.

27 RFC 3031, Multiprotocol Label Switching Architecture, IETF January 2001.

28 ITU-T Recommendation Y.1291, An Architectural Framework for Support of Quality of Service in Packet Networks, May 2004.

29 RFC 2205, Resource ReSerVation Protocol (RSVP), IETF, September 1997.

30 RFC 2475, An Architecture for Differentiated Services, IETF, 1998.

31 RFC 3209, RSVP-TE: Extensions to RSVP for LSP Tunnels, IETF, December 2001.

32 RFC 3212, Constraint-Based LSP Setup Using LDP, IETF, January 2002.

33 RFC 4124, Protocol Extensions for Support of DiffServ-aware MPLS Traffic Engineering, IETF, June 2005.

34 RFC 4364, BGP/MPLS IP Virtual Private Networks (VPNs), IETF, February 2006.

35 ITU-T Recommendation Y.2770, Requirements For Deep Packet Inspection In Next Generation Networks, November 2012.

36 RFC 5102, Information Model for IP Flow Information Export, IETF, January 2008.

37 ITU-T Recommendation Y.2771, Framework for Deep Packet Inspection, July 2014.

38 ITU-T Supplement 8 to E.800 series, Guidelines for Inter-provider Quality of Service, November 2009.

39 ITU-T Recommendation Y.1545, Roadmap for the Quality of Service of Interconnected Networks That Use the Internet Protocol, May 2013.

40 Janevski, T. (2003). *Traffic Analysis and Design of Wireless IP Networks*. Norwood, MA: Artech House.

41 ITU-T Technical Paper, How to Increase QoS/QoE of IP-based Platform(s) to Regionally Agreed Standards, March 2013.

3

QoS in NGN and Future Networks

The quality of service is always defined end-to-end. The traditional approach in ICT worlds is based on support of end-to-end QoS in the networks by using globally accepted standards, such as ITU standards for telephony. However, the Internet has changed the telecom game by separating network on to one side and services on the other side, which provided the possibility to invent new services on the run without the need to change the underlying telecommunication network infrastructure. That was also one of the reasons for Internet technologies to become main or only packet-based networks of the present time, including fixed and mobile access networks. However, many services and applications over the same IP-based networks make the QoS provisioning more difficult. There are millions of different applications in different ecosystems (e.g. in Google's Play Store, Apple's iStore), so theoretically it is not possible to guarantee the QoS per individual application, which come and go. However, some of the services and applications are legacy ones (e.g. those transferred to IP environments from legacy or traditional telecom networks, such as voice and television), while other applications and services can be considered as Internet-born ones. However, not all of them get the massive number of end-users because end-users (as humans) are influenced by different factors, including trends, price, affiliation, experience, region, personal philosophy, business requirements, time in which we are talking about services, and so on. Further, besides humans as end-users of services (either human-human or human-machine communication), machine-type communications are continuously emerging and spreading, also referred to as the Internet of Things (IoT) concept, which is based and will be based on the same IP network infrastructures. This is also referred to as the ICT world going into different verticals by connecting different things to the Internet, for the sake of humanity, to make life better and more pleasant as the main target (smart homes, smart cars, smart cities, etc.) [1].

There are standardized umbrella specifications for such a transition of the telecom world to all-IP environments and to different verticals; these are NGNs and future networks umbrella by the ITU-T [2].

3.1 ITU's Next Generation Networks

The QoS in telecommunications networks at the beginning of the twenty-first century is directly linked to the standardization of the NGN by the ITU-T [1]. Why? Because QoS end-to-end is directly related to signaling, and signaling end-to-end needs to be

QoS for Fixed and Mobile Ultra-Broadband, First Edition. Toni Janevski.
© 2019 John Wiley & Sons Ltd. Published 2019 by John Wiley & Sons Ltd.

standardized globally to be the same in all telecommunication networks of all telecom operators, fixed or mobile, in each country. For example, end-to-end signaling is needed for services such as voice (to locate the end-user device to deliver a call to it) or television. Although the signaling for voice was created by the IETF as SIP (Session Initiation Protocol), there were other options such as the H.323 standard from the ITU-T for call signaling. However, if every vendor and every operator implements some standardized signaling (or other) protocols and mechanisms, there can be problems with interoperability for signaling between different operator networks, which are autonomous regarding each other. End-to-end signaling in a standardized manner, and also end-to-end QoS provisioning in a standardized manner (which is based largely on signaling as a necessary "tool"), are accomplished with NGN standardization on a global scale. Hence, NGN networks implemented by telecom operators provide the base for interoperability between networks based on standards and offer the possibility of having end-to-end QoS for those services that need it.

What have NGNs brought to the telecom world? They have introduced the Internet concept (the winning concept in packet-based networks) of separation of services and underlying networks (including all types of networks and services). That provides the possibility of having different types of access networks, such as fixed access (copper or fiber based) accompanied with Ethernet or WiFi access in the last meters, and mobile access networks (e.g. from 2G to 5G and beyond).

NGNs are all-IP networks, something that is realized as a convergence of PSTN/ PLMN toward the Internet as a single networking platform for all types of services, the existing ones as well as future ones. However, one may note that traditional telecommunication networks have influenced the NGN network concept, particularly for QoS-enabled end-to-end VoIP and IPTV.

Traditional telecommunication networks (before the appearance of NGNs) were designed primarily to carry circuit-switched digital voice by using unified 64 kbit/s bitstream in each direction for a single voice connection (ITU-T G.711 voice codec standard). Such networks were based on simple telephone devices (without any computational capabilities), circuit-switched transport networks based on time division multiplexing (TDM) with 64 kbit/s "capacity slots" (e.g. SDH/SONET (synchronous optical networking) transport networks), complex and expensive digital exchanges located in telecom operators' networks, and usage of globally unified Signaling System No. 7 (SS7, standardized by ITU-T). As shown in Figure 3.1, the evolution from PSTN and PLMN toward NGNs is performed by gradual replacement of TDM-based transport systems and interfaces, circuit-switching network nodes (e.g. telephone exchanges), and SS7 signaling (which is packet-based, separated from the user traffic, but is not IP-based) with IP/MPLS transport networks, media gateways, and IP-based signaling (e.g. SIP), respectively.

NGN specifies IP on network layer end-to-end, and such requirement defines the all-IP network concept. That means that all access, core (or backbone), and transit networks in NGN are IP based. On one side, IP hides the lower protocol layers (i.e. OSI-1 and OSI-2) from the upper layers, so an NGN can have single core and transport network infrastructure for different fixed and wireless access networks. In the all-IP network principle, the interconnection between networks is also done via IP links established between pairs of gateway routers or controllers.

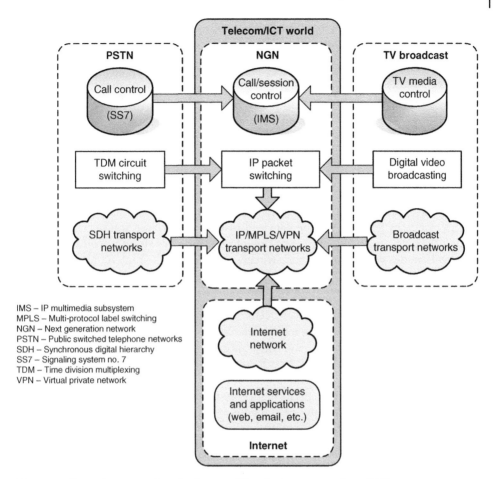

Figure 3.1 Network concept changing from traditional telecom networks to NGN.

On the other side, IP hides the network interfaces from the upper protocol layers (e.g. the transport and application layers), thus providing single network abstraction (through the socket interface) to the applications running on the top of the IP stack (even in cases when there are several network interfaces on a given host). So, while the Internet was designed with separated transport and service parts, with IP (including both IPv4 and IPv6) as the main protocol between them in each host and each network node connected to the Internet, NGN puts that into a well-defined framework which specifies the functionalities in given network nodes and end-user hosts.

3.2 Transport and Service Stratum of NGNs

Following the Internet philosophy of separation of services/applications and transport networks, the NGN provides the same for the legacy services (voice, TV) over IP networks and some emerging services (e.g. IoT services) [1]. This separation is done in the NGN by definition of two distinct blocks of functionalities (called stratums), which define the basic reference model for NGNs [3]:

- NGN service stratum: It provides user functions which relate to transfer of service-related data (e.g. signaling and control information related to voice, video, data, and all other services) to network-based service functions which manage service resource and network services with a goal of enabling user applications and services. So, the service stratum provides originating and terminating calls/session between end peers (i.e. hosts, network nodes). It typically consists of servers, databases, and service functions on different network elements (e.g. routers, switches, gateways) and user devices.
- NGN transport stratum: It provides functions that transfer various user data. On the network's side, transport stratum provides functions that control and manage transport resources for carrying the data end-to-end. The related transferred data may carry user, management, and/or control information. This stratum typically includes network nodes such as routers and switches and links for their interconnections, with different sets of functions to provide the transfer of all kinds of information (user data, control data, or management data).

Figure 3.2 shows the separation of NGN service and transport stratums. The separation includes all planes in the telecommunication networks: the user plane (refers to data generated or received by the end-user device, such as voice, video, web pages, emails), the control plane (refers to call/session control, such as signaling), and the management plane (refers to communication between entities related to configuration, accounting, fault management, performance, and security management). Overall, both stratums consist of different types of functions.

The main target of the NGN functional architecture is real-time services, including conversational multimedia (e.g. VoIP with video or multimedia telephony) and content delivery services (e.g. IPTV). The NGN functional architecture is shown in Figure 3.3.

The central position in the NGN architecture for control functions is reserved for the network access control functions (NACFs). These interact with other functional entities in the transport stratum (resource and admission control functions (RACFs), mobility management control functions (MMCFs), and several transport functions), with the

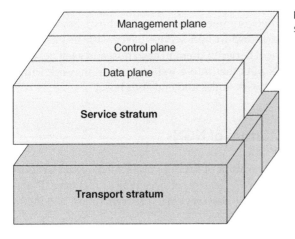

Figure 3.2 NGN transport and service stratums.

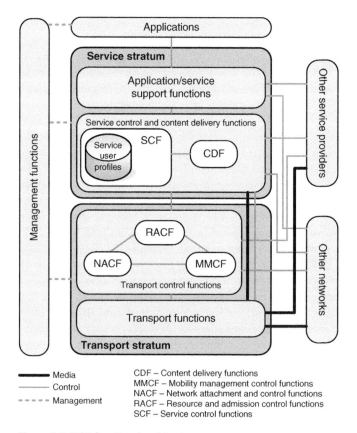

Figure 3.3 NGN functional architecture.

service stratum (service control functions (SCFs)), and with customer equipment. A single NACF can be used for multiple access networks connected to the same NGN core, but it is not required in the NGN architecture (it is a network design choice).

3.3   Service Architecture in NGN

The main part of the service architecture in NGN is targeted at standardized signaling systems as a replacement for SS7 (Signaling System 7), which was the legacy in digital telephone networks (PSTNs). Although SS7 was a packet-based system, it was not an IP-based one. With the transition to all-IP networks, the SS7 traffic started to be tunneled over IP-based core networks (e.g. in many mobile operators' core networks in the 2010s). The standardized signaling in the all-IP environment for legacy as well as for innovative services is based on the Internet Protocol multimedia subsystem (IMS).

IMS was initially standardized with 3GPP Release 5 in a form of an application development environment. 3GPP Release 7 has further directed IMS usage focus toward telephony replacement in IP-based networks, regardless of access type (i.e. fixed or mobile). The common IMS version was standardized in 3GPP Release 8, which implemented different requirements from all other bodies for standardization of mobile and fixed networks (ITU, 3GPP2, etc.). Although IMS is standardized by the 3GPP, it is included as an integral part of the NGN service stratum [4]. Also, it has become a legacy approach for fixed-mobile convergence (FMC) deployments [5] because the IMS is access independent, which means that the same systems can be used for both main types of access networks, fixed access and mobile access. Also, the 3GPP continues with the development of the IMS and its functionalities in further releases after Release 8, which defined the common IMS version [6].

The IMS uses standardized IETF protocols, of which the most important (for the IMS) is SIP [3], as the signaling and control protocol for different services, including multimedia session services (e.g. voice and video telephony services provided by telecom operators) and some non-session services (e.g. message exchange services provided by telecom operators). Overall, the IMS uses SIP for signaling in the same manner as PSTN/ISDN (Integrated Services Digital Network) were using SS7. However, SS7 (in the past) and the NGN framework (at the present time) are standardized by the ITU, SIP and the Diameter protocol are standardized by the IETF, while IMS itself is a 3GPP standard. So, we have three different SDOs with synergy in provision of a common IP networking environment for legacy telecommunication services such as voice services.

In practice, the IMS was initially used for voice services in fixed networks in the first half of the 2010s, and then in mobile networks in the second part of the 2010s. Although the IMS can support a wide range of services and there were efforts for such innovations by major telecom operators during the 2010s, it appears that telecom operators (telcos) cannot match the pace of OTT service providers (e.g. Google, Facebook, Amazon) in the innovation of Internet-based services for end-users (human users). So, although the IMS provides an environment for many new services, it has arrived on the market later and telecom operators usually have longer time from innovation to the market (due to different reasons, such as legislation in a given country, the need to have a sustainable business case, having a big corporate hierarchy, having limited national markets, and so on). The following sections define the main aspects of the IMS functions and its two most important accompanying protocols, SIP and Diameter.

3.3.1 IMS Architecture

The IMS is the main signaling system in NGN, but it has been standardized by 3GPP (3G Partnership Project) and also accepted by the other SDOs in its "common" version (3GPP Release 8). In general, the IMS is access independent, so the same system can be included for both fixed and mobile access networks. It is primarily targeted at services that require end-to-end signaling, which include QoS-enabled VoIP and IPTV.

Figure 3.4 shows the IMS architecture with its functional entities. Note that the words "function" (in 3GPP terminology) and "functional entity" (in NGN terminology) refer to the same functional entities of the IMS.

The main functional entities in the IMS architecture are three types of call session control functions (CSCFs):

- *Proxy-call session control function (P-CSCF)*. This is the first contact point on the IMS side for the user equipment (e.g. mobile terminal, fixed terminal). P-CSCF acts as a SIP proxy node and it interacts with the admission control subsystem to provide authorization of media components that can be provided with appropriate QoS. In

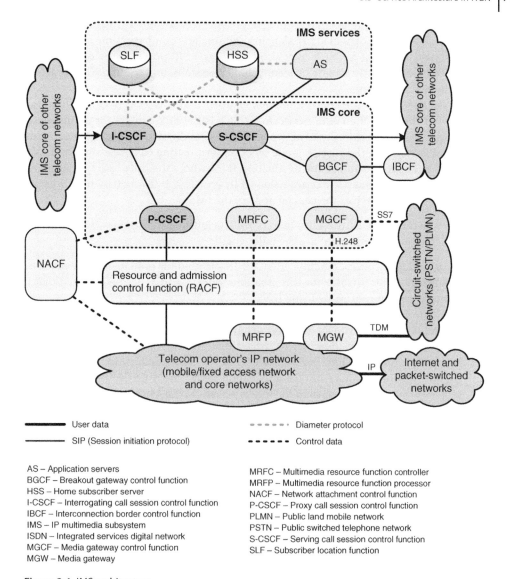

Figure 3.4 IMS architecture.

AS – Application servers
BGCF – Breakout gateway control function
HSS – Home subscriber server
I-CSCF – Interrogating call session control function
IBCF – Interconnection border control function
IMS – IP multimedia subsystem
ISDN – Integrated services digital network
MGCF – Media gateway control function
MGW – Media gateway

MRFC – Multimedia resource function controller
MRFP – Multimedia resource function processor
NACF – Network attachment control function
P-CSCF – Proxy call session control function
PLMN – Public land mobile network
PSTN – Public switched telephone network
S-CSCF – Serving call session control function
SLF – Subscriber location function

many implementations P-CSCF also behaves as a user agent (UA), with the aim of hiding IMS functions from end-user terminals.

- *Serving-call session control function (S-CSCF).* This is the central node in the IMS architecture which has SIP server functionalities. The main function of this node is session/call control. S-CSCF behaves also as a SIP registrar because it accepts registration requests and then makes the registration information available via the location server (that is, the home subscriber server (HSS)). Also, it binds the public IP address (allocated to the network interface of the end-user equipment) with the SIP address. Unlike P-CSCF, S-CSCF is always located in the home network of the subscriber and it uses Diameter protocol to access the HSS database with the aim of obtaining user profiles. A given telecom operator may implement several S-CSCF nodes in the network for the purposes of load balancing and smaller round-trip time for signaling. One should note that S-CSCF may also behave as a SIP proxy server or a SIP user agent to independently initiate or terminate SIP sessions, and to support interaction with service platforms.

- *Interrogating-call session control function (I-CSCF).* This is a SIP location function located at the edge of an IMS network. I-CSCF is the first contact for all incoming messages from other IMS-based networks, so the IP address of the I-CSCF node is the one that is stored in the carrier-grade DNS, so terminating calls/sessions from IMS in another network can find the destination IMS (where the called party is located). However, all SIP requests that are received by an I-CSCF from other networks are routed toward S-CSCF as the central node of the IMS.

Besides the three CSCFs, the "core" part of IMS includes additional functional entities, from which the most important are the following:

- *Breakout gateway control function (BGCF).* This is used for processing user requests for routing from an S-CSCF for the situations when S-CSCF has determined that it cannot use session routing by using DNS.
- *Media gateway controller function (MGCF).* This performs signaling translation between SIP and ISUP (ISDN user part, from the SS7 signaling system in PSTN).
- *Multimedia resource function (MRF).* This is split into two parts:
  - Multimedia resource function controller (MRFC) is a signaling node that interprets the information coming from the application server and S-CSCF, and further uses such information for control of the media stream in the MRFP.
  - Multimedia resource function processor (MRFP) is a user plane node that provides mixing of media streams.

To cover the complete architecture shown in Figure 3.4 one should also note the following IMS entities:

- *Home subscriber server (HSS).* This is the main user database in the IMS architecture that contains home subscriber profiles, subscription information, and the user's network location (i.e. IP address). Also, HSS provides authentication and authorization of users, based on requests from CSCF entities.
- *Subscriber location function (SLF).* This is a resolution function of the IMS that provides information about the HSS which has information on a given IMS user (e.g. upon requests from I-CSCF, S-CSCF, or ASs).
- *Application server.* This hosts and executes a given service, and is using SIP for communication with S-CSCF.
- *Interconnection border control function (IBCF).* This is used as a gateway to external networks, and provides firewall and NAT (network address translation) functionalities in the IMS architecture.

The IMS is used for different services and different access networks which are all-IP based. It can be used in fixed networks such as xDSL networks, passive optical networks (PONs), QoS-enabled Ethernet, cable networks, as well as mobile networks (e.g. Universal Mobile Telecommunication System (UMTS), LTE/LTE-Advanced), and wireless networks (e.g. WiMAX, Worldwide Interoperability for Microwave Access). In telecom

NGN networks with deployed IMS, QoS-enabled services from telecom operators are provided by using signaling through the IMS, which further communicates with RACF for the resource reservations in the NGN architecture. However, the most important signaling protocols used in the IMS and overall in the NGN are SIP and Diameter.

3.3.2 Session Initiation Protocol (SIP)

SIP is a signaling protocol for IP networks, which is standardized by the IETF [6]. SIP appeared in the second half of the 1990s (when IP dominance in the data world became evident) and was finalized at the beginning of the twenty-first century. Although it is already a well-known and mature protocol, its major impact on the telecommunication world becomes more evident with the transition of legacy telecommunication networks to all-IP networks, i.e. to NGN, which has already started to happen in fixed networks and will continue to happen in mobile networks with existing 4G and 5G developments.

SIP can be used as a component with other IETF standardized protocols for multimedia, including HTTP, Real-Time Streaming Protocol (RTSP) [7], and Session Description Protocol (SDP) for describing multimedia sessions [8]. For transmission of real-time multimedia (e.g. voice, video streaming) SIP typically uses RTP [9].

SIP is a client-server protocol, based on SIP requests from the client to the server and responses in the opposite direction. For example, in the case of operator-grade IP telephony, the caller acts as a client and the called party acts as a server. A single call may involve several clients and servers, and a single host can be addressed as a client and as a server for a given call. In general, a SIP caller also may directly contact the called party by using its IP address (previously resolved from the SIP address), which is a peer-to-peer SIP communication. So, SIP can be used either in a client-server or in a peer-to-peer manner.

Generally, SIP is a text-based protocol similar to HTTP (and HTTP was similar to SMTP for email communication). SIP uses messages in a request/response manner similar to HTTP. For such a purpose SIP defines six main messages used for session setup, management, and termination, as given in Table 3.1.

The responses for the SIP messages are used in the same format as responses for HTTP (which appeared before SIP). So, similar to the tradition already established by the IETF

Table 3.1 Main SIP messages.

SIP message	Description
INVITE	SIP session initialization message.
ACK	Acknowledgment response upon receipt of successful SIP request message (e.g. INVITE message).
BYE	SIP session termination message.
CANCEL	To cancel initialization process already started (e.g. it is used when a client sends an INVITE and then changes its decision to call).
REGISTER	Used by SIP user agent (in user equipment) to register its current IP address and the SIP URIs for which the user wants to receive calls.
OPTIONS	Used to request information about capabilities of the caller, while this message does not set up a session by itself (a session is set up only with the INVITE message).

for its standardized protocols on the application layer, SIP does not reinvent the wheel when it is not needed (e.g. for SIP response codes and messages). All response codes are grouped into six groups, where each group of codes is identified by the first digit (response code groups are given in Table 3.2).

To invite a given party to a session, SIP uses its naming/addressing schemes. Since the most used addressing form on the Internet at the time of SIP standardization (i.e. year 2002) was the form of email addresses (e.g. user@FullyQualifiedDomainName) and URI (Uniform Resource Identifier) schemes (e.g. http://www.example.com for WWW), SIP also adopts the URI scheme and user addressing scheme based on general standard syntax (as used for WWW and email). The host name in the URI contains either a fully qualified domain name (FQDN) or an IP address (IPv4 or IPv6 address). The port number is used where necessary. If it is omitted, the default SIP port numbers are used, which are ports 5060 and/or 5061 (for both, TCP and UDP as transport layer protocols). The general form of the SIP URI is the following:

```
sip:{user[:password]@}host[:port][;uri-parameters][?headers]
sips:{user[:password]@}host[:port][;uri-parameters][?headers]
```

Generally, there are two types of SIP network elements, i.e. two types of user agents (UA):

- *User agent client (UAC).* This makes SIP requests and sends them to servers.
- *User agent server (UAS).* This receives the requests, processes them, and returns SIP responses.

Regarding SIP UAs, a single UA can function as both, UAC and UAS, where the roles of either UAC or UAS last only during a given SIP transaction. SIP UA is a logical endpoint in the SIP network architecture, which is used to create or receive SIP messages, used for management of the SIP sessions. On the SIP servers' side, there are several server types defined by the IETF, which include redirect server (used for redirection of SIP client

Table 3.2 SIP response codes and messages.

SIP response code	Description
1xx	Provisional response, used by SIP servers to indicate a progress, but it does not terminate a SIP transaction (e.g. 100 Trying).
2xx	Success response, which means that action was successfully received, understood, and accepted (e.g. 200 OK).
3xx	Redirection response, which indicates that further action is needed for completion of a given request (e.g. 302 Moved Temporarily).
4xx	Client error, which usually means that the request contains bad syntax or cannot be completed by the server (e.g. 404 Not Found).
5xx	Server error, which is server-side error on a given valid request from the client side (e.g. 504 Server Time-out).
6xx	Global failure, which means that the request cannot be fulfilled by any server (e.g. 603 Decline).

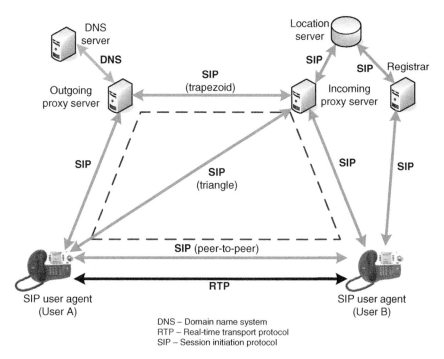

Figure 3.5 SIP network architecture.

requests to corresponding SIP servers), proxy server (used to enforce network policy), registrar server (provides binding IP address of the client with one or more SIP URIs), and location service (database that stores information about user location), as shown in Figure 3.5.

3.3.3 Diameter

Diameter is standardized by the IETF to overtake its predecessor, the RADIUS protocol. It is developed as a framework protocol for authentication, authorization, and accounting (AAA), which works on the application layer. Diameter has been accepted by the 3GPP for the IMS, and with that also in the NGN.

The Diameter protocol was initially specified by the IETF in RFC 3588, but a new version appeared in 2012 with the RFC 6733 [10]. Its name is influenced by the name of its predecessor, RADIUS – the diameter of a circle is twice larger than the radius, which points to the idea that Diameter is created to function "two times" better than RADIUS.

Diameter is a message-based protocol (every such message is transported in a separate IP packet). There are defined two types of Diameter messages: request messages and answer messages [10]. Each Diameter message contains a header followed by the packet body, which consists of one or more attribute-value pairs (AVPs). An AVP is used to encapsulate protocol-specific data, which may include (but is not limited to) QoS information, traffic information, routing information, AAA information, etc.

The Diameter is defined as a framework for different applications to use for different purposes (so it is an extensible protocol over time). For differentiation between applications, IANA allocates an Application ID (32 bits) which is placed in the Diameter packet (i.e. message) header.

The Diameter network architecture consists of the end-host peers (clients and servers) and Diameter agents. Overall, one may distinguish among four types of Diameter agents:

- *Relay agent.* It is used to route a message to other Diameter nodes using the routing information in the message (e.g. Destination-Realm AVP). This node does not change the Diameter message, only advertises its Application ID.
- *Proxy agent.* It performs the same Diameter messages routing function as the relay agent, but this node can also change Diameter messages. Proxies must maintain the states of their downstream peers (i.e. devices), while enforcing of given policies by proxy agents is usually application dependent.
- *Redirect agent.* It just replies to requests with responses, without routing or forwarding Diameter messages to other nodes.
- *Translation agent.* It translates RADIUS messages to Diameter messages and vice versa.

An example of the Diameter architecture is shown in Figure 3.6, with all four types of nodes. However, when a relay agent or a proxy agent can route the Diameter messages by using their own Diameter routing table, a redirect agent is not needed.

Diameter includes a number of AVPs related to QoS. For example, the Traffic Model (TMOD) AVP [11] is used to describe the traffic source based on the following parameters: token rate, bucket depth, peak traffic rate, minimum policed unit, and maximum packet size.

Diameter also has a standardized QoS application [12], targeted to allow network entities to interact with Diameter servers for the purposes of allocation of QoS resources in the given network. There are two modes of operation defined:

- *Pull mode.* In this mode the network element requests QoS authorization from the Diameter server which is based on certain trigger information (e.g. from a QoS signaling protocol).
- *Push mode.* In this mode the Diameter server proactively sends a command to the given network elements with the aim of installing the QoS authorization state. The trigger for this mode can be done by off-path signaling protocol, where SIP is a typical example.

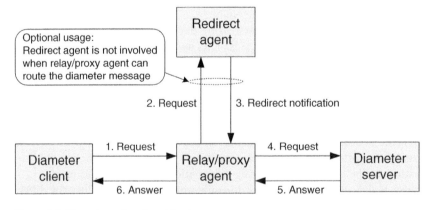

Figure 3.6 Diameter network architecture.

There is a set of Diameter codes defined for the QoS application which intends to support both push and pull modes. The QoS application of Diameter runs in a given QoS framework, as shown in Figure 3.7. The Diameter QoS application functions between network elements (e.g. QoS-aware router that acts as a Diameter client, which triggers QoS information with Pull mode and receives QoS information in Push mode) and application entity (that is a Diameter server that supports the QoS application, and it can authorize QoS requests for a flow or aggregate of a given application). The AAA cloud in Figure 3.7 represents a cloud of Diameter servers, which communicate with each other. Network elements request authorization through the AAA cloud based on the incoming QoS requests. The authorizing entity (AE) selects the appropriate mode (push or pull) after receiving the QoS request. The AAA cloud (Figure 3.7) may also include business relations needed in cases of interconnections between network providers or between the given network operators and application providers as third parties.

Generally, NGN uses the Diameter as the main protocol for AAA and user subscription management (i.e. it is the main protocol for communication with the HSS, the users' database in NGNs).

Finally, SIP and Diameter are standardized by the IETF, and have become crucial elements in the IMS developed by 3GPP and in NGN developed by the ITU. So, they can be noted as examples of the synergy between SDOs which started in the twenty-first century and is targeted at having unified standards on a global scale, especially important for the control plane in the telecommunication systems where SIP and Diameter belong. From the NGN perspective, both protocols belong in the service stratum, together with the IMS as their system umbrella. The control plane is crucial for end-to-end QoS provisioning because it is directly related to service profiles of users in the system and signaling in the control plane regarding resource allocation for a given call or session (or simply a flow, either treated individually or aggregated with the same class flows).

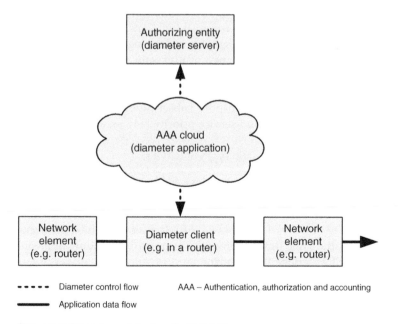

Figure 3.7 Diameter architecture for QoS AAA.

3.4 QoS Architectures for NGN

How does NGN provide QoS? It has two stratums (service and transport), which perfectly corresponds to the Internet approach where applications/services are independent of the underlying transport infrastructure. With such an approach, QoS provision must be made via certain functions between the two stratums because the service stratum maintains QoS requirements from different services, while the transport stratum applies QoS requirements (coming from the service stratum) by using certain Internet QoS techniques in network nodes (e.g. switches, routers). The main entity for QoS in NGN is the resource and admission control function.

3.4.1 Resource and Admission Control Function

QoS functions in NGN are defined through the standardization of the RACF [13]. The RACF is the central functional entity in NGN regarding QOS, with further communication with other functional entities dependent upon the type of access network (e.g. Ethernet, cable, xDSL, Fiber To The x – FTTx, mobile access networks).

The RACF is a "link" between the SCFs and transport functions in NGN. Its main goal is to provide admission control regarding given calls/sessions to/from end-users. So, a user connection can be admitted in the transport network or rejected (or modified) according to transport subscription information, service level agreements, different policies and service priorities, etc. The RACF hides network topology from the service stratum, thus providing an abstract view of transport architecture to SCFs. The RACF is not mandatory in NGN, such as in cases in which admission control is not needed (e.g. best-effort Internet traffic, such as the Web). On the other side, the RACF interacts with SCFs for applications that require certain QoS support, which results in a need for admission control (e.g. QoS-enabled VoIP or IPTV). Besides that, RACF also provides network address and port translation (NAPT) traversal functions which are necessary when an NGN operator uses private IP addresses for addressing end-user terminals. Finally, RACF interacts with NACF for exchange of user subscription information, as well as with RACF in other NGN for delivering services over multiple network or service providers.

RACF policy decisions are based on transport subscription information, SLA, network policy rules, admission control, and service priority [14], as well as transport resource status and policy information.

Functional architecture consists of two main types of RACFs:

- *Policy decision functional entity (PD-FE).* It provides a single contact point to the SCF (i.e. the service stratum) and RACF hides all details of the transport infrastructure and transport functions.
- *Transport resource control functional entity (TRC-FE).* It deals with different transport technologies and at the same time provides the resource-based admission control decision. The TRC-FE is service-independent.

The RACF is connected via its PD-FE with the NACF. Only optionally, when QoS support for mobile nodes is required (e.g. in mobile networks), the RACF is also connected with MMCF.

When QoS is a matter of interest, as in the case of RACF, the performance measurements become crucial, with aim of following QoS provisioning by measuring the network performance of different network segments (e.g. access network, core network, transit network). For that purpose there is an umbrella standard for management for performance measurement (MPM) in NGN, which is also connected to the RACF.

The PD-FE is connected to SCFs in the service stratum and to transport functions in the transport stratum, in particular to the policy enforcement functional entity (PE-FE). The PE-FE is packet-to-packet gateway located at the boundary of different networks, or between the core and access networks.

The TRC-FE is used in the transport stratum only. It instructs the transport resource enforcement functional entity (TRE-FE) to enforce the transport resource policy rules, which are technology-dependent. Hence, the TRE-FE differs for different transport technologies, including access and core networks.

3.4.2   Ethernet QoS for NGN

Ethernet has already become the dominant access technology in all corporate and residential environments. Note also that WiFi is the wireless extension of Ethernet.

However, Ethernet is developing even further, from the access to the core and transport networks (e.g. metro Ethernet or, in other words, carrier-grade Ethernet). In such cases the all-Ethernet network (on layer 2) with all-IP network (on layer 3) is becoming the reality on the two endpoints of the communication sessions, end-users on one end (either humans or machines) and data centers (where typically servers and data bases are located) on the other. Of course, communication can also be peer-to-peer (e.g. between users connected via Ethernet on both ends of the communication end-to-end session). Ethernet is used for all fixed access networks in the last meters (at home or office or at public places). The mobile access networks are based on completely different technologies than Ethernet or WiFi on the lowest two layers of the protocol stack (i.e. physical and data-link layers, below the network layer where IP is positioned). Due to the widespread use of Ethernet access networks in LAN or MAN environments, Ethernet-based NGN is becoming increasingly important as a network architecture. In practice, main network nodes such as routers and gateways (which are routers with additional functionalities) have primarily Ethernet physical ports for interconnection with other network elements (other routers, multiplexers, etc.).

The main idea for Ethernet in NGN is to use the Ethernet frame format throughout the NGN networks end-to-end, without frame conversion. However, for the purpose of QoS provisioning Ethernet frames (in the header) must contain information about the QoS, that is to have implemented IEEE 802.1Q standard. Ethernet-based NGN is intended for use of Ethernet technology over any physical media, including fixed and wireless (e.g. WiFi) environments (in the transport stratum of the NGN) in access, core, and transport networks [15], as shown in Figure 3.8.

One should note that there are requirements to provide end-to-end Ethernet, such as user equipment needing to support the same Ethernet format for various existing technologies, including capabilities for operation and maintenance, load sharing, protection and restoration; VPNs; virtual LANs; auto-configuration capability, security control; QoS mapping between the core and access networks (and vice versa); as well as the possibility for traffic management over Ethernet (needed to guarantee QoS).

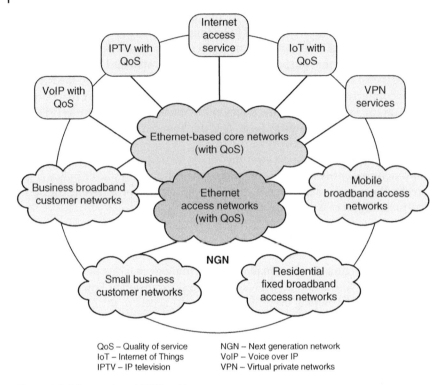

QoS – Quality of service   NGN – Next generation network
IoT – Internet of Things   VoIP – Voice over IP
IPTV – IP television   VPN – Virtual private networks

Figure 3.8 Ethernet-based NGN architecture.

The interface between different operator networks is the Ethernet Network-Network Interface (E-NNI). The Ethernet User to Network Interface (E-UNI) includes the IEEE 802.1Q standard. The Carrier Ethernet uses Ethernet virtual connections (EVCs) in one of three possible service types:

- *Ethernet line (E-line)*: point-to-point EVC
- *Ethernet local area network (E-LAN)*: multipoint-to-multipoint EVC
- *Ethernet tree (E-tree)*: routed multipoint EVC.

In Ethernet-based NGNs the EVC extends between two E-UNIs. Overall, the EVC-based services are provisioned by using several IEEE standards, from which the most important for Carrier Ethernet are the following three:

- IEEE 802.1Q for virtual LAN [16]: each virtual local area network (VLAN) is uniquely identified within the given Ethernet network by a Q-tag (it is also referred to as VLAN ID). The Q-tag field (i.e. VLAN ID) has a length of 12 bits, which results in 4096 different values (from which two are reserved for network administration).
- IEEE 802.1ad for provider bridges [17]: these add a tag in the Ethernet frame, called an S-tag. Similar to the Q-tag, its length is 12 bits, which results in 4094 possible instances (2 of 4096 values for the S-tag are reserved also for administration).
- IEEE 802.1ah for provider backbone bridges (PBBs) [18]: they give additional source and destination medium access control (MAC) addresses in Ethernet frames (at the UNI) when they enter the PBB network to transfer from one provider's network to

another provider's network via a backbone network. A PBB uses an additional tag, called an I-tag (it has a length of 24 bits), to identify the services (i.e. Q-in-Q networks), thus separating the PBB network (as a backbone Ethernet network) and service providers' networks.

Besides the EVC-based services, the typical approach for providing Ethernet end-to-end connectivity is by using VPN. The VPN provides multipoint-to-multipoint Ethernet service that can span multiple metro Ethernet areas by providing connectivity between such multiple sites as if they were attached to the same Ethernet LAN. Simply, VPN encapsulates Ethernet frames (i.e. MAC frames) into IP packets and tunnels them from one Ethernet LAN or metro Ethernet network on one site to all other sites (and other VPN tunnels are established for the traffic in the opposite direction). With this approach, when a telecom operator uses MPLS in its transport network, VPN goes over MPLS [19]. The reference model for using VPNs for the end-to-end Ethernet service is shown in Figure 3.9.

3.4.2.1   QoS Services in Ethernet-based NGN

Ethernet QoS services for NGN include premium service, gold service, and best-effort service. They are created according to the traffic parameters, provider edge rules, transfer capability (regarding the bandwidth for the service), and ITU QoS class for IP-based services.

Providing QoS for Ethernet-based NGN requires connection admission control (CAC); however, this is operator specific [15]. Once a given end-to-end connection (over Ethernet-based NGN) has been accepted, the network operator's policies

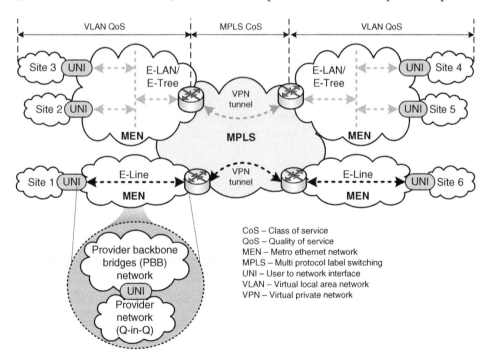

Figure 3.9 Reference model for VPN use for end-to-end Ethernet service.

determine further the settings of the CAC values and QoS parameters at E-UNIs and E-NNIs. The edge nodes at these interfaces implement the mechanisms for traffic management of Ethernet QoS-enabled services by specifying the network edge rules, traffic parameters, QoS class mappings between interconnected networks, and other QoS parameters in use. The mapping between Ethernet QoS services and ITU QoS classes is given in Table 3.3.

When a telecom operator has implemented the NGN, then Ethernet traffic management functions are based on Ethernet-based NGN services. In such cases SCFs determine the requirements for a given request for Ethernet-based NGN service and then inform the RACF. The RACF then does the policy-based resource control in the transport stratum (for the given request). Further, the transport functions in the access and core networks perform the traffic conditioning (as part of the QoS control) based on the policy rules given by the RACF. The architecture for Ethernet traffic management is shown in Figure 3.10.

Based on the policy rules received by the RACF, the transport nodes (e.g. switches), which belong to the NGN transport stratum, are performing the following functions:

- *Metering*. This is used to ensure conformance of Ethernet frames at the interfaces (e.g. E-UNI).
- *Classification*. For Ethernet services this is based on layer 2 ID information, such as different tags used in Ethernet MAC frames (e.g. VLAN ID, i.e. Q-tag, S-tag, or I-tag, as well as use of higher layer IDs such as IP addresses on protocol layer 3). For flow classification Ethernet tag values (e.g. VLAN IDs) are mapped onto labels (used as IDs between layer 2 and layer 3 in the MPLS networks) when the traffic is transported over MPLS transport networks.
- *Marking*. Based on the metering Ethernet frames can be colored, recolored, or dropped. So, frames are marked with the appropriate class in the network to which they are being forwarded.
- *Dropping/shaping*. Non-conformant Ethernet frames are either marked or dropped.

However, the Internet as a network consists of interconnected IP networks, where the main network elements are routers and the packets are being routed based on their IP addresses (i.e. protocol layer 3 IDs), while the layer 2 IDs are LAN (or sometimes MAN)

Table 3.3 Ethernet QoS services mapping.

Ethernet QoS service type	Traffic parameters	Edge rule	Bandwidth	ITU QoS class for IP traffic
Premium	Constant bit rate and committed burst size, no excess bitrate or bursts	Non-conformant frames are dropped	Dedicated	Classes 0, 1, 6, and 7
Gold	Constant bit rate and committed burst size, excess bitrate and bursts allowed	Excess frames are admitted with high discard precedence	Delay-sensitive statistical	Classes 2, 3, and 4
Best-effort	No guaranteed bitrate, no guaranteed burst size	All frames are admitted, and the first is dropped at congestion	Best-effort	Class 5

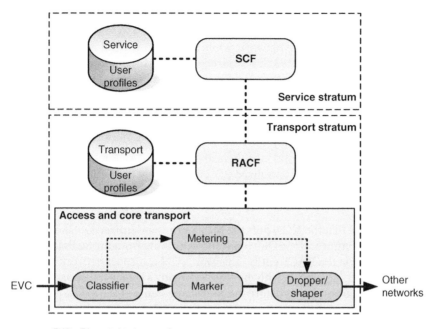

EVC – Ethernet virtual connections
RACF – Resource and admission control functions
SCF – Service control function

Figure 3.10 Architecture for traffic management of Ethernet-based services in NGN.

limited. Each router has its own scheduling and buffer management schemes which are also referred to as nodal behavior. Although theoretically IP networks are based on per hop behaviors (PHBs), that is, each router routes the packets on the next hop (toward the next router) based on certain policies or metrics (e.g. least cost routing), in practice (in telecom operators' networks) that is not really the case in the 2010s. Why? Because each telecom operator does its own network planning and traffic engineering (which influence the capital investments, as well as operational costs for operation and maintenance of the network equipment) and hence puts predefined routes in the network and traffic is aggregated (e.g. in VPNs) and after that is routed (or switched, in the case of MPLS or Ethernet-based NGN) through the network. So, in practice we have pre-established tunnels (over IP networks) or paths (over MPLS-based transport networks) between pairs of nodes in each core and transport network of the telecom operators.

3.4.3 Multi-Protocol Label Switching (MPLS)

The main transport network technology from the beginning of the twenty-first century has been IP/MPLS in different variants (or shortly MPLS). NGN is the definitive standardized approach for QoS provision which is based on the RACF between the service and transport stratums (to transfer service requests from the service stratum into concrete traffic management policies and triggered mechanisms in networks nodes in the transport stratum). A combination of MPLS and NGN (and with it the RACF) results in two main architectures:

- Centralized RACF architecture [20]: In this case there is a single centralized RACF in the core network, which typically consists of PD-FE and TRC-FE. TRC-FE monitors all traffic in the MPLS core network, while admission control is carried by the PD-FE. The flows are mapped into LSPs at the edge of the MPLS network. Bandwidth reservation in LSPs or tunnels is performed in two modes:
  - *Static mode*: reservation of predetermined bandwidth associated with a given LSP or tunnel.
  - *Dynamic mode*: aggregate bandwidth in a given LSP is adjusted with each new admitted or released media flow.
- Distributed RACF architecture [21]: In this case the two RACF functional entities are separated. There is a single centralized PD-FE and many distributed TRC-FEs at the edges of the MPLS core network. Each flow is mapped into an LSP or a tunnel in the same manner as in centralized RACF architecture.

Overall, one may note that the NGN puts the MPLS in a standardized QoS framework within the transport stratum and the QoS requests are coming from the SCFs (e.g. IMS) in the service stratum via the RACF entity. This way NGN transport network provides the possibility for higher integration with the MPLS (already a legacy transport technology), which was on the ground even before the NGN appeared, but the MPLS continues to be the main QoS technology for transport networks of telecom operators, combined with Ethernet technologies in local and metropolitan areas.

3.5 Management of Performance Measurements in NGN

Performance measurements in NGNs are important, with the aim of providing QoS support in heterogeneous network environments consisting of many interconnected autonomous domains, as well as heterogeneous services with different QoS requirements that are provided using the same IP-based transport networks. For that purpose there is a defined general reference network model for performance management in NGNs, which is in line with the basic network model for QoS analysis (see Figure 2.4 in Chapter 2).

So, performance measurements in NGNs on the path end-to-end involve two customer premises networks (CPNs), the access networks, and one or multiple core networks, as well as zero or multiple transit networks. In such a model, customer equipment is connected to the access networks, then access networks are connected to core networks, and different core networks are connected via transit networks. In this manner, on the end-to-end path, the networks are partitioned into several network segments and domains, where each of them is responsible for maintaining QoS support in the given segment. Overall, NGNs support three types of QoS-based delivery of services regarding the endpoints:

- *Edge-to-edge*. These services extend to the network provider's edge nodes (e.g. gateways). The demarcation points for performance measurements are access or core network edge nodes.
- *Site-to-site*. These services extend to customer premises (e.g. home, office, public place). The demarcation points are typically the edge nodes (e.g. home gateways for residential users) at the customer premises.

- *TE-to-TE (TE stands for terminal equipment).* These services extend to end-user terminals within the given customer premises network. The demarcation points are typically the terminal equipment of the customer.

All three services with assured delivery in NGNs have to be supported by a measurement model. So, the points in the network in which performance measurements are performed are referred to as demarcation points, which are typically located on the edge nodes in each of the given network segments in the NGN general model for performance management, as shown in Figure 3.11.

In the edge-to-edge model, the service delivery is assured between the edge nodes in the access networks to which CPNs are connected. For this model the measurements provide performance information for all segments between the egress and ingress demarcation points, including all network segments between them. The transit segment has the demarcation point for the performance measurement at the ingress NGN core network and the egress NGN core network. A given transit segment may include different transit network providers, so it can span larger geographical areas (cities, countries, regions, as well as globally). However, the choice of the demarcation points for the measurements is done autonomously by the NGN provider. But such choice may be influenced by national legislation on the ICTs, i.e. electronic communications in such a

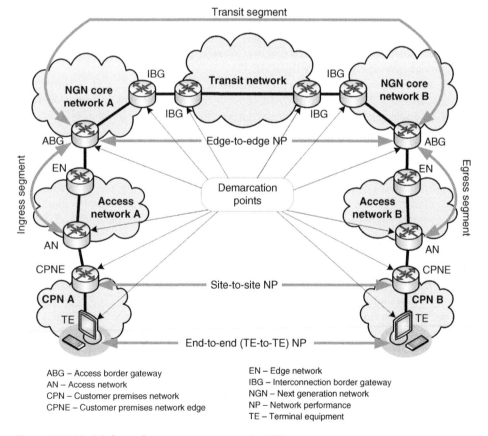

ABG – Access border gateway
AN – Access network
CPN – Customer premises network
CPNE – Customer premises network edge

EN – Edge network
IBG – Interconnection border gateway
NGN – Next generation network
NP – Network performance
TE – Terminal equipment

Figure 3.11 Models for performance management in NGN.

way that certain demarcation points in certain cases should be implemented (e.g. at the Internet Protocol eXchanges (IPXs)).

The second model for NGN performance measurements is the site-to-site model, in which the service is assured between the customer premises edge nodes. In this case the ingress and egress segments for the performance measurements may include access links toward the customers, such as xDSL, PON, cable networks, Ethernet, WiFi, mobile access network (e.g. 4G, 5G), etc.

In the third model, TE-to-TE, the NGN performance characteristics are the aggregate of performance characteristics of all network segments between the two terminals (the terminals can be used by humans or by machines). This approach may cover performance measurements for all different services, including legacy voice and TV services, smart services (smart cars, smart homes, etc.), services for corporations (i.e. business services), as well as a variety of terminals.

The MPM system is logically separated from the service and transport stratums in NGN. As one may expect, it is connected to both such NGN stratums, as well as to end-user functions and to MPM systems in other NGNs. MPM functional architecture consists of three functional entities:

- *Performance measurement execution functional entity (PME-FE).* It performs functionalities of active probe initiations (i.e. measurements initiation), active probe termination, and passive measurements (i.e. measurements on ongoing data and signaling traffic, without generation of any additional probing traffic in the network).
- *Performance measurement processing functional entity (PMP-FE).* It collects all measurement reports generated by the PME-FEs.
- *Performance measurement reporting functional entity (PMR-FE).* It is positioned highest in the hierarchy of MPM, so collects the measurement information from PMP-FE and reports it to MPM applications or to other MPMs in other NGNs.

Overall, implementation of MPM in NGN aims to provide the desired level of QoS support, per flow or aggregate, including all types of end-users. Also, performance measurements give inputs to the business management system and have a direct influence on different aspects of NGN operators, such as network planning and dimensioning, services marketing, and SLAs. Overall, MPM supports business processes to plan, provision, operate, and maintain NGN networks and services, with the desired level of QoS provisioning.

3.6 DPI Performance Models and Metrics

Deep packet inspection (DPI) performs analysis on different packet headers from layer 2 at the bottom up to the application layer on the top, as we discussed in Chapter 2. The output of a given DPI function is usually used in subsequent functions which may include reporting (on the data carried by the packet) or actions on the packet (e.g. for the purposes of QoS classification, selective charging approach for different types of services, security control to locate and stop security threats for the network or the end-users, and so on). So, DPI is becoming an important and integral part of telecom networks.

The general aspects for applying DPI technologies inside the network are typically one or several of the following [22]:

- monitoring the status of the networks
- network optimization and redesign
- improving network performance.

Besides the general objectives for the DPI deployment in the network, there are certain specific objectives, which include monitoring different types of traffic, identifying invalid traffic according to the policy rules, analyzing network performances, reallocation of network resources, or rebuilding the network based on network performance monitoring, as well as improving the satisfaction of end-users.

The deployment of DPI nodes should not interrupt the performance aspects of services and applications. In theory, out-of-path DPI nodes (which analyze copies of the packets) can be implemented without interruption of services and applications. If DPI nodes are inserted on the path of packets, services and applications interruption time should be less than 50 ms. However, requirements may be stricter for certain mission-critical services (e.g. control of driverless cars or drones via a mobile network).

Since DPI influences the performance of services, a certain number of DPI performance metrics need to be defined [23]. Such metrics should comply with national (or regional) laws, policies, and regulations. The DPI performance metrics are based on the defined DPI functions and measured at defined measurement points. The DPI performance model is shown in Figure 3.12.

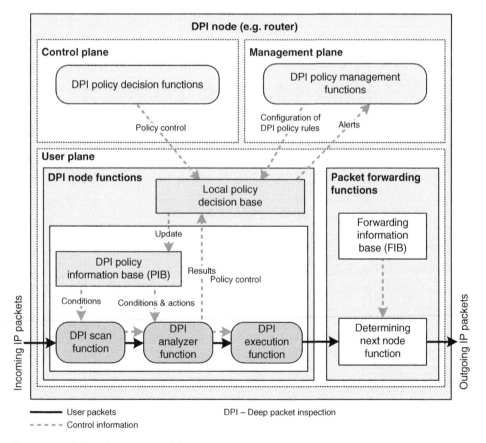

Figure 3.12 DPI performance model.

DPI performance is defined as a matrix, which includes three DPI functions: application identification, PIB (policy information base) maintenance, and system status. Performance is measured with respect to four metrics: speed, accuracy, dependability, and resources [24]. Certain DPI performance metrics are defined for the user plane traffic, and others are defined for the control or management plane traffic, as given in Table 3.4.

One of the key requirements of DPI is identification of the application, with the aim of ensuring QoS and QoE (when applicable). Such typical applications over public Internet access are HTTP (i.e. Web), email protocols, peer-to-peer applications (e.g. BitTorrent), video streaming (e.g. "tube" video sharing sites such as YouTube), etc. However, one should note that network neutrality applied to Internet access service does not allow prioritization of one application (through such Internet access) over the other, although

Table 3.4 DPI performance metrics.

DPI function	Performance metric	Speed	Accuracy	Dependability	Resources	Plane
Application identification	Node-internal transfer delay	X				User plane
	Packet processing rate	X				User plane
	Error rate		X			User plane
	False-positive error rate		X			User plane
	False-negative error rate		X			User plane
	Rate of successfully identified packets	X				User plane
System status	DPI inspection depth				X	User plane
	TCP successful connection establishment rate	X				User plane
	Number of concurrent TCP connections				X	User plane
	TCP connection establishment success rate	X				User plane
	Packet loss rate		X			User plane
	Failover time	X				Control plane
	Network management system (NMS) response time	X				Management plane
	Supported concurrent NMS number			X		Management plane
	Energy per bit				X	DPI node
	Energy per packet				X	DPI node
PIB maintenance	Number of supported application types			X		User plane
	DPI PIB size at the line rate				X	User plane
	Rule take effect time	X				Control plane
	NMS DPI rule age report time				X	Control plane

certain traffic management schemes (with or without DPI in use) can be beneficial to all applications served in a network neutral and best-effort manner.

For example, there can be defined application-specific DPI performance metrics for HTTP, such as HTTP application transaction identification rate. It is defined as the number of HTTP transactions per second (through the given DPI node) that have been successfully established and identified by the DPI node. One may note here that an HTTP connection is successfully established, which means that the response code "HTTP 200 OK" is sent from the HTTP server to the HTTP client after the connection has been initiated with the HTTP GET request from the client to the server. The measurement unit of this metric is a reciprocal second, so the result of 50 means that 50 HTTP connections were identified (by the DPI node) as successfully established in the time period of one second. However, this is not a key performance indicator (KPI) for the DPI node, while most of the performance metrics given in Table 3.4 can be considered as KPIs for the DPI.

3.7 QoS in Future Networks

The further evolution of NGN within the ITU is referred to as future networks. In fact, it is a continuation of the work on NGN, targeted to smart and more advanced services, considering different types of awareness, including:

- *Service awareness.* Future networks should provide new service deployments without significant increase in network deployment and operational costs.
- *Data awareness.* Future networks are expected to carry constantly increasing amounts of data in all parts (access, core, and transit network segments), which needs to be accessed easily, quickly, accurately, safely, and with desired QoS, regardless of the access network, fixed or mobile.
- *Environmental awareness.* Future networks should be energy efficient and environmentally friendly.
- *Social and economic awareness.* Future networks should be developed with awareness of the costs and competition, so that services can be accessible by all players in the service ecosystems.

The question is how to provide these types of awareness in future networks. Well, the answer is by adding more flexibility in the network regarding the allocation of its resources (e.g. bandwidth, memory or processing power of the nodes, etc.) regarding all planes (user plane, control plane, and management plane). Which are the available "tools" for such a goal? They are software-defined networking (SDN) and network function virtualization (NFV) technologies. They are seen as enablers to reduce operational costs and provide potential revenue streams by expanding the telecom world into different verticals (e.g. machine-type communications, the IoT, used for different services in different verticals, such as smart cities, smart industry, smart homes). However, telecom operators have a large amount of equipment which does not have SDN or NFV capabilities. So it is very probable that many telecom operators in the future will consider the benefits of SDN and NFV in hybrid implementations, which include legacy (non-SDN and non-NFV) network elements together with SDN- and NFV-enabled network devices.

Generally, future networks are expected to be more complex (with many new services in different verticals) and more efficient (higher utilization of the same network infrastructure for all services), considering all given awareness aspects. The standardized form of NFV is called network slicing (e.g. network slices are defined in 5G mobile networks). However, there are several other terms associated with the same issue. For example, in the series of recommendations for NGN and future networks, ITU-T has standardized logically isolated network partitions (LINPs), which is in fact network slicing.

QoS is important for all networks; however, when a physical network is separated into logical network slices, the QoS may be more important in certain slices (i.e. isolated network partitions) than in others.

3.7.1 Network Virtualization and QoS

Network slicing and virtualization were defined as early as the end of the 2000s [24], set as the target for the development of the future Internet, so-called management and service-aware networking architectures (MANA). Among some of the main slicing capabilities are the following:

- *Infrastructure capabilities*. These include the introduction of virtualization and programmability in network nodes, such as core nodes, edge nodes (e.g. switches, routers, gateways), as well as wireless and mobile nodes.
- *Control and elastic capabilities*. They include (but are not limited to) cognitive control (by decoupling of the user plane and control plane in telecommunication systems), self-configurability of the network components and the networks, as well as flexible and cost-effective operations of different service platforms.
- *Accountability capabilities*. They include, for example, cross-layer optimization, including physical resources, the network as a whole, transport and service layers (or stratums) to provide application-enabled QoS approaches. Also, required are mechanisms for handling non-technical issues of accountability, which includes ethical, legal, and governance aspects.
- *Virtualization*. This refers to having virtual resources (e.g. resource allocation to slices of the network infrastructure) and virtual infrastructures (e.g. dynamic creation of slices for different services), and programmability of virtual clouds (including network clouds, service clouds, and virtual infrastructures).
- *Self-management capabilities*. These include self-functionality mechanisms (in a given domain and cross-domain) as well as self-functional infrastructure and systems (self-adaptation and self-composition of resources, context awareness, automated auditing, with increased cost-effectiveness of the networks and systems).
- *Service-enabling capabilities*. This is targeted toward creation of service-aware networks, where networks can discover services and have self-negotiation and self-contraction capabilities regarding SLAs.
- *Orchestration capabilities*. These capabilities govern the integrated activities and performances of different systems of systems and dynamically adapt to the changing contextual information, and based on that optimize usage of the network resources (switches, routers, etc.) and service resources (e.g. servers, service cloud resources).

The ITU-T has defined network slicing via the definition of LINPs [25], which may be considered as the trigger for further standardization of network virtualization. By

ITU-T's definition [25], network virtualization is a technology that allows multiple virtual networks called LINPs to coexist on a single physical network infrastructure. Virtual resources are created over the physical objects in the given network, which includes (in IP networks) nodes such as switches, routers, servers, and end-user terminals. LINP can use virtual resources based on their programmability features, which are required with the aim of having the virtual networks. Regarding legacy end-users (e.g. residential users, enterprises), LINPs are able to provide to them the same services as provided by networks without applied virtualization. This approach is in fact the basic concept for network softwarization, which is in fact a convenient short description of network virtualization. So, a network slice is a LINP (or vice versa), which can be considered as a unit of programmable resources, where the resources can be network, storage, or computation.

ITU's architecture for network virtualization is shown in Figure 3.13, the LINP architecture. Each LINP is built from multiple virtual resources, which are mapped on the

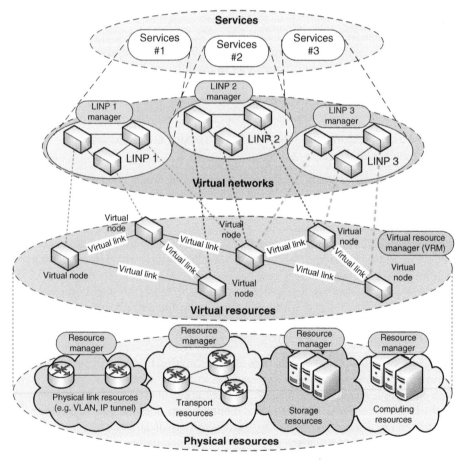

LINP – Logically isolated network partitions
VLAN – Virtual local area network

Figure 3.13 Architecture for network virtualization.

physical resources (e.g. switches, routers, servers). However, each physical resource may have multiple virtual instances, in other words it can be shared among multiple LINPs. The sharing of physical resources between virtual networks is provided by using virtual resource managers (VRMs). On one hand, one of the main targets of network virtualization is to provide higher utilization of physical resources (through their sharing between LINPs) and hence better cost-efficiency. On the other hand, virtualization helps toward better adaptation of the networks to the requirements from existing services and end-users, as well as easy introduction of new services with given QoS and security requirements without a need to build a parallel network infrastructure or to change the existing network infrastructure. Overall, there are four parts in the LINP architecture (going from physical resources toward services, Figure 3.13):

- *Physical resources.* These are all physical hardware, including physical resources management and logical separation of the physical resources [26], where logical resources have the same characteristics as physical resources.
- *Virtual resources.* They abstract the given physical resources.
- *LINPs.* They are built on multiple virtual network resources.
- *Services.* In virtualized (i.e. softwarized) networks, the services can be composed of large numbers of software components, which include appropriate combinations of application programs and virtual resources on different LINPs. The service management is targeted to enable the network operator, which has network virtualized resources (LINPs), to support service provisioning with the required QoS.

Achieving efficient usage of LINPs or network slicing in future networks requires standardized functional architecture [26]. The following four user roles are defined in the LINP functional architecture:

- *Service developer.* It designs, develops, then tests and finally carries out operation and maintenance of services based on service developer functions. Such services are developed on one or more LINPs (when more LINPs are used for a given service it is referred to as a LINP federation).
- *LINP operator.* This is a network operator which aims to design and program network services over different LINPs. A further LINP operator configures, manages, and terminates such network services.
- *Network operator.* It designs, deploys, and manages physical resources in the network as well as their abstracted sets within the operator's administrative domain.
- *End-user.* When an authorized user has access to a LINP which carries a given service, that user can utilize the service.

The QoS in virtualized networks is becoming decoupled from the physical network infrastructure. So, one may not talk about QoS in the network but QoS on a given network slice (in a given LINP) because services are provided via different network slices which use different virtual resources mapped onto different physical resources. The same QoS metrics can be applied to services in virtualized networks as applied to the same services in non-virtualized networks. However, different slices can be isolated regarding the QoS metrics (e.g. dedicated bandwidth, i.e. bitrate) or not. QoS isolation is traditionally provided in telecom networks by use of VPNs, which are based on tunneling of IP packets between pairs of nodes within a network (or between networks at the interconnections), where each VPN carries a certain type of traffic (aggregated traffic of a given type such as voice, video, or data, or individual traffic such as traffic for a given corporation). However, QoS is always end-to-end. So, for a given flow (from terminal A to terminal B), the packets may travel via several networks on the path (the networks belonging to different administrative domains are autonomous by default), which can carry that traffic by using different network slices (i.e. different LINPs). A complete slice is composed of not only various network functions which are based on virtualization in the access networks or core networks, but also transport network resources assigned to have a complete end-to-end path.

Overall, various ICT/telecom businesses in future networks require different QoS parameters, such as different values for throughput, delay, and losses. Certain services may need guaranteed throughput and very low delay (e.g. mission-critical services, such as control in real time of various transportation vehicles, as one example). For such services the transport networks need to provide QoS isolation and flexible network operation and management, and improve the network utilization among different services in different verticals (i.e. different business).

So, traditionally QoS isolation in transport networks is provided by VPN technology, which can give physical network resource isolation across multiple network segments. However, VPN is less capable of supporting hard QoS isolation because QoS isolation on the user and control planes requires better coordination with the management plane. The flexibility behind the network slicing approach needs to address QoS guarantees in the transport networks and at the same time enable transport network openness because future network design is targeted to provide end-to-end QoS guarantees per network slice.

3.7.2 Software-Defined Networking and QoS

According to the ITU [27] and also the IETF [28], SDN is a set of techniques that enables the direct programming, orchestration, control, and management of network resources which facilitate the design, delivery, and operation of network services in a deterministic, dynamic, and scalable manner.

SDN is a framework which already has a defined layering framework (Figure 3.14) [27] – convenient for interpretation and standardization of different SDN techniques that exist today and those that will appear in the future.

According to the SDN layered architecture shown in Figure 3.14, there are three main layers:

- *SDN resource layer.* This includes data transport and data processing by network devices [28], and hence includes forwarding and operational planes [29].
- *SDN control layer.* This does resources abstraction, control layer services, and application layer services, including management support and orchestrations in this SDN control layer [28]. The control abstraction is done in the control plane while the management abstraction layer belongs to the management plane [29].
- *SDN application layer.* This includes all SDN applications, where the term "SDN application" refers to applications that control and manage network resources and not applications or services used by end-users. One example of an SDN application is user application-specific routing in the SDN-enabled network.

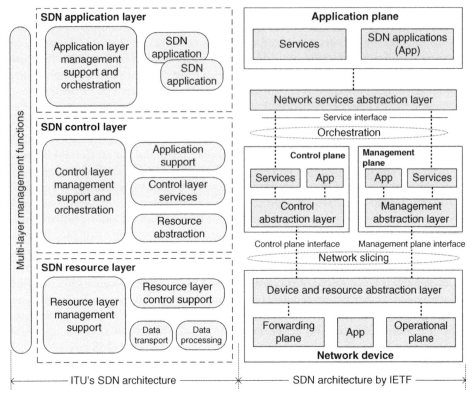

Figure 3.14 SDN layered architecture.

Overall, SDN spans multiple planes [29]. The forwarding plane is responsible for handling the packets based on instructions from the control plane. The control plane includes all signaling messages, which are pushed to the network devices with the aim of implementing decisions about how the packets should be forwarded (in the routers, switches) in the forwarding plane. While the control plane focuses mainly on the forwarding plane of network nodes, the management plane focuses on operational aspects of the network (i.e. on the operational plane), which include monitoring, configuring, and maintaining various SDN-capable network devices. For example, one of the most popular protocols in the SDN research phase, the OpenFlow, is intended for use in the control and management planes [29]. Finally, the application plane is where SDN applications and services reside, where such SDN applications/services directly or indirectly define network behavior. Such applications can be developed in a modular manner and can span multiple planes in the SDN architecture.

Generally, SDN is based on separation of user plane (i.e. data plane) and control plane in the network. That also provides the possibility to enhance QoS provision in the network via use of SDN. Typically QoS is measured via parameters such as bandwidth, delay, jitter, and loss (the most frequently used ones). QoS is provided to given sets of services by giving precedence on QoS-enabled services over other services (e.g. best effort ones). The specification of the QoS targets between the end-users and service providers is traditionally done by SLAs. There are two types of QoS [30]:

- *Hard QoS*. This method guarantees the QoS requirements of a given call/session, but may suffer in the case of limited resources. Typical example of hard QoS is constant bitrate bandwidth reservation (e.g. guaranteed bandwidth).
- *Soft QoS*. This method is not strict QoS provisioning, i.e. the QoS is not strictly guaranteed for the entire duration of the connection (i.e. call/session), so it is more flexible in adaptation to the available resources. A typical example of soft QoS is class-based routing (e.g. the DiffServ approach).

What is the relationship between SDN and QoS? Well, SDN improves the QoS with QoS-motivated routing, for example with per-flow routing (instead of class-based aggregate routing) and with simpler mechanisms for that purpose than legacy telecom network infrastructures. Further, SDN may help telecom operators to create more a powerful and more flexible QoS management system framework, with well-defined control mechanisms to adapt to the constantly changing environment in the network regarding the volume of traffic or number of supported end-user services and applications. SDN also allows QoS monitoring on a fine-grained level, such as monitoring per queue, per packet, etc. Finally, SDN is inseparably connected with network virtualization (many times the terms NFV and SDN are used together), and it can be used for virtualization-based QoS provisioning and QoS policy management.

3.8 Business and Regulatory Aspects

The move toward NGNs in the 2010s started the transformation of the legacy telecommunication sector. The main approach of the NGN is having a software-based platform with various functional entities (e.g. IMS) that uses standardized Internet technologies to enable the provision of all services, including Internet data services, QoS-enabled voice, video, or TV streaming, via any access network infrastructure, including fixed access (copper or fiber) and mobile access (e.g. 4G, 5G). The NGN provides a transition from separate networks and functions for separate services into a single network and single market for all services, including those that require certain strict QoS provisioning (e.g. real-time services such as voice, critical machine type communication) and those that are more "relaxed" regarding QoS requirements (e.g. email, file download), and at the same time facilitates FMC.

3.8.1 NGN Policies

The development of new networks and new services is dependent upon broadband and ultra-broadband deployments, which are dependent upon national and regional strategies and policies. In the absence of these, regulators, investors, consumers, and service providers can develop their own approaches. Although in developed ICT national markets the ICT networks and services are left to free market forces, even in such economies there is a need for certain policy, regulation, and directions. What should the policy reflect? It needs to consider the overall national needs regarding the ICT/telecom supply side and the demand side. The supply side is most usually considered by national strategies and policies, and it typically covers licenses, funding, and investments in broadband, spectrum policy for mobile access, and so on. The demand side is considered less often, but it is highly important (it includes content and applications to use the NGN, use of government services, education, and facilities to exploit available NGN broadband).

The main policy factors for the telecom world include the following [31]:

- *Spectrum policy*. It is critical because spectrum is a limited resource and mobile broadband access has higher penetration than fixed broadband access (especially in countries with developing ICT markets). Also, in mobile networks (similar to the fixed networks) the focus moves from traditional voice services (provided by mobile telecom operators) to data services via mobile broadband/ultra-broadband Internet access.
- *Services*. Broadband Internet-based services are mainly developed by private companies, which are different from telecom operators that provide access to the Internet and overall provide broadband/ultra-broadband access. Other significant benefits come from the delivery of government services, such as education, health-care, public transportation, and government services. This also may increase the need for QoS in certain cases. Regulatory policies should facilitate supply of end-user services by private companies. Also, regulatory policies should allow competition as deep as possible into the layers of service provisioning, including broadband network infrastructure services, ultra-broadband access services, various content, and services via OTT applications as third-party applications provided based on network neutral Internet access services.
- *Legal frameworks*. Sometimes beyond the telecom regulator's scope, certain policies need to consider enabling legal frameworks and structures for services and networks (e.g. for data protection, security, and monitoring).
- *Education and contents*. Education is using more and more broadband Internet access and various OTT services. ICT literacy (how to access and appropriately use Internet services and application for different purposes) is also required to benefit from certain NGN services (e.g. various smart services).

3.8.2 NGN Regulation Aspects

With the transition of telecom operators from legacy digital networks to NGN, the telecom regulation does not change significantly. However, there are a number of issues regarding NGN regulation. For example, the regulator needs to regulate access to the non-competitive wholesale markets (e.g. controlled by one or two NGN providers). Similarly, regulation can be targeted to ensure that ISPs have wholesale access to the NGN provider.

Further, various diverse services may share the same NGN. NGN fosters mobile–fixed convergence (e.g. the same service functions are used for both types of access networks), so mobile and fixed services should be treated as interlinked. Also, regulation must focus more on the key bottleneck of the access network technologies.

However, NGN adds some new factors into the regulation sphere. One of the specific approaches to regulation with the deployment of NGN is separation, which has two main forms [31], structural and functional.

Structural separation creates a separate access business which is strongly regulated, while service providers are then lightly regulated. Such separate businesses provide services to the other players in the ICT ecosystem (in different verticals) on a wholesale basis. Meanwhile, functional separation does not require separate businesses, so in this case the broadband access service provider remains part of a larger business. The access business is only functionally separated (and not legally as a certain separate business), so in this case it can deliver equivalent services to its own services as well as to other service providers.

Regarding regulation, the targets with the NGN remain the same as legacy networks, that is, encouragement of competition between different providers where it is possible. When it is not possible to have more than one core NGN, the access rules and the prices for using the NGN-capable network become critical. How can the regulator be certain that a given NGN is built efficiently and at lowest cost, without the encouragement of competition? Regulators typically had to focus in most cases on the telecom/ICT areas where there was the least competition, which is the highest barrier to market entry by new players. So, it appears that mobile and fixed access networks are the most critical areas to regulate, particularly mobile access where the resources (e.g. available spectrum) are limited and hence there can be only several competing telecom operators on a national level. Deploying fixed access (which will be only fiber-based in the fixed networks, looking toward the future) has high costs per customer and therefore it is less cost-effective for the telecom business (especially when it does not bring return of investments or increasing revenues from customers), hence fixed access is a natural bottleneck, particularly because it is not practical to have many fiber cables everywhere from different service suppliers (although that is also possible and also existing practice in certain areas). The competitive supply of ultra-broadband fixed access networks can be increased if infrastructure sharing is permitted or even enforced (e.g. by regulatory support). In such cases, the ducts and cabinets can be reused to reduce applicable costs for alternative providers to provide fixed fiber access to their customers.

With the development of NFV and SDN, and their application in NGN telecom operator networks, there will be even greater possibilities for resource sharing by network and service providers based on network virtualization (e.g. LINP, NFV) and softwarization (e.g. SDN). In developed ICT markets this process of further development of NGNs toward future networks can be left to free market forces; however, that should be monitored regarding QoS and even enforced when necessary. In all countries the network virtualization (e.g. network slicing) may be a subject for regulation regarding QoS and QoE. In some countries, such as those with developing telecom/ICT markets, it can be enforced, while in countries with developed telecom/ICT markets, QoS/QoE should be monitored from the end-users' side.

3.8.3 NGN Business Aspects

The policies, regulation, and business aspects in NGN broadband are related. However, it is also important that regulation is kept as separate as possible because the main role of regulation is to ensure the best outcomes and to assist with competition among NGN business players. So, regulation is not directly concerned with creating the funding possibilities for deployment of broadband/ultra-broadband and NGN, but it regulates what is built. However, regulation should encourage investments in NGN and broadband.

Regarding the funding options for NGN and broadband [31], they are mostly related to deployment of access (fiber-based) networks, but the same funding approach can be used for core NGN as well as international broadband links. There are generally four funding options:

- *Government funding.* The standpoint here is that fiber access is essential for broadband infrastructure based on NGN and it is needed to enable the service ecosystems to function. The justification for this type of funding is that there cannot be multiple fiber cables to the same house or office. Furthermore, it is justified by the fact that such investment cannot be paid for at rates that private companies require. In this case the broadband network may be funded centrally (by the government) and then it can be leased to other operators. Then the services can be used by different service providers, which may have different core networks and separate service platforms over the deployed physical (fiber) infrastructures.
- *Localized government funding.* This is similar to government funding but on a smaller scale. In this localized case the funding addresses smaller or selected areas. The positive side is that it can help to attract other industry and investors in the given area. The negative side may be negative outcomes with objections from network operators on the basis of competition, which may leave the locality with no NGN at all because local regions are usually too small for investments by private telecom operators.
- *Public-private funding.* Private network operators can bid for public broadband and NGN funds, for example to help telecom/ICT investments in some rural areas. This is also most commonly targeted at deployment of fiber access. Similar approaches can be applied for backbone networks and international links, especially needed in emerging economies where there is not enough market demand that can sustain private-only investment with justified business logic for investments.
- *Private funding.* In this case the funding comes from the telecom operators themselves (based on debt funding) and/or from other private investors. However, every such investor typically considers the risk to the investment and the return of the investment (i.e. payback rewards). The question is, what will be the revenues versus costs over time (for longer and shorter time periods)? Another problem comes from the uncertainty of both future prices (which influence the revenues) and number of customers. With the saturation of the number of customers connected to broadband and to NGN, the motivation for private (commercial) funding drops, which is a natural business reaction to having no growth potential. However, ICTs going into different verticals by using NGN and broadband access (where broadband does not equal NGN) gives more options for growth in new sectors, hence attracting new funding sources into the telecom private business, which may give very positive future expectations and thus motivation in the telecom (and NGN) business.

Overall, the need for a consolidated network supply becomes critical as network operators (telecom operators) tend to deliver broadband services to their customers. The convergence given with the NGN transformation helps operators to reduce costs in the longer term (one network fits all services, including those that require strict QoS provisioning and those that have higher flexibility regarding QoS). The NGN helps the traditional telecom operators (which also have become the main ISPs) to deliver new services to their customers with less capital investment and reduced operational costs, better regulation, and higher QoS. In conclusion, there are many types of business and regulation challenges facing NGNs, but business opportunities are varied, unlimited, reliable, and promising in the long term, especially with the continuous development of NGN toward future networks based on network virtualization and slicing, with QoS provided per network slice, service need, or connection request.

References

1 Janevski, T. (2014). *NGN Architectures, Protocols and Services*. Chichester: Wiley.
2 ITU-T Recommendation Y.3001, Future Networks: Objectives and Design Goals, May 2011.
3 J. Rosenberg, H. Schulzrinne, G. Camarillo, A. Johnston, J. Peterson, R. Sparks, M. Handley, E. Schooler, SIP: Session Initiation Protocol, RFC 3261, June 2002.
4 ITU-T Recommendation Y.2021, IMS for Next Generation Networks, September 2006.
5 ITU-T Recommendation Y.2808, Fixed Mobile Convergence with a Common IMS Session Control Domain, June 2009.
6 3GPP TS 23.228, IP Multimedia Subsystem (IMS); Stage 2 (Release 12), June 2013.
7 IETF RFC 7826, Real-Time Streaming Protocol Version 2.0, December 2016.
8 IETF RFC 4566, SDP: Session Description Protocol, July 2006.
9 IETF RFC 3550, RTP: A Transport Protocol for Real-Time Applications, July 2003.
10 IETF RFC 6733, Diameter Base Protocol, October 2012.
11 IETF RFC 5624, Quality of Service Parameters for Usage with Diameter, August 2009.
12 IETF RFC 5866, Diameter Quality-of-Service Application, May 2010.
13 ITU-T Recommendation Y.2111, Resource and Admission Control Functions in Next Generation Networks, November 2011.
14 ITU-T Recommendation Y.2171, Admission Control Priority Levels in Next Generation Networks, September 2006.
15 ITU-T Recommendation Y.2113, Ethernet QoS Control for Next Generation Networks, January 2009.
16 IEEE 802.1Q – 2005, Virtual Bridged Local Area Networks, May 2006.
17 IEEE 802.1ad – 2005, Virtual Bridged Local Area Networks – Amendment 4: Provider Bridges, May 2006.
18 IEEE 802.1ah – 2008, Virtual Bridged Local 14 Area Networks – Amendment 6: Provider Backbone Bridges, August 2008.
19 IETF RFC 2917, A Core MPLS IP VPN Architecture, September 2000.
20 ITU-T Recommendation Y.2175, Centralized RACF Architecture for MPLS Core Networks, November 2008.
21 ITU-T Recommendation Y.2174, Distributed RACF Architecture for MPLS Networks, June 2008.
22 ITU-T Recommendation Y.2772, Mechanisms for the Network Elements with Support of Deep Packet Inspection, April 2016.
23 ITU-T Recommendation Y.2773, Performance Models and Metrics for Deep Packet Inspection, February 2017.

24 A. Galis, H. Abramowicz, Brunner, M., et al., Management and Service-aware Networking Architectures (MANA) for Future Internet, Communications and Networking in China – ChinaCOM 2009.

25 ITU-T Recommendation Y.3011, Framework of Network Virtualization for Future Networks, January 2012.

26 ITU-T Recommendation Y.3012, Requirements of Network Virtualization for Future Networks, April 2014.

27 ITU-T Recommendation Y.3302, Functional Architecture of Software-defined Networking, January 2017.

28 IETF RFC 7149, Software-Defined Networking: A Perspective from within a Service Provider Environment, March 2014.

29 IRTF RFC 7426, Software-Defined Networking (SDN): Layers and Architecture Terminology, January 2015.

30 Karakus, M. and Durresi, A. (2017). Quality of Service (QoS) in software defined networking (SDN): A Survey. *Journal of Network and Computer Applications* 80.

31 Telecommunication Development Sector, ITU, Strategies for the Deployment of NGN in a Broadband Environment – Regulatory and Economic Aspects, March 2013.

4

QoS for Fixed Ultra-Broadband

Quality of service is equally important for both main types of access networks, fixed and mobile. Both have converged on the all-IP principles, meaning that the user traffic end-to-end is IP traffic, so the QoS is directly related to end-to-end QoS in IP networks. In this chapter we cover the fixed access network technologies which are developing toward ultra-broadband access as the first wave of broadband access. However, the term "ultra-broadband" is a relative term, which refers to higher bitrates than previous broadband bitrates. Any bitrates in the access part (which directly influence the end-to-end bitrates if the access part is considered as a major possible bottleneck) are directly related to the time when a certain book is written.

So, if we are writing at the end of the second decade of the twenty-first century about tens, hundreds, Mbit/s, and even Gbit/s achievable per individual users (including fixed and mobile access parts), in a couple of decades the numbers will develop further toward higher and higher bitrates, and that is a normal expectation. Of course, no one can really predict the access technologies in three, four, five, or more decades in the future because if that were the case then such technologies would be developed sooner. But, as is typical in telecommunications, one may look into the past and predict the development of the technology in a certain future. We do that for signals on the physical layer, so copper or fiber links are designed to have certain capacity and certain characteristics (based also on their physical characteristics, such as diameter of the wire – higher diameter is better for copper wires, while smaller diameter of fiber core is better for fiber links). We have Moore's law regarding the development of computer technologies and the processing power of computers, which doubles approximately every two years. Hence, someone can predict with a certain degree of accuracy what processing capabilities computers will have in a decade from now – that is, according to Moore's law they will be $2^5 = 32$ times more powerful (if one assumes that processing power doubles every two years), or more than $2^6 = 64$ times (if one assumes that processing power doubles every 18 months). In practice, Moore's law refers to a number of components on integrated circuits, which directly influence the processing power of computers. However, different factors also may influence trends, such as prices.

In the twenty-first century most telecommunication end-user devices are computers because they have hardware and software (which includes the operating system of the device that contains protocols above the physical layer and applications on top of the protocol layering stack). So, one may apply Moore's law directly to such telecommunication devices. But what happens to the bitrates in the access networks?

QoS for Fixed and Mobile Ultra-Broadband, First Edition. Toni Janevski.
© 2019 John Wiley & Sons Ltd. Published 2019 by John Wiley & Sons Ltd.

First, let's discuss a question: since when can we apply a law on the bitrates in the access networks? The answer is: since globalization of the Internet – that is, the middle of the 1990s, or more accurately, from 1993 when the WWW was allowed to spread "free of charge" outside the CERN labs in Switzerland where it was invented by Tim Berners-Lee. Considering that, let's start in the first half of the 1990s and continue until the present day. So, in the 1990s the main type of access to the Internet for residential users was 56 Kbit/s dial-up via digital telephone networks (i.e. via twisted pair copper lines). The introduction of ADSL in 1999 brought bitrates of 8 Mbit/s in downlink and 1 Mbit/s in uplink. ADSL spread during the 2000s, reaching bitrates over 10 Mbit/s in the downlink and even over 20 Mbit/s with ADSL2+ in the 2010s. At the same time fiber technologies started to spread in the last mile by the end of the 2000s and further increased in the 2010s. Typical access rates with shared passive optical networks are in the range of 50 Mbit/s (although dedicated fiber links in the access part may provide Gbit/s speeds). So, regarding fixed access per individual user (not aggregate bitrates, which normally are many times higher at any time), one may notice that in two decades, there was an increase of 1000 times (in individual user bitrates), from 56 Kbit/s up to 50 Mbit/s (on average). That corresponds to doubling the bitrates for individual users in fixed access networks (either copper or fiber) every two years, which gives 2^{10} ($\sim$1000) times higher bitrates in two decades. We may refer to this as some telecommunication law, which is similar to Moore's law for integrated circuits (i.e. processing power of computers). Also, one may note that for higher bitrates we also need higher processing power in end-user devices (computers) to be able to process all the data (all bits that are sent or received). Computing (i.e. processing) power and bitrates increase at a similar pace, and that is expected to continue in the future.

If we apply this telecommunication law on bitrates, we may expect the individual average access bitrates in the 2030s to be in the range of 50 Gbit/s, which is an increase of 1000 times. However, different factors influence bitrates, such as saturation of broadband markets for human end-users in the 2010s, which may decrease the expectations for bitrates in the longer term. If such a decrease is in the range of two times, i.e. doubling of bitrates to occur every four years (instead of every two years, as in past decades), the expected increase can be 2^5 times ($\sim$30 times) for a period of two decades in the future (or 5–6 times increase of the bitrates per decade). The bitrate is the parameter that most influences QoS and the user experience.

The fixed access networks can be copper-based or fiber-based. Copper-based access networks can use twisted pairs (used for analogue and digital telephony in the twentieth century, and still in the twenty-first century), coaxial cables (for cable access networks), and fiber access. What is broadband and what is ultra-broadband in all these access networks? Well, broadband first refers to high bitrates, which provide the possibility to use all services, including video as most demanding (on average) for humans as end-users. Typically such bitrates are in the range of Mbit/s and tens of Mbit/s. The speeds in the range of 100 Mbit/s and even Gbit/s per individual access (i.e. fixed end-user) can be referred to as ultra-broadband (around 2020). Considering such a view in the longer term, the only ultra-broadband technology in the access part is fiber access. However, there are a lot of deployed copper access networks (including twisted pairs and coaxial cables) which will be used until they can be concurrent with fiber access in both bitrates (copper access loses the battle with fiber access gradually) and price (copper-based access has been in a better starting position because it was deployed during the twentieth century in many parts of the world and is still present). So, copper-based access also plays a role in the ultra-broadband story in the 2010s and 2020s.

4.1 Ultra-broadband DSL and Cable Access

For provision of ultra-broadband fixed access, telecom operators require a wide range of technology solutions. The fiber solution is the best regarding provided capacities (i.e. bitrates) per cable; however, copper infrastructure such as DSL and cable access networks (based on coaxial cable) can still be used in the last mile, or more accurately in the last kilometer (or less) where fiber installation is not feasible or practical.

4.1.1 DSL Ultra-Broadband Access

The initial technology for DSL access (which goes over twisted pair lines) was ADSL, starting with the first standard in 1999 from ANSI [1] and following with standardization by the ITU-T [2], also known as G.DMT (because it is using Discrete Multi-Tone, i.e. DMT modulation). For longer lengths of the subscriber's local loop there is also the ITU-T standard [3] known as G.Lite (due to its lower data rates, up to 1.5 Mbit/s in downlink, as a tradeoff for longer lengths of the subscriber's line).

ADSL was fitted to average telephone lines between user premises and telephone exchanges (at telecom operators' premises) which are typically in the range up to 5–6 km (of course, there are also shorter lines in the local loop, depending on the concrete location of the user). ADSL was the main broadband fixed access technology in the 2000s because it reused existing copper lines (i.e. twisted pairs) deployed for telephony in the twentieth century. The standards ADSL2 [4] and later ADSL2+ [5] provided over 10 Mbit/s and 20 Mbit/s in the downlink direction, respectively (Table 4.1), and hence have prolonged the ADSL lifetime until the 2010s.

4.1.1.1 ADSL (Asymmetric DSL)

Figure 4.1 shows a typical deployment of the ADSL network architecture. The access consists of DSLAM (digital subscriber line access multiplexer) nodes, to which are connected digital local loops. Typically several local loops end into a single DSLAM, which further aggregates the IP traffic from users, and vice versa in the opposite direction. The

Table 4.1 Main ADSL broadband standards and bitrates.

ADSL name	Downstream bitrate (max) (Mbit/s)	Upstream bitrate (max) (Mbit/s)	Standard name
ADSL	8	1	ANSI T1.413-1998
ADSL (G.DMT)	12	1.3	ITU G.992.1
Splitterless ADSL (G.Lite)	1.5	0.5	ITU G.992.2
ADSL2	12	1.3	ITU G.992.3
Splitterless ADSL2	1.5	0.5	ITU G.992.4
ADSL2+	24	3.3	ITU G.992.5

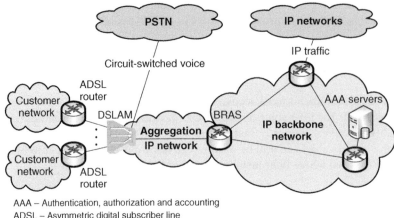

AAA – Authentication, authorization and accounting
ADSL – Asymmetric digital subscriber line
BRAS – Broadband remote access server
DSLAM – Digital subscriber line access multiplexer
PSTN – Public switched telephone network

Figure 4.1 ADSL network architecture.

main gateway node between the ADSL access network and the IP core network of the telecom operator is a broadband remote access server (BRAS), a router node which carries out AAA functions (by communication with a centralized AAA server, such as the RADIUS server). The BRAS performs traffic aggregation from the access network (from DSLAMs) toward the core IP network, and vice versa.

However, ADSL as a technology cannot provide bitrates over 100 Mbit/s, which is referred to as ultra-broadband speed. But speeds higher than 100 Mbit/s can be provided by very-high-bit-rate DSL (VDSL2) (the initial VDSL was not compatible with ADSL2+, but VDSL2 was further developed with such a goal); however, the crosstalk between the two copper wires in the twisted pair (i.e. local loop) meant it was unrealistic to expect ultra-broadband speeds of 100 Mbit/s using the initial VDSL2 [6]. Therefore vectoring solutions were added to VDSL2 to reduce crosstalk and hence boost the speed of VDSL2 over 100 Mbit/s on medium local loop lengths (Table 4.2).

Regarding DSL, bitrates are several times lower at the approximate maximum local loop length shown in Table 4.2 (in general, that refers to every DSL type, including VDSL, ADSL, and others), i.e. they decrease by increasing the length of the subscriber line.

Table 4.2 Main VDSL standards for broadband and ultra-broadband access.

DSL type	Maximum bitrate downstream	Maximum bitrate upstream	Approximately maximum local loop length (m)
VDSL	52 Mbit/s	16 Mbit/s	1200
VDSL2	100–300 Mbit/s	100–300 Mbit/s	500
G.fast	Maximum aggregated UL and DL 1 Gbit/s		250

VDSL, Very high bit rate DSL.

Considering Table 4.2, VDSL2 [7], besides vectoring that cancels crosstalk to provide bitrates up to 100 Mbit/s, can provide higher bitrates up to 300 Mbit/s with use of a larger spectrum (on the copper wire) of up to 35 MHz. Additionally, copper local loop may achieve very high bitrates of up to 1 Gbit/s, although only on a short length of the twisted pairs (or coaxial) wires, while the rest of the infrastructure toward the core network is based on fiber. So, with all DSL solutions, whether it is ADSL2+, VDSL2 Vplus, or G.fast [8], the copper phone line (the local loop) is in the last hundreds of meters or last couple of kilometers. This entails using FTTdp (fiber to the distribution point) as a broadband access solution for taking fiber very close to the customer premises, as close as 250 m or less for G.fast implementations. Regarding G.fast as the real ultra-broadband solution (with bitrates up to 1 Gbit/s), the practical lengths of the copper wires are in the range of 30–50 m (i.e. existing wiring inside buildings), where on 30 m loops it should support at least 0.5 Gbit/s on a single twisted pair. Meanwhile, VDSL2 Vplus (with its 35 MHz frequency spectrum) supports longer loop lengths (such as 200 Mbit/s over 500 m) and also up to a couple of hundred subscribers (that is, denser solution) compared with G.fast. This makes VDSL2 Vplus the best possibility for deployment of ultra-broadband FTTN/FTTcurb (fiber to the node/fiber to the curb) over medium local loop lengths. Overall, Vplus bitrates are comparable to bitrates that can be achieved by many FTTH (fiber to the home) deployments.

4.1.2 Cable Ultra-Broadband Access

Another copper access technology that has been used massively since the end of the twentieth century is coaxial cable access (or shortly, cable access). However, where cable access is used it is typically implemented in a hybrid optical-coaxial (HFC) manner, that is, only the last mile (or meters) toward the end-user are implemented via coaxial cable, while the rest of the network toward the operator core network is optical. For Internet access via the cable access networks, DOCSIS (Data Over Cable Service Interface Specification) is the CableLabs standard. In particular, CableLabs developed architectures for integrated cable networks for different services, including initially analog TV and radio, then digital TV, HDTV, video-on-demand (VoD), broadband Internet access, and telephone services provided via VoIP. A typical architecture for cable access networks is given in Figure 4.2. It consists of several network segments, which include home IP network, DOCSIS, PacketCable, and the core IP network.

DOCSIS exists in three main versions and several subversions, all listed in Table 4.3. Initially appearing as a standard DOCSIS 1.0 in 1997, it became the first cable standard for Internet broadband access. Further, DOCSIS 2.0 enhanced the upstream direction by providing higher bit rates up to 30 Mbit/s and adding VoIP services to DOCSIS, necessary for emulation of PSTN such as voice services. The great enhancement of capacity (i.e. bitrates) for IP-based access over cable access networks came with DOCSIS 3.0 (reaching aggregate speeds of up to 1 Gbit/s in downlink), which included support for IPv6. Further capacity and efficiency progression came with DOCSIS 3.1 in 2013, which evolved to the full duplex version (with up to 10 Gbit/s in both directions, downlink and uplink) in 2017 [9].

High bitrates with DOCSIS are obtained with channel bonding, which uses multiple physical TV channels for provision of broadband access (either in the downlink or the uplink direction). The bitrates given in Table 4.3 are aggregate bitrates on a single coaxial

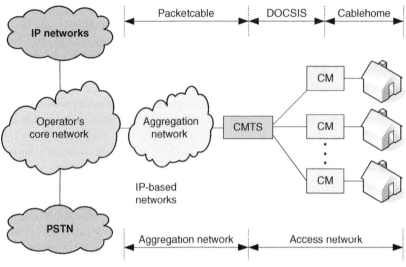

CM – Cable modem
CMTS – Cable modem termination system
DOCSIS – Data-over-cable service interface specifications

Figure 4.2 Cable access network architecture.

Table 4.3 DOCSIS standards.

Version of DOCSIS	Downlink	Uplink	Release date
DOCSIS 1.0	40 Mbit/s	10 Mbit/s	1997
DOCSIS 1.1	40 Mbit/s	10 Mbit/s	2001
DOCSIS 2.0	40 Mbit/s	30 Mbit/s	2002
DOCSIS 3.0	1 Gbit/s	100 Mbit/s	2006
DOCSIS 3.1	10 Gbit/s	1–2 Gbit/s	2013
DOCSIS 3.1 full duplex	10 Gbit/s	10 Gbit/s	2017

cable, so they are shared between all those using the same cable in the DOCSIS access network. However, with DOCSIS 3.1, full duplex can provide ultra-broadband speeds of over 100 Mbit/s per cable modem (CM) in both directions.

The PacketCable specification defines the interface used to enable interoperability of equipment for the transmission of packet-based voice, video, and other broadband multimedia services over an HFC network using DOCSIS. The main reason for the development of PacketCable was provisioning of packet-based voice communication for users connected to cable networks. Its architecture includes call signaling, accounting, configuration management, security, as well as PSTN interconnection. To guarantee QoS over the DOCSIS access network (which is also used for access to best-effort Internet services, such as web, email, etc.), PacketCable services are delivered with guaranteed priority in the DOCSIS access part and that ensures guaranteed bit rates and controlled latency (i.e. packet delay).

4.2 Ultra-Broadband Optical Access

Optical access is the long-term way to achieve ultra-broadband speeds per subscriber. Although copper-based access networks are still on the ground and with VDSL2 and similar access technologies have prolonged their lifetime in the telecommunication world, future access on the physical layer is targeted at all optical networks, due to the better transmission of optic compared with copper access. But let's first look through the history of fiber communication [10].

Although the invention of the laser and its demonstration for optical communications goes back to the 1960s, the real research on fiber optic communication started around 1975 [10]. Since then one may distinguish five different phases in the development of lightwave telecommunication systems:

- *Phase 1.* The first fiber systems operated near 850 nm (wavelength), with the first commercial products appearing around 1980. The bitrates were in the range 34–45 Mbit/s and repeaters placed on a distance of 10 km. This was also the era of plesiochronous digital hierarchy (PDH), which in Europe was based on digital hierarchies of 2 Mbit/s, 34 Mbit/s, and later 140 Mbit/s (8 Mbit/s was also specified but rarely used in practice), while in the United States PDH bitrates of 1.5 Mbit/s, 6 Mbit/s, 45 Mbit/s, and 140 Mbit/s (the latter was the same for both Europe and the U.S.) were available.
- *Phase 2.* In the 1980s wavelength region was moved toward 1300 nm, which provided the possibility to extend the distance between the repeaters. By 1988 commercial fiber systems operating up to 1.7 Gbit/s with a repeater distance of 50 km had appeared. In this phase came ITU-T recommendation G.957, which specified transmitting bitrates of up to 2.5 Gbit/s targeted for use with SDH and STM 16. As a useful note, STM-N has a bitrate of exactly $N \times 155.52$ Mbit/s, where 155.52 Mbit/s $= 2430 \times 64$ Kbit/s (i.e. 2430 time slots, each with bitrate of the ITU-T G.711 digital voice, the global standard in digital telephony in the 1980s and 1990s).
- *Phase 3.* In this phase the target was wavelength at 1550 nm over single mode fibers, to extend the distance between repeaters in excess of 100 km. Third-generation optical systems became operational at speeds of 2.5 Gbit/s in 1992. These systems were capable of operating at bitrates up to 10 Gbit/s (i.e. to carry STM-64 in SDH).
- *Phase 4.* Next came wavelength division multiplexing (WDM) systems. With WDM, optical systems' capacity started to double every six months, which was a kind of revolution at that time. By 1996 there was demonstrated transmission over optical networks of a distance of 11 600 km (with the possibility to go over 21 000 km, which is near to half of the perimeter length of the equator), which brought the possibility of using submarine cables. In this phase commercial transatlantic and transpacific optical cable systems were deployed, thus providing terrestrial communication between almost all countries on the planet (which introduced the possibility to decrease delays, compared with satellite communications which travel over longer distances from Earth to the satellites in a given orbit and then backwards). This also influenced the increase of QoS due to lower end-to-end delays on a global scale.
- *Phase 5.* This is the current phase of fiber systems' development (in the 2010s and around 2020). The main target in this generation is to further increase transmission capacity over optical cables, reaching speeds of multiple Tbits/s per optical cable (by using WDM). Each wavelength in WDM is called an optical channel (or simply,

a channel). With this notation, if one channel carries, for example, 10 Gbit/s and WDM multiplexes 160 channels on a single fiber cable, then the available bitrate is 1.6 Tbit/s (= 160 channels × 10 Gbit/s/channel). In this phase bitrates of 40 Gbit/s and 100 Gbit/s per channel (i.e. per wavelength) are also being developed, which further boosts the bitrates over a single cable to tens and even hundreds of Tbit/s (with dense WDM schemes). The higher bitrates per cable offer the possibility to use the already installed fiber cables infrastructures and to provide higher bitrates by changing the operating equipment (at the cables' ends), which is a prerequisite to increase bitrates also in the access parts and with that to improve QoS (because the bitrates are one of the most important QoS parameters in all types of networks).

Terabit per second speeds are targeted for transport networks, not for access networks, at least in the 2010s and 2020s. However, with the evolution of fiber-based technologies, they are expanding from the transport networks toward the access networks, including the last mile. There are different architectures for fiber implementations and design in the access part. The general name for fiber access networks is FTTx (fiber-to-the x), where x stands for home (FTTH), curb or cabinet (FTTC), building (FTTB), premises (FTTP), desk (FTTD), etc. With FTTH, optical connection reaches the end-user's home premises (Figure 4.3), while FTTB or FTTC can be used for different solutions, including HFC access networks. However, deployment of fiber requires larger capital investments in the network, therefore the penetration of fiber is gradual, starting from hot traffic zones such as urban areas (e.g. major cities and their downtowns) and then going toward suburban areas. One of the challenges for cost-effective FTTx deployments is minimization of the number of fibers (e.g. by using WDM over existing deployed fibers) as well as of the number of optical-electronic conversions

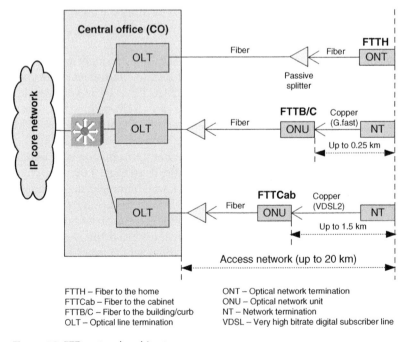

FTTH – Fiber to the home
FTTCab – Fiber to the cabinet
FTTB/C – Fiber to the building/curb
OLT – Optical line termination

ONT – Optical network termination
ONU – Optical network unit
NT – Network termination
VDSL – Very high bitrate digital subscriber line

Figure 4.3 FTTx network architecture.

on the way of the signals. All data processing, including buffering, scheduling, etc., at network nodes is done in the electronic domain, while transmission is carried over fiber.

In FTTx network architecture electronics is located in the optical line termination (OLT) on the telecom operator's side and in the optical network termination (ONT) on the end-user's side (where the optical link ends, in the home, cabinet, etc.). In the case of FTTC, the termination node is the optical network unit (ONU). The difference between ONT and ONU disappeared over time, so they will be used interchangeably.

Depending whether the fiber architecture is point-to-point or point-to-multipoint (P2MP), four basic architectures for the FTTx access network can be distinguished, which are also referred to as the optical distribution network (ODN):

- *Point-to-point architecture.* It uses a separate fiber link between the telecom operator and the user's home. So, the number of fiber links needed in this case equals the number of homes being connected. Obviously this architecture can provide maximum capacity because there is dedicated optical cable per home; however, it is the least cost-effective.
- *Passive optical network architecture.* This is P2MP architecture which uses a shared fiber link between the OLT and passive splitters toward the ONTs on the end-user's side (they perform signal splitting in the optical domain, without any electric power connection to the unit) between the shared fiber and ONTs.
- *Active optical network (AON) architecture.* It uses a shared P2P fiber link between the OLT on the telecom operator's side and an active remote switch (sometimes referred to as a curb switch), and P2P links between such active remote switch and ONTs. The active switches do optical-electronic conversion to process data in the switch and vice versa (on the next fiber link).
- *WDM PON architecture.* It is P2MP passive optical architecture with applied WDM (which multiplexes several wavelengths per single fiber).

The optical networks, including PON architectures, are based on TDM, which is necessary to have strict QoS provisioning over the optical network part. Legacy TDM-PON standards (e.g. gigabit PON – GPON) define line bitrates of up to 2.5 Gbit/s and a maximum length of links of 20 km. Commonly, use splitting for PON is 1 : 32 or 1 : 64 (1 : 64 means that feeder fiber splits to 64 fiber links on the way to end-users); however, up to 1 : 128 can be found in the products, but that is limited by optical power available at the ONT.

In general, FTTH can carry voice, video (including TV), and Internet data services to end-users. AON and PON multiplex the data on a single wavelength in a given direction, while different wavelengths are used for uplink (i.e. upstream) and downlink (i.e. downstream). With applied WDM, there is the possibility to carry different types of traffic on different wavelengths (e.g. voice on one wavelength, TV on another). WDM implementations can be based on PON or AON, where initially the more cost-effective solution is WDM-PON because in such cases less active equipment is needed to be deployed in the access network. Also, WDM-PON can be implemented as an upgrade of legacy PON by changing transceivers on both ends of the fiber, OLTs and ONTs.

Table 4.4 gives the main PON standards. The PONs with higher bitrates than GPON are referred to as next-generation PON (NG-PON). The standardization of NG-PON is carried out by ITU. The ITU standardization of 10 Gbit/s capable PON by ITU-T is called XG-PON, ITU-T G.987.1 [11], with 10 Gbit/s in downlink and 1 Gbit/s in uplink,

Table 4.4 PON standards.

	Maximum bitrate (Gbit/s)	Transmission	Standard
GPON	2.5	Ethernet, ATM, TDM	ITU-T G.984
NG-PON1	10	TDM, TWDM	ITU-T G.987 ITU-T G.9807
NG-PON2	40	TDM, TWDM	ITU-T G.989
1G-EPON	1	Ethernet	IEEE 802.3ah
10G-EPON	10	Ethernet	IEEE 802.3av

ATM, Asynchronous Transfer Mode; EPON, Ethernet PON; NG-PON, Next Generation PON; PON, Passive Optical Network; TDM, Time Division Multiplexing; TWDM, Time and Wavelength Division Multiplexing.

and XGS-PON with symmetrical 10 Gbit/s access in both directions (these standards are referred to as NG-PON1), and further toward the next-generation 40 Gbit/s capable PON (i.e. XLG-PON, also referred to as NG-PON2), ITU-T G.989 standard [12]. The NG-PON3 is targeted to provide bitrates up to a maximum 100 Gbit/s.

Besides the ITU, the IEEE also standardizes the PON, based on use of Ethernet technology. Hence, its standards are called Ethernet PONs (EPONs). The first such IEEE standard is 1G-EPON (or shortly, EPON), with symmetrical access of 1 Gbit/s in both directions. The next IEEE PON standard is 10G-EPON, which provides bitrates of up to 10 Gbit/s [13]. Similar to the ITU's work on PON, the third PON standard in the 2020s will be 25/50/100G-EPON (IEEE 802.3ca), aiming to provide bitrates gradually up to 100 Gbit/s [14].

4.3 QoS for Fixed Ultra-Broadband Access

Fixed broadband and ultra-broadband access networks are based on three main deployment approaches: DSL, cable, or optical access. In the previous subsection we introduced the main characteristic regarding network architectures as well as theoretical bitrates (in downlink and uplink) for each of the three main fixed access technology "groups." In this subsection the focus is more on the QOS solutions applied in the access part of each of the three types of fixed broadband access networks.

4.3.1 QoS for DSL Access

DSL technologies appeared in the 1990s and made progress in the 2000s, at a time in the packet-switching telecom world when ATM and Internet technologies coexisted. Many implementations of DSL access networks at the beginning were based on ATM QoS mechanisms. In ATM access nodes, every ATM permanent virtual circuit (PVC) is assigned a traffic class and a traffic profile. Typically, ATM traffic classes are scheduled by using strict priority. Internet fixed access is inseparably connected to using Ethernet as LAN technology, and in such cases QoS support is not mandatory for all Ethernet implementations. So, specific amendments to the Ethernet standard (IEEE 802.3) should be implemented with the aim to provide QoS traffic differentiation (via its separation into a limited number of classes). In Ethernet, the traffic class is determined on a frame by frame approach (on protocol layer 2) by examining the VLAN tag priority field, which provides eight different priority values. Such values can be mapped into lower number of classes (e.g. 2, 3, or 4), however. One important difference between ATM and Ethernet lies in the possibility of multiplexing multiple Ethernet traffic classes over one VLAN. Meanwhile, single ATM traffic class can be used per single ATM virtual circuit.

The QoS in DSL access networks is defined by the lower three layers in the protocol stack: (from the top to the bottom of the stack) the network layer, data-link layer, and physical layer. The network layer is Internet protocol, either IPv4 (as the dominant one in the DSL era) or IPv6 (which is typically considered as an optional for DSL) [15]. The data link layer for the IP traffic is Ethernet or ATM. On the physical layer, physical ports can include Ethernet, as well as ports for PDH and/or SDH interfaces. Considering the QoS approaches in the access part of the DSL network, one may distinguish among the following [15]:

- *ATM-based QoS (on layer 2).* This approach was the first phase of QoS for DSL access networks, based on ATM. However, with the aim to transfer IP traffic over ATM, IP QoS (e.g. DiffServ) should be mapped on the underlying ATM QoS classes.
- *L2 Tunneling protocol (L2TP).* The L2TP is typically used over IP as opposed to L2TP over ATM. A given L2TP tunnel can be associated with a certain traffic engineering specification, but the constituent flows in that L2TP tunnel do not receive differentiated service. So, the flows aggregated in a single L2TP tunnel receive the same aggregate QoS treatment.
- *IP QoS.* This is a network layer QoS provisioning, which includes standardized approaches, such as DiffServ (from the IETF), as well as customized approaches (e.g. chosen scheduling schemes and their combinations). In this case the main QoS control is done at BRAS as the gateway router which interconnects the IP core network of the telecom operator and the DSL access network.

Regarding the DSL access network, three models for QoS support can be distinguished, which include the following (they may be used separately or jointly, in the same access network):

- *Partitioning of bandwidth.* In such a case rate limits are established on broadband network gateways, based on business or application characteristics of the gateways that control each partition.
- *Distributed precedence and scheduling.* This marks services (based on layer 2 precedence) and lowest traffic classes will be dropped in the case of congestion appearance. This model provides weighted fairness among traffic classes of the same precedence, but it cannot provide fairness between users within the same class. Priority in this case is used to enhance the option with bandwidth partitioning (illustrated in Figure 4.4). However, if traffic bursts of higher-priority traffic above its partitioned rate are allowed, then it may discard the bandwidth portion for other traffic. So, the burst traffic will be delivered only in cases when there is excess bandwidth that is not being utilized by other traffic

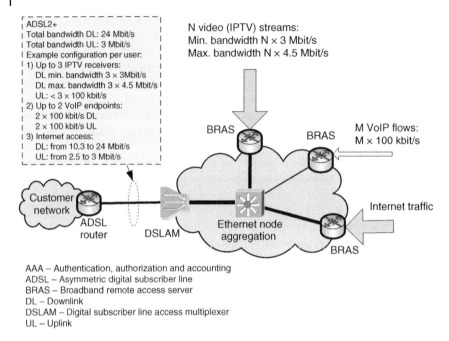

AAA – Authentication, authorization and accounting
ADSL – Asymmetric digital subscriber line
BRAS – Broadband remote access server
DL – Downlink
DSLAM – Digital subscriber line access multiplexer
UL – Uplink

Figure 4.4 Use of priority and scheduling for QoS provisioning in a DSL network.

- *Hierarchical scheduling (HS).* It must be supported on BRAS gateways. In that manner, to preserve IP QoS downstream toward customer premises, it uses packet classification, traffic shaping (which limits the peaks in the bitrate from the bursty traffic such as video or the Web), and hierarchical scheduling that is implemented as a logical tree-based network between the BRAS and the residential (i.e. home) gateways. While for ATM there are five levels of hierarchy (corresponding to session, VC, group of VCs, virtual path, and port), in Ethernet-based DSL there is a minimum of three levels of hierarchy (BRAS or broadband network gateway port, access node uplink, and synch rate between the network and residential/home gateways in the access local loop).

4.3.2 QoS for Cable Access

The DOCSIS standards 3.0 and 3.1 define a set of QoS mechanisms on MAC and upper layer protocols based on a fundamental network resource management construct: service flow. Service flow is defined as a MAC-layer transport service which is created to provide unidirectional transport of packets from the upper layer service entity to the physical layer of the cable (the RF), as well as to shape, police, and prioritize traffic according to a set of QoS traffic parameters defined for the given flow [16].

As typical for all fixed access networks with a shared medium (a single cable in this case), the uplink and downlink directions are treated differently regarding QoS at the end node of the cable access segment, which is the CMTS (cable modem termination system). This is in fact a consequence of the fact that uplink RF channels are contentious and shared-access mediums, based on a topology with many CMs connected to a single

CMTS. On the other side, the downlink RF channels behave similarly to a traditional IP router where IP packets arrive and are queued and then forwarded to one (for unicast) or multiple (for multicast traffic, such as TV) destinations. So, applied QoS mechanisms are different for service flows in uplink (i.e. upstream) and downlink (i.e. downstream) directions. There are five service flow scheduling mechanisms for the uplink and one scheduling mechanism for the downlink. Upstream service flows may be defined with one of five service flow scheduling types:

- *Best-effort.* This is a standard contention-based resource management which is based on a FCFS (First Come First Served) principle, although it is being coordinated by the CMTS scheduler. For this scheduling type it is possible to set a limit on the maximum bitrate for a given service flow.
- *Non-real-time polling.* This is a reservation-based resource management approach in which the CMTS polls CMs on a fixed time interval basis to determine whether service flow data has been queued for transmission, and, if so, to provide a transmission grant for the given service flow.
- *Real-time polling.* This is similar to non-real-time-polling, with the main difference in the duration of the fixed polling interval, which in this case is usually very short (e.g. shorter than 500 ms). Both polling scheduling types are more suitable for traffic which has variable bit rates and is more tolerant to delay and throughput.
- *Unsolicited grant.* This is reservation-based resource management based on a fixed-size grant that is allocated to a given service flow at almost fixed time intervals, which is done without polling of the CMs by the CMTS. This type is convenient for traffic with constant bitrate such as voice traffic.
- *Unsolicited grant with activity detection.* This is a reservation-based resource management which is in fact a hybrid of the polling and unsolicited grant scheduling types. Here CMTS provides CMs with fixed grants at nearly fixed time intervals as long as data is queued for transmission in the upstream. While the CM is non-active, the schedulers go into the polling mode of operation.

In the downstream direction (from CMTS toward the CMs), service flows are defined with the same QoS parameters that are used for the best-effort scheme in the upstream direction.

For all scheduling types, all dynamic service flows may be in one of the three logical states: authorized state (network policy rules are applied, resulting in an authorization), admitted state (resources are reserved by the scheduler for inactive service flows, which assures that subsequent activation requests will succeed), and active state (here the service flow is already activated and QoS-marked packets can traverse). Speaking literally, one should note that DOCSIS in fact does not define states, but they are logically constructed to describe the resource management (i.e. QoS management) which is performed by the CMTS for the cable data traffic.

Optionally, DOCSIS 3.1 provides hierarchical QoS (HQoS) [17]. The HQoS aims to provide an intermediate level in the scheduling hierarchy between service flows on one side (upper protocol side) and physical channels (in the cables) on the other side, by introducing aggregate traffic QoS treatment (via aggregated service flows – ASFs). There are two options for HQoS (for implementations). One is to provide aggregation of unicast service flows which are associated with a single CM, and the second option is to aggregate service flows coming from multiple CMs but sharing the same cable toward the CMTS. In this manner, the HQoS enables telecom operators (which use access networks with DOCSIS) to define policies on aggregated service flows. It is based on a strict hierarchical approach, where the available total channel capacity is the root and each "child" node can have only one parent node (that means the term "strict" hierarchy). So, one may note that HQoS is in fact a CMTS-only QoS feature because no CMs will be aware of the existence of the HQoS.

Regarding the PacketCable part of the cable network architecture, PacketCable 2.0 from CableLabs introduces an IP multimedia subsystem, which is standardized by the 3GPP (3G Partnership Project) [18]. The PacketCable QoS architecture (high-level view) is shown in Figure 4.5.

The PacketCable application manager (PAM) receives session-level QoS requests from the IMS. PacketCable 2.0 is designed to perform policy control for access to services by end-users (via their CMs) in the same manner as implemented by the 3GPP the architecture, with PCRF (policy and charging rules function) node between the CMTS system on one side (the cable network access) and the IMS on the other (the telecom operator's core network).

4.3.3   QoS for PON Access

For QoS support the same approaches which are network access agnostic are used for PON and GPON as they have been previously defined for DSL access networks and triple play services (voice, TV, and Internet access) by the Broadband Forum [19]. However, GPON technology on the network interface layer is different that of DSL, so mapping of GPON to Ethernet is required and vice versa. GPON supports two layers of traffic encapsulation [20]:

- Encapsulation of Ethernet frames into GEM (GPON encapsulation method) frames.
- Encapsulation of GEM into the GPON transmission convergence (GTC) frame, which includes pure TDM traffic as well as ATM cells which are 53 bytes fixed-length

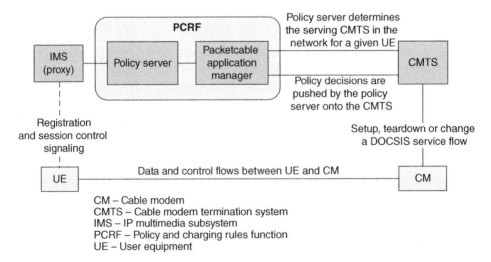

Figure 4.5 PacketCable QoS architecture.

packets. Although ATM lost the "battle" with the Internet technologies in the 1990s, certain QoS solutions in ATM are well defined and they are still being used on a link level for QoS provisioning in broadband access links. The downstream GTC frame is broadcast to every ONT with the aim of ONT using the synchronization information carried in the physical control block downstream (PBCd) field.

For the purposes of synchronization between the OLT and ONTs, the downstream frames are transmitted continuously even when there is no user data to carry in them. Meanwhile, the upstream traffic contains multiple transmission bursts coming from multiple ONTs. Each of the upstream burst frames contains physical layer overhead upstream (PLOu), which is in fact bandwidth allocation interval that contains the dynamic bandwidth report upstream (DBRu) field as well as allocation identifiers (Alloc-IDs). When the upstream traffic from ONTs arrives at the OLT, it is scheduled based on the class of service (CoS) dependent on the type of transmission containers (T-CONTs) specified in the Alloc-ID. The T-CONT frames are used for establishing a virtual connection between ONTs and the OLT. In GPON there are five defined types of T-CONT frames (Table 4.5) which one may find equal to five QoS classes in the upstream direction. Single T-CONT may have one or multiple GEM ports. Then, a GPON interface on an ONT may contain one or multiple T-CONTs.

Because different ONTs may be deployed on different distances of the same OLT to which they are connected, the OLT also has a ranging function to avoid overlapping of frames coming from different ONTs in the uplink direction. The ranging is based on calculating a specific delay time (also called equalization delay) for each of the connected ONTs. Another tool for improving QoS in the uplink direction is the dynamic bandwidth assignment (DBA), which enhances the uplink utilization of PON ports by providing the possibility for higher-bandwidth services for users who require greater change in the allocated bandwidth. DBA is based on service priorities, where the priorities are set separately for each ONT, as well as minimum and maximum bandwidth for each ONT. Typically, higher-priority service is voice, then video, and after that data services (data services typically refer to network-neutral Internet access service). The QoS specifics for different services are specified in the service level agreements. The OLT on

Table 4.5 QoS types for GPON.

QoS type (T-CONT type)	QoS requirements	Bandwidth type	Traffic type
T-CONT type 1	Jitter = 0, very low delay	Fixed	Constant Bit Rate (CBR) traffic such as voice
T-CONT type 2	Higher delay than T-CONT 1	Assured	Committed Information Rate (CIR) traffic such as TV
T-CONT type 3	Does not guarantee delay	Non-assured	Variable Bit Rate (VBR) traffic such as video
T-CONT type 4	Best effort approach	Best-effort	Web browsing, FTP, email, etc. (Internet access)
T-CONT type 5	Mix of the previous four types of T-CONT types	Best-effort	General traffic

the operator's side grants the bandwidth based on the SLA, service priorities, and the current state of the ONT. Higher priority refers primarily to lower delay (necessary for voice services, for example) or higher assured bandwidth (e.g. for IPTV, which is TV broadcast over the IP network with guaranteed QoS).

The first NG-PON1 refers to 10 Gbit/s PON, i.e. XG-PON. It was created to support multiple services across multiple market segments, which included the consumer segment, business services, and the mobile backhaul network. Such services in G-PON and XG-PON include Internet access, IPTV, and VoIP, as well as certain legacy services, such as digital telephony (i.e. plain old telephone systems – POTS). In addition, an XG-PON provides access to carrier-grade Metro Ethernet services, such as P2P, P2MP, and rooted-multipoint Ethernet virtual connections (EVCs) services, also known as E-Line, Ethernet LAN (E-LAN), and E-Tree [11], as defined by the Metro Ethernet Forum (MEF). The heterogeneous services provided via optical access networks such as PON require QoS support and hence certain traffic management mechanisms.

Regarding QoS, XG-PON (i.e. NG-PON1) and NG-PON2 support POTS voice quality with guaranteed fixed bandwidth, low delay, and low jitter required for voice services. In that manner, NG-PON technologies are developed to provide TDM services (e.g. for business customers), and mobile backhauling applications with low delay and jitter as well as guaranteed bandwidth.

Because the typical telecom operators' offers to both individual (i.e. residential) and business customers are targeted at provision of VoIP (as a PSTN replacement), IPTV (as a TV replacement), ultra-broadband Internet access (based on network neutrality and best effort principles), and VPNs as leased lines replacement for business users, NG-PON needs also to support at least four classes of services (and should support six classes) and to appropriately map each of the user-to-network interface (UNI) flows. The QoS requirements regarding the number of classes and their mapping are set initially for GPON [21], and they continue to be supported in the NG-PONs, which go up to 10 and on longer-term 100 times higher bitrates than GPON, respectively.

Further, NG-PON1 and NG-PON2 also support different mixes of both consumer and business users within a multiple subscriber ONT/ONU, that is, a mix of bitrate-based (e.g. with traffic policing, traffic shaping, etc.) and priority-based traffic management mechanisms within a single ONT/ONU on the same optical network. NG-PON technologies also provide support for VLAN services, including N : 1 and 1 : 1 VLANs, as well as business Ethernet services on the PON.

Initially only one priority queue for each T-CONT was specified. However, NG-PON 1 and 2 (for different types of use cases) require multiple queues per T-CONT and thus scheduling between such queues (in the same T-CONT) is needed [22, 23]. One use case of such an approach is when a service provider assigns a VLAN to a TCONT, but at the same time it needs to give different priorities to different flows in the given T-CONT (illustrated in Figure 4.6). This is achieved by use of mapping between specific P-bit values within the given VLAN to separate queues, and further applying strict priority as the scheduling scheme for the traffic from the queues into the T-CONT for the given VLAN.

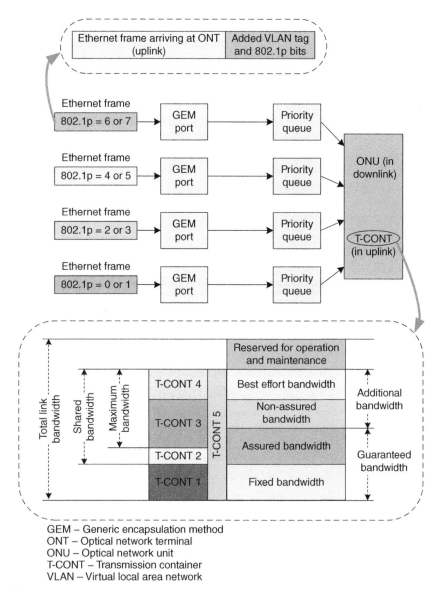

GEM – Generic encapsulation method
ONT – Optical network terminal
ONU – Optical network unit
T-CONT – Transmission container
VLAN – Virtual local area network

Figure 4.6 Priority scheduling for multiple queues in T-CONT in GPON/NG-PONs.

4.4 QoS in Ethernet and Metro Ethernet

In the past, telecom operators' core and transport networks were primarily built on SDH technologies. However, Ethernet as a unified local area network technology since the 1990s has spread to metro and to transport environments, including providers', corporate, and home networks. The organization which supports the expansion of

Ethernet from LAN to MANs (metropolitan area networks) and further to WAN and international networks is MEF. In general, Ethernet (IEEE 802.3 standard) has become unified LAN due to lower cost of the equipment, lower operation and maintenance costs, and dedication to transport IP packets on the network layer. In general, IEEE is developing Ethernet standards on physical and data-link protocol layers. Initially Ethernet has no QoS built-in and no TDM support such as legacy telecommunication networks for transport of voice services (e.g. PSTN/ISDN). With moving to Carrier Ethernet (that is, Ethernet for telecom operators), certain end-users are demanding the same level of performance that they had in the past with TDM-based WAN [24].

The evolution of Ethernet toward service providers' requirements and telecom operators' networks has led to the MEF's definition of "Metro Ethernet network" or "Carrier Ethernet" (the names are interchangeable). For the development of Carrier Ethernet, the requirements were to provide network scalability, QoS for different user types (the same performance as in TDM-based access and transport networks), reliability (in a similar manner as found in SDH transport networks, e.g. 50 ms link recovery after a failure), the possibility to guarantee the SLA, and provision of TDM emulation services [25].

There are two main types of interfaces in Carrier Ethernet:

- *User-network interface (UNI)*. This is the interface between the customer network and the network operator, which separates the responsibilities between the end-user and the operator as network provider.
- *NNI (network-network interface)*. This is the interface between two Carrier Ethernet networks.

The bitrates on the physical interfaces are the same as those defined for the LAN Ethernet standards, which include 10, 100 Mbit/s, 1, 10, and further up to 100 Gbit/s in the 2020s. Further, each of the Ethernet services can be provided as port-based or as a VLAN service.

The MEF defines Ethernet-based services in metropolitan areas by using Ethernet virtual connections (EVC). There are three main types of EVCs, also referred to as service types [26]:

- *Ethernet line (E-Line)*. P2P EVC.
- *Ethernet LAN (E-LAN)*. Multipoint-to-multipoint EVC.
- *Ethernet tree (E-Tree)*. Routed multipoint EVC.

There is also an operator virtual connection (OVC) when an EVC spans multiple Carrier Ethernet networks, so end-to-end appears as a composition of segments (from different Carrier Ethernet networks) concatenated with the aim of forming an EVC. Such segments are in fact OVCs, the association of UNIs and ENNIs (external network to network interfaces) within a single Carrier Ethernet network.

The QoS for EVC-based services is implemented by using several IEEE standards, which include IEEE 802.1Q for VLAN [27], IEEE 802.1ad for provider bridges [28], and IEEE 802.1ah for provider backbone bridges (PBBs) [29].

- *Virtual LAN (VLAN)*. The E-Line service uses the IEEE 802.1Q standard for VLANs. They are created to provide virtualization of the Ethernet infrastructure, so the different traffic can be separated into different VLANs. Each VLAN is identified by a Q-tag (also called a VLAN ID), which has a length of 12 bits, resulting in 4096 different values (from which two are reserved for network administration). Such an approach works well in a single enterprise. However, when multiple customer networks are connected to the Internet via a shared Ethernet infrastructure by using virtual LANs, the VLANs are separated by using different Q-tags for different VLANs. For example, VLANs can be a convenient solution for virtual separation of the same physical Ethernet infrastructure within a single enterprise.

- *Provider bridges (IEEE 802.1ad)*. Provider bridges are introduced with IEEE 802.1ad [28]. They put an additional tag in the Ethernet frame (besides the Q-tag), called the S-tag (or provider ID). Its length is 12 bits (the same length as the Q-tag). Due to two same-length tags (S-tag, i.e. Provide ID, and Q-tag, i.e. VLAN ID), this approach is also called Q-in-Q.
- *Provider backbone bridges (IEEE 802.1ah)*. PBBs are standardized with IEEE 802.1ah [29], which adds source and destination MAC addresses in Ethernet frames (at UNI) when they enter the PBB network. For this purpose PBB uses an additional tag, called I-tag (with length of 24 bits), to identify the services (such as Q-in-Q networks being interconnected via the PBB). For the network deployment, only switches on the edge of the PBB network should be PBB-enabled, while other Ethernet switches in the core PBB network can be only provider bridges.

Figure 4.7 illustrates the use of S-tags, Q-tags, and I-tags for deployment of Carrier Ethernet networks.

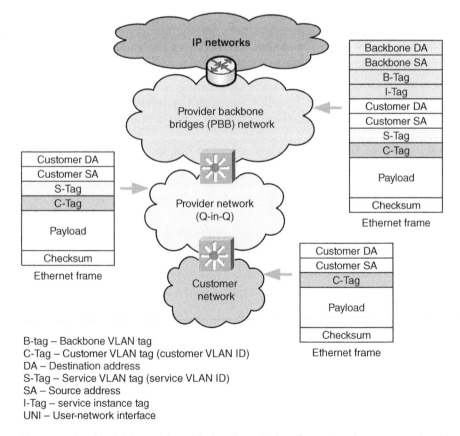

B-tag – Backbone VLAN tag
C-Tag – Customer VLAN tag (customer VLAN ID)
DA – Destination address
S-Tag – Service VLAN tag (service VLAN ID)
SA – Source address
I-Tag – service instance tag
UNI – User-network interface

Figure 4.7 Provider bridges and provider backbone bridges for carrier ethernet network architecture.

In some network environments synchronization is highly important for certain events, such as handovers in mobile networks. So, when Carrier Ethernet is used for mobile backhaul networks, there is a need for synchronization for mobile applications and services in a similar manner as with TDM-based transport technologies. That is particularly required to minimize interference in the radio access network, as well as to support handovers in the cellular radio access network (handing over calls/sessions between neighboring cells), as well as to satisfy certain regulatory requirements on QoS. In general, synchronization refers to distribution of common reference for time and/or frequency to all networks (or network segments) to align their frequency and/or time scales. For example, different generations of mobile technologies have different demands regarding synchronization. However, most synchronization requirements are related to distribution of a common frequency reference within the mobile network, expressed in unit ppb (parts per billion). The ppb for global systems for mobile communication (GSM) (2G), universal mobile telecommunication system (UMTS) (3G), long-term evolution (LTE)/LTE-Advanced (4G), should be less than ± 50 ppb for wide-area base station/eNodeB with cells sizes in range of kilometers, ± 100 ppb for local area base station/eNodeB, and ± 250 ppb for home base station/eNodeB [30].

4.4.1 Class of Service for the Carrier Ethernet

The main approach of QoS provisioning in the Carrier Ethernet network is related to CoS. Bandwidth profiles per EVC and per CoS are governed by six parameters:

- *CIR (committed information rate).* This defines assured bandwidth, which is assured via bandwidth reservation based on traffic engineering.
- *EIR (excess information rate).* The EIR bandwidth is considered "excess" and it improves the network's goodput. The excess traffic is dropped at congestion nodes in the network.
- *CBS/EBS (committed burst size/excess burst size, two parameters which are expressed in bytes).* Typically, higher burst size gives better performance.
- *Color mode.* The carrier Ethernet can be "Color Aware" or "Color Blind." When set as "Color Aware" this parameter governs discard eligibility, with marking usually done at ingress network nodes. The defined colors are:
  - *Green:* used for forwarded frames and CIR conforming traffic.
  - *Yellow:* used to mark for discarding of eligible frames, which are over CIR but within the given EIR.
  - *Red:* used for marking of discarded frames which exceed the given EIR.
- *Coupling flag (it is set to either 1 or 0).* This parameter governs which frames are classed as yellow.

Bandwidth profiles are used to divide the available bandwidth per EVC over a single UNI. There are multiple services over the same port (at UNI), and CoS markings provide the possibility for the network to determine the required QoS.

Overall, there are several phases in the development of Metro Ethernet (i.e. Carrier Ethernet) CoS. In CoS phase 1 the MEF specifies a three-CoS model and allows for subsets and extensions. Also, it provides possibilities for interconnections of Carrier Ethernet networks with implementation of CoS models. In that manner priority code point/differentiated services code point (PCP/DSCP) values are used as part of the CoS ID. The PCP values are mandatory at ENNI (for the purposes of interconnection), while they are only optional at UNIs. The further CoS phases (e.g. phase 2 and phase 3) further extend and specify the QoS details via CoS performance objectives (CPOs). They also include new performance tiers. In general, the performance objectives apply to qualified frames in EVC or OVC. An Ethernet frame which arrives at an external interface (EI) may have different delay in cases with different geographic distance between different pairs of EIs and that has a significant influence on the frame delay (the longer the distance between the EIs, the larger the delay). The defined implementation agreement (IA), according to the MEF terminology, is created to provide guidance to service providers, telecom operators, and subscribers by specifying five sets of CPOs called performance tiers (PTs), given in Table 4.6. Each set includes objectives for seven performance metrics for P2P and multi-point CPOs.

An example of bandwidth profiles for mobile backhaul (e.g. for 4G) is given in Figure 4.8. The extension of the Carrier Ethernet QoS is via the usage of bandwidth profiles and traffic management techniques, based on quantification of delay, delay variation, Ethernet frame loss ratio, network availability, and so on. The CoS models give the possibility to define the SLA at the UNI.

The MEF defines three particular CoS names, which are also noted as CoS labels:

- *H (High) label.* This is targeted to applications which are very sensitive to delay, delay variation, and losses, such as mobile backhaul control traffic and VoIP.
- *M (Medium) label.* This is targeted to applications which are less sensitive (i.e. more tolerant) to delay and/or delay variation, but are sensitive to losses. These are critical data applications and near real-time ones.
- *L (Low) label.* This is targeted to non-critical applications, which are more tolerant to delay, delay variation, and losses.

In Carrier Ethernet there is a CoS frame set (CoS FS) which is defined as association of a CoS name and a set of ordered end-point pairs (OEPPs). The CoS EF represents all Ethernet frames which traverse between a particular set of EVC/OVC endpoints within a particular CoS. Each of the CoS FSs within a given EVC or OVC should be associated

Table 4.6 Performance tiers for Carrier Ethernet.

Performance tier (PT) number	Performance tier (PT) name	Distance, delay
PT0.3	City PT	<75 km, 0.6 ms
PT1	Metro PT	<250 km, 2 ms
PT2	Regional PT	<1200 km, 8 ms
PT3	Continental PT	<7000 km, 44 ms
PT4	Global PT	<27 500 km, 172 ms

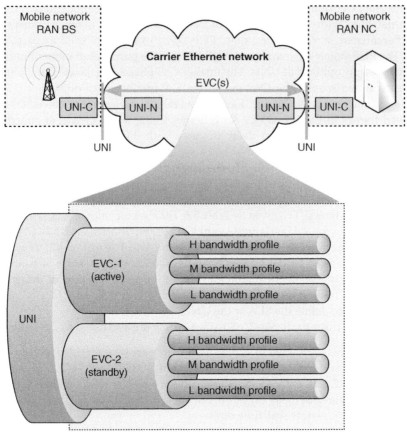

EVC – Ethernet virtual connection
RAN BS – Radio access network base stations
RAN NC – Radio access network network controllers

UNI-C – User-network interface-customer
UNI-N – User-network interface-network

Figure 4.8 Metro Ethernet bandwidth profiles for mobile backhaul.

with one of the defined performance tiers. There are certain requirements regarding the flows associated with different CoS labels. For example, CoS labels H and M must be satisfied CBS ≥ MFS. Flows associated with CoS label L must have CBS ≥ MFS or EBS ≥ MFS or both. Overall, the performance objectives specified by MEF [31] must apply to CoS FS.

The CPOs for Carrier Ethernet (defined per performance tier) are given in Table 4.7. There are several defined performance metrics for the CoS IA [31]. For example, the IA requires support for at least one of the two defined delay parameters, one-way frame delay (FD) and one-way mean frame delay (MFD) [31]. Also, IA requires support for at least one of the two performance metrics on delay variation to be included in SLS: one-way inter-frame delay variation (IFDV) and one-way frame delay range (FDR).

Table 4.7 Carrier Ethernet CoS performance objectives (CPO) per PT for point-to-point EVC.

	CoS label	FD (ms)	MFD (ms)	IFDV (ms)	FDR (ms)	FLR (percent)
	Label H	≤ 3	≤ 2	≤ 1	≤ 1.25	≤ 0.00001 i.e. 10^{-5}
PT0.3	Label M	≤ 6	≤ 4	≤ 2.5 or N/S	≤ 3 or N/S	≤ 0.00001 i.e. 10^{-5}
	Label L	≤ 11	≤ 9	N/S	N/S	≤ 0.001 i.e. 10^{-3}
	Label H	≤ 10	≤ 7	≤ 3	≤ 5	≤ 0.0001 i.e. 10^{-4}
PT1	Label M	≤ 20	≤ 13	≤ 8 or N/S	≤ 10 or N/S	≤ 0.0001 i.e. 10^{-4}
	Label L	≤ 37	≤ 28	N/S	N/S	≤ 0.001 i.e. 10^{-3}
	Label H	≤ 25	≤ 18	≤ 8	≤ 10	≤ 0.0001 i.e. 10^{-4}
PT2	Label M	≤ 75	≤ 30	≤ 40 or N/S	≤ 50 or N/S	≤ 0.0001 i.e. 10^{-4}
	Label L	≤ 125	≤ 50	N/S	N/S	≤ 0.001 i.e. 10^{-3}
	Label H	≤ 77	≤ 70	≤ 10	≤ 12	≤ 0.00025 i.e. 2.5×10^{-4}
PT3	Label M	≤ 115	≤ 80	≤ 40 or N/S	≤ 50 or N/S	≤ 0.00025 i.e. 2.5×10^{-4}
	Label L	≤ 230	≤ 125	N/S	N/S	≤ 0.001 i.e. 10^{-3}
	Label H	≤ 230	≤ 200	≤ 32	≤ 40	≤ 0.0005 i.e. 5×10^{-4}
PT4	Label M	≤ 250	≤ 220	≤ 40 or N/S	≤ 50 or N/S	≤ 0.0005 i.e. 5×10^{-4}
	Label L	≤ 390	≤ 240	N/S	N/S	≤ 0.001 i.e. 10^{-3}

FD, one-way frame delay; MFD, mean frame delay; N/S, not specified; IFDV, inter-frame delay variation; FDR, frame delay range; FLR, frame loss ratio.

4.5 End-to-End QoS Network Design

QoS is always an end-to-end requirement. For the end-user, it is important that the QoS satisfies their needs and provides high user experience. Of course, that refers to humans as end-users (regarding the end-user experience). For machine-to-machine (M2M) communication, the QoE has no meaning, but QoS has an end-to-end meaning and it is equally important as in the cases of human-to-human (e.g. telephony) or human-to-machine (e.g. Web browsing) communication.

In general, services in telecommunications including IP-based platforms and NGNs are performed using two models [32]:

- *Vertical model.* This consists of three layers: lower protocol layers (physical and data-link layers as part of the network interfaces), the IP layer, and higher protocol layers (transport protocols such as UDP and TCP, up to the applications on the top).
- *Horizontal model.* This refers to horizontal configuration and interconnection of IP-based networks, which is primarily composed of two types of sections: network section and exchange links. The IP network components include source and destination hosts, routers, and links (link is a P2P physical or virtual connection used for transporting IP packets between a pair of end hosts, which operates below the IP protocol layer).

Understanding the performance features of both the horizontal and vertical model aspects is required for identifying service performance. The vertical model reflects the design of the application and the OS of the given host or router (as network node) as well as connection of OS with applications through the API via the sockets and connection to the network interfaces on the given machine below. Regarding the vertical model, QoS depends also on the capabilities of the host or router regarding the processing power of the machine, operational memory, and optimization of the applications and OS as well as OS and network interfaces in use for communication. The horizontal model reflects the network architecture, the positioning of network nodes (routers, gateways), the nodes' functionalities (e.g. in which nodes and where the signaling functions are implemented, for example in mobile networks), and server hosts (data centers, CDNs), and their connection (e.g. connection of the CDN to a certain network provider) and interconnection (each IP network must be connected to other neighboring IP networks, which may belong to different telecom operators and also to different countries and regions around the globe).

4.5.1 End-to-End Network Performance Parameters for IP-based Services

The transition from PSTN/ISDN to all-IP networks and Internet technologies-based services which run on the application layer over TCP/IP or UDP/IP protocol stack requires definition of end-to-end network performance parameters. Such parameters are defined by the ITU (performance parameters are also being defined by other organizations, but we will focus on those from the ITU as the largest organization in the world for standardization of ICT/telecommunications), to be used in specifying and assessing IP network performance. In that regard, the most important performance parameters are the following [33]:

- *IPTD (IP packet transfer delay).* This is calculated over all successful and errored packets and refers to time difference between the occurrences of two corresponding IP packet reference events (an IP packet reference event is a packet transmission via a given measurement point in the network). There are several possible definitions or uses of this parameter. Minimum IPTD is the smallest IP packet delay, median IPTD refers to 50th percentile of the frequency distribution of IPTDs, while average IPTD is calculated as an average of all measured IPTDs.
- *IPDV (IP packet delay variation, also referred to as jitter)*: This is the difference between the one-way delay of IP packet and reference IPTD (e.g. average IPTD as a reference delay).
- *IPLR (IP packet loss ratio).* This is calculated as the ratio of the total number of lost IP packets to the total number of transmitted IP packets.
- *IPER (IP packet error ratio).* This is calculated as the ratio of the total number of IP packets with some errored bit(s) to the total number of transmitted IP packets.

The values of these performance parameters can be different for different traffic types, which are mapped to network QoS classes. Of course, the use of the classes is not mandatory in all network environments, and may differ from one network to another.

One may note that the transfer capacity (typically expressed in bit/s, i.e. Kbit/s, Mbit/s, Gbit/s, etc.) is the network performance parameter that has certainly the highest impact on the performance perceived by the end-user. Why? Because higher bitrates end-to-end provide a better user experience due to better performances for most of the services (e.g. higher transfer capacity as well as lower delay and delay variation). Note that the theoretical "net" capacity of a given system (e.g. cable, optical, wireless/mobile) is always larger than the bitrates that can obtained by individual end-user applications.

Regarding end-to-end communication, the network model is also referred to as UNI-to-UNI, where UNI is the interface that connects the user premises network such as home network or enterprise network on one side and the telecom operator network on the other side [32]. End-to-end interconnected IP networks may support user-to-user connections, user-to-host connections, and other endpoint variations.

In the network presentation, the network sections (NSs) can be represented as clouds with edge routers on their borders and a certain number of interior routers which may have different functions (e.g. routers on the network edges, with additional functions, are referred to as gateways). What influences the number of network sections on the end-to-end path? Well, that depends upon the CoS offered, the geographic area through which the given network section is deployed, the data servers' location within the CDN of the service providers (e.g. OTT service provides), and interconnection agreements between telecom operators (which are also influenced by business relations). In general, there can be one network section (e.g. in the case of local traffic, which has both source and destination hosts in the same network section) or more network sections on the path. However, Internet architectures allow networks to come and go on the run, so the network sections used for transport of the packets may change over time. It is less likely that the network path will change during the lifetime of a given flow (e.g. during seconds or minutes), although in some cases it is possible or even mandatory (e.g. at handovers in mobile networks); however, transport networks (e.g. international interconnection networks) which interconnect core networks of telecom operators are less likely to change the architectures and routing principles (implemented via routing tables and policies in network routers and gateways) in a short period of time. In general, most of the routes do not change in short time intervals (e.g. seconds, minutes, or even hours). As with circuit switched communication in the past, the IP connectivity spans international boundaries, but it does not follow and does not need to follow the circuit-switched conventions, simply because the principles of IP communication end-to-end differ from circuit-switching at least in two aspects: (i) there is no mandatory need for reservation of resources end-to-end in the case of IP communication; and (ii) it is not mandatory to establish the path over the public Internet network between the two (or more) end-hosts by using signaling prior to the communication (e.g. any client host connected to the public Internet via any access network may contact any publicly available server without any signaling prior to the session/call or any reservation of network resources on the end-to-end path).

The end-to-end performance of a path can be estimated if the performances of all subsections on the path are known (Figure 4.9). For that purpose there can be different parameters; however, if one needs to distinguish the most important standardized QoS parameters for IP-based services (from the end-to-end point of view, which is the only point of view for any QoS observation), they will be the following [32]:

- *Mean transfer delay*. This refers to the sum of the mean delays (IPTDs) contributed by network sections on the path UNI-UNI.

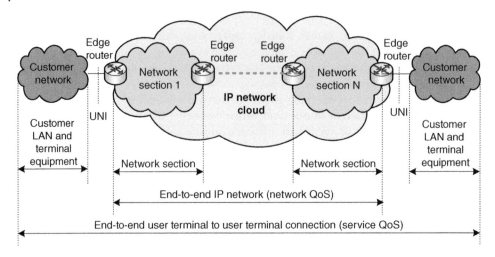

Figure 4.9 End-to-end IP network QoS.

- *Delay variation.* This is UNI-UNI delay variation which is influenced by IPDV from the network section on the path between the two ends. However, each network section influences this parameter via its traffic management techniques, which may differ from one network section to another. Therefore, it is difficult to estimate accurately the delay variation without having considerable information about individual delay distributions in different network sections. Such detailed information can only occasionally be shared among telecom operators (it is not a rule). Therefore, the network end-to-end delay variation (i.e. UNI-UNI IPDV) estimation may have certain limitations regarding accuracy.
- *Error packet ratio.* For estimation of IPER across network sections (which have their own IPER) to obtain UNI-UNI performance one should proceed by inverting the probability of error-free packet transfer across n network sections, given as follows:

$$\text{IPER}_{\text{UNI-UNI}} = 1 - \{(1 - \text{IPER}_{\text{NS1}}).(1 - \text{IPER}_{\text{NS2}}).(1 - \text{IPER}_{\text{NS3}}).\ldots.(1 - \text{IPER}_{\text{NSn}})\} \tag{4.1}$$

- *Loss ratio.* The UNI-UNI performance regarding the IPLR parameter can be estimated by inverting the probability of successful packet transfer across n network sections on the path end-to-end, given as follows:

$$\text{IPLR}_{\text{UNI-UNI}} = 1 - \{(1 - \text{IPLR}_{\text{NS1}}).(1 - \text{IPLR}_{\text{NS2}}).(1 - \text{IPLR}_{\text{NS3}}).\ldots.(1 - \text{IPLR}_{\text{NSn}})\} \tag{4.2}$$

The units of IPLR values should have a resolution of at least 10^{-9} to include different types of transport technologies, including the optical transport networks as dominant ones, which have the lowest loss ratio (in the range of 10^{-9}).

4.5.2 QoS Classes by the ITU

Network QoS classes are specified by ITU-T [32], based on the requirements of the main applications/services such as telephony, digital TV, and video streaming on demand,

Table 4.8 ITU QoS classes.

QoS Class	Upper bound on IPTD (milliseconds)	Upper bound on IPDV (milliseconds)	Upper bound on IPLR	Upper bound on IPER
Class-0	100	50	10^{-3}	10^{-4}
Class-1	400	50	10^{-3}	10^{-4}
Class-2	100	Unspecified	10^{-3}	10^{-4}
Class-3	400	Unspecified	10^{-3}	10^{-4}
Class-4	1000	Unspecified	10^{-3}	10^{-4}
Class-5	Unspecified	Unspecified	Unspecified	Unspecified
Class-6	100	50	10^{-5}	10^{-6}
Class-7	400	50	10^{-5}	10^{-6}

data applications based on use of reliable TCP (e.g. Web, email), and others. The QoS classes and their upper bounds on the main QoS network parameters are shown in Table 4.8.

In general, one may use the defined QoS classes for certain types of application considering their QoS requirements, given as follows:

- Class-0 and Class-1 are appropriate for real-time applications that are sensitive to mean delay and delay variations (e.g. VoIP, video telephony services) where Class-0 has a lower bound on the IPTD parameter and hence it can be used for voice (without video), gaming, or some time-critical smart services (e.g. driverless vehicles).
- Class-2 and Class-3 are targeted at transaction data, from which class-2 is intended for signaling traffic, while class-3 is for interactive applications.
- Class-4 is targeted at short transactions, video streaming, or bulk data.
- Class-5 is unspecified (regarding all performance parameters) and is targeted at traditional best-effort Internet applications.
- Class-6 and Class-7 are only provisionally specified in the table. However, such QoS classes are needed for certain new or emerging applications with strict performance parameters. They also introduce in their definition a new parameter named IP packet reordering ratio (IPRR), which has the same upper bound as IPER for these two classes.

The specified upper bounds on IP-network performance parameters refer to end-to-end QoS provisioning, which is not an easy task to implement in the existing Internet. On the other side (the users' side), the concept of QoS is attractive if the presentation of services in any type of media (audio, video, data, or multimedia as a combination) at the user interface satisfies the needs and expectations of users. However, such expectations are driven not only by users but also by network providers (which provide broadband access with certain QoS guarantees) and service providers (e.g. OTT service providers which use Internet access service on the basis of the network neutrality). Since QoS is an end-to-end concept by definition, this means that an overall approach is necessary to provide QoS in practice (i.e. end-to-end QoS provisioning).

4.5.3 End-to-End QoS Considerations for Network Design

Generally, providing end-to-end QoS in IP-based platforms and networks is difficult due to their heterogeneity. Although the telecom world converges on all-IP principles and IP-based networks are using IP protocols on the network protocol layer, that does not mean that all networks and platforms are exactly the same or compatible with each other. This particularly refers to QoS. As discussed in Chapter 1, the Internet is built of interconnected autonomous systems, which are owned by different public and private organizations (e.g. telecom operators). However, each AS may have different traffic management concepts, which can be completely different from those found in neighboring ASs, to which the given AS is connected. More ASs on the path of the IP traffic add more delay (each AS adds its own delay to the transferred IP packets) and with that a higher probability of bottlenecks appearing somewhere on the path between the two endpoints of a given call/session. However, the number of ASs that the IP traffic passes end-to-end remains almost constant, although the total number of active ASs increases either exponentially or linearly over time. Of course, this positively influences QoS (as an end-to-end feature) and shows that the Internet has good scalability on a global scale.

So, regarding QoS from the end-to-end point of view, the extreme case will be a scenario where IP networks have different traffic management mechanisms, different levels of performance control, and different QoS provision (e.g. telecom operator 1 uses IntServ, backbone provider 1 uses DiffServ, backbone provider 2 uses MPLS traffic engineering, telecom operator 2 uses bandwidth over-provisioning, as given in Figure 4.10) [34]. In such a case it is hard to provide services with guaranteed end-to-end QoS level when each of the networks (and network sections) on the end-to-end path uses a different QoS mechanism.

The basic network model introduced in Chapter 1 of this book (which consists of access, backbone, and transit networks) should be applied commonly by network providers, but the technologies that compose such a network model would be different depending on the providers and their geographic regions or countries (Figure 4.10). Additionally, differences exist between national regulatory frameworks regarding addressing SLAs between service providers (typically national telecom operators) and customers (typically residential and business customers) and accompanied QoS and QoE frameworks.

Overall, end-to-end network design is essential in order to provide end-to-end QoS for both fixed and mobile access networks, which however requires QoS to be part of the peering or transit agreements between the telecom operators and the transit network providers.

Certain QoS issues at interconnection points are shown and discussed in Figures 4.11 and 4.12. As illustrated, IP interconnection points are possible bottlenecks for end-to-end IP traffic. This includes also best-effort traffic from the Internet access service (access to the public Internet) and other IP traffic which is provided to customers by telecom operators with QoS guarantees (e.g. carrier-grade VoIP as a replacement for PSTN/ISDN). The interconnection points of a given telecom operator should be able to transfer the demanded traffic from/to all access networks during peak time or busy hours if one uses the traditional telecom terminology (that is, the time period in the usual weekday with the highest volume of traffic in both directions, downlink and uplink). So, peering and transit agreements of a given telecom operator (which

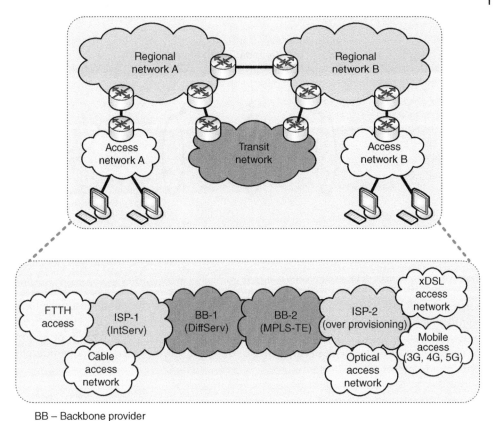

BB – Backbone provider
ISP – Internet service provider

Figure 4.10 End-to-end QoS provision.

FTTH access network has capacity, for example: 100 Mbit/s in DL per user, and 100 Mbit/s in UL per user, and there are 10 000 users connected to it, from which a maximum of 80% are online at the same time, then the needed capacity in the access network will be 10000 × 100 Mbit/s × 0.8 = 800 Gbit/s in each direction (UL and DL).

At this interconnection point, if we assume that 90% of the traffic in both UL and DL goes out of the ISP-1 network, it requires capacity of 800 Gbit/s × 0.9 = 720 Gbit/s only for users from FTTH access network in this example. If the capacity at the interconnection is significantly below 720 Gbit/s, then the advertised speeds cannot be reached and we have **QoS degradation**.

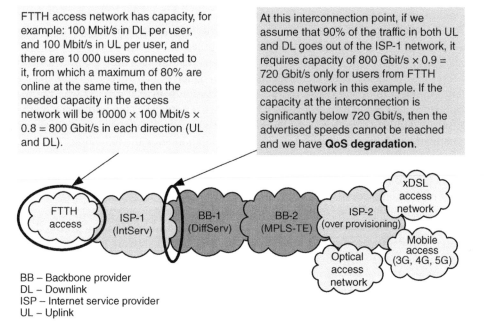

BB – Backbone provider
DL – Downlink
ISP – Internet service provider
UL – Uplink

Figure 4.11 Dependence of network design on the interconnection points and access network.

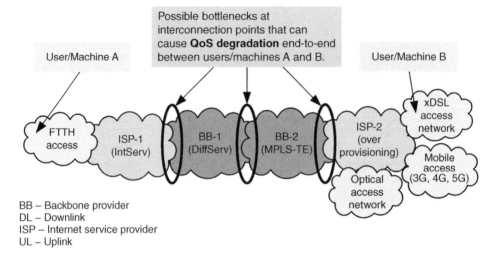

BB – Backbone provider
DL – Downlink
ISP – Internet service provider
UL – Uplink

Figure 4.12 Bottleneck at interconnection points in end-to-end network design.

serves also as ISP, where ISP refers to provision of access to the public Internet) have significant influence on end-to-end QoS, and therefore interconnection between IP networks belonging to different operators may require regulatory attention or action to ensure QoS as end-to-end performance of different services with generally different QoS requirements.

4.6 Strategic Aspects for Ultra-Broadband

The ultra-broadband access gives additional power to telecommunications/ICT in the second and third decades of twenty-first century. The deployment of ultra-broadband has multiple effects on all aspects, including business and regulatory aspects, as well as on society in general. Ultra-broadband coupled with QoS provisioning per traffic class or per flow increases further the possibilities for spread of ICT into different verticals.

It is generally accepted that broadband has a beneficial impact on economic growth [35]. One may note that broadband speed matters, which is a trivial conclusion. Such a conclusion refers to different speeds at different times; however, it matters whether the broadband access speed is higher (typically, the access network is the bottleneck on the way end-to-end). From an economic point of view, one may distinguish several impacts of broadband and ultra-broadband deployment, which include the following:

- *Direct effects.* Large-scale ultra-broadband infrastructure investments lead to increased economic activity in the investment area, such as an increase in employment in different sectors (e.g. telecom sector for ICT deployments, civil engineering sector for building terrestrial physical infrastructure, financing services).
- *Indirect effects.* These are long-term effects which boost innovation and productivity via increased affordability and broadband speeds. People can work at a distance, can collaborate more easily, can have virtual meetings, etc. So, an employee does not need to go to an office to access a document or have a meeting. All these possibilities save time for completion of the same tasks, which in other words is increased productivity.

- *Induced effects.* These include spillovers into other economic sectors (in different verticals) by providing access to education, healthcare, online banking services, e-commerce and the overall digital economy, entertainment, smart cities, smart industry, smart transportation systems, and so on.

So, the general benefits of ultra-broadband include economic effects such as increased productivity and innovation, then social effects via improved access to different services as well as improved public services via broadband access such as healthcare, or improved energy efficiency via creation of smart homes or cities.

However, at the end of the second decade of the twenty-first century there are billions of people still not connected via broadband. The question that arises is, why are all people not yet connected? Overall there are several reasons for this, which include the following [36]:

- *Lack of infrastructure.* Absence of sufficient physical infrastructure means no provision of meaningful access.
- *Non-affordability.* This happens when the cost of broadband Internet access is too high for the given people in a given country or region.
- *Lack of skills.* This refers to Internet applications and services unawareness, as well as lack of digital literacy and digital skills.
- *Lack of digital content.* This is a problem when there is no relevant content in a given national language, or there is no content required for certain economic sectors (e.g. buying and selling agriculture products).

In this chapter we have focused on fixed broadband and ultra-broadband access. The technologies are continuously being updated. While you are reading this book a new standard is probably being developed and research is going on for each of the technologies. For example, DSL networks enter the ultra-broadband arena with VDSL2 and G.fast. One may expect that in the 2020s the number of gigabit/s connections per customer will reach hundreds of millions on a global scale [37].

Additionally, the long-term future regarding fixed ultra-broadband access is fiber (e.g. FTTH); however, the copper (i.e. metallic) access networks will continue to exist in the near future because the infrastructure has already been deployed and minimizes the cost and time needed for ultra-broadband delivery to customers. Nevertheless, in some markets on national levels, the tendency for fixed-mobile substitution is occurring. In other markets, with initially very low penetration of copper-based fixed access (e.g. regions of Africa), deployment of broadband access is focused on mobile networks, not fixed ones.

Overall, the ultra-broadband deployments require investments in infrastructure on a long term. The costs may differ depending on the types of technologies as well as geographic area. Also, the costs are influenced by other factors such as number of ultra-broadband subscribers or connected households and businesses, coverage areas, and so on. The calculation of costs is different for fixed access than for mobile or satellite technology investment, which is based on coverage area per base station and satellite, respectively. Calculation of investments for fixed fiber-based connectivity is performed per household. Also the type of geographic area (e.g. dense urban area, sparse populated area) influences different technologies to be used in different cases to provide connectivity. In such cases, combinations of different cost bases for different technologies are more likely to be used in practice [37].

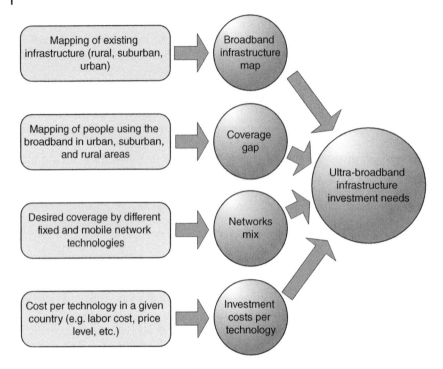

Figure 4.13 Framework for ultra-broadband infrastructure investment needs.

A framework for determining total ultra-broadband infrastructure needs is given in Figure 4.13. Based on such frameworks are developed and continuously updated national broadband and ultra-broadband plans and strategies, which by default include the investments required to achieve desired levels of broadband/ultra-broadband access, bearing in mind the costs, the technology mix per geographic area, and/or the targeted types of customers.

In parallel with economic aspects of ultra-broadband, the regulatory environment (for the telecommunication/ICT sector) is the one that can provide guarantees that customers and citizens in general can exploit the new possibilities and services through ultra-broadband access, having also greater choice of end-user devices (for human end-users), services, and applications, and at the same time developing the appropriate framework for consumer protection, as well as innovations and investments in the ICTs.

Typically, regulators on the global scene use a set of best practice guidelines [37] to protect consumer interests, and at the same time they use a light-touch regulatory approach to ensure a working field for traditional and new market players.

Regarding the development of the broadband targets toward ultra-broadband in the 2020s, an example from Europe is given in Table 4.9, based on the targets set by the European Commission, by broadband category and access speed [38].

But who is leading the investments in the broadband and ultra-broadband sector? Well, they are led by the telecom operators, and in some places also supported by governments. So, the telecom operators are leading the direct investment game while they "live" in a free market on one hand and in the regulated market on the other. In other words, they compete among themselves in a given country, but they also are obliged to

Table 4.9 European targets regarding broadband and ultra-broadband by 2020.

Broadband targets	Broadband category	Access speed
Target I: Basic broadband for all by 2013	Basic broadband	150 Kbit/s – 30 Mbit/s
Target II: High or very high access speed for all by 2020 (above 30 Mbit/s)	High speed Very high speed	30–50 Mbit/s 50–100 Mbit/s
Target III: 50% or more of households subscribe to Internet access above 100 Mbit/s by 2020	Ultra-broadband	100 Mbit/s – 1 Gbit/s

comply with all national regulations, which may differ from country to country. There are OTT service providers which are generally global (Google, Amazon, Facebook, etc.). They do not invest directly in the broadband and ultra-broadband infrastructures. But do they have a certain impact on broadband and ultra-broadband? Well, the OTT service providers have the highest possible innovation pace, which is the main driver of the telecommunication/ICT sector. If one compares the ICT sector in the current century with the ICT/telecommunication sector during the twentieth century, there is a noticeable difference. This comes in the "freedom" for innovation (which in fact appeared with public Internet access), which provides the possibility for plenty of different services and applications to be offered to end-users via different application ecosystems (Play Store, iStore, and others). Of course, some services and applications are more successful than others, but that is usual in every business segment. What is the benefit for end-users? They are getting more choice and freedom to choose which services they want to use. As usual, with every choice comes the responsibility for having such choice.

So, OTT providers also impact broadband and ultra-broadband development. In fact, they have the highest impact on the end-user's needs for broadband and ultra-broadband access. Why? Because without the OTT services, broadband or ultra-broadband Internet access would be useless. Of course, we would have telephony and television (with much higher picture resolution with ultra-broadband access) and other QoS-enabled services provided by telecom operators (e.g. VPN services for business users), typically on a national level. So, one may state that the OTT service providers and the telecom operators are complementary. Overall, considering broadband and ultra-broadband access, the telecom operators are providing the high-quality "infrastructures," while OTT service providers are producing the Internet "vehicles" to be used on such ultra-broadband "highways."

References

1 ANSI T1.413 issue 2 (1998). *Network and Customer Installation Interfaces – Asymmetric Digital Subscriber Line (ADSL) Metallic Interface*. Ansi.
2 ITU-T Recommendation G.992.1, Asymmetric Digital Subscriber Line (ADSL) transceivers, 1999.
3 ITU-T Recommendation G.992.2, Splitterless Asymmetric Digital Subscriber Line (ADSL) transceivers, June 1999.

4 ITU-T Recommendation G.992.3, Asymmetric Digital Subscriber Line Transceivers 2 (ADSL2), 2009.

5 ITU-T Recommendation G.993.5, Asymmetric Digital Subscriber Line 2 Transceivers (ADSL2) – Extended Bandwidth ADSL2 (ADSL2plus), 2009.

6 ITU-T Recommendation G.9700, Fast Access to Subscriber Terminals (G.fast) – Power Spectral Density Specification, 2014.

7 ITU-T Recommendation G.993.2, Very High Speed Digital Subscriber Line Transceivers 2 (VDSL2), 2015.

8 ITU-T Recommendation G.9701, Fast Access to Subscriber Terminals (G.fast) – Physical Layer Specification, 2014.

9 CableLabs, The Evolution of DOCSIS, https://www.cablelabs.com/full-duplex-docsis, accessed in March 2018.

10 ITU-T Series G Supplement 42, Guide on the Use of the ITU-T Recommendations Related to Optical Fibers and Systems Technology, 2014.

11 ITU-T Recommendation G.987.1, 10-Gigabit-capable Passive Optical Networks (XG-PON): General Requirements, 2016.

12 ITU-T Recommendation G.989.3, 40-Gigabit-capable Passive Optical Networks (NG-PON2): Transmission Convergence Layer Specification, 2013.

13 IEEE 802.3av-2009, Local and Metropolitan Area Networks – Specific Requirements Part 3: Carrier Sense Multiple Access with Collision Detection (CSMA/CD) Access Method and Physical Layer Specifications Amendment 1: Physical Layer Specifications and Management Parameters for 10 Gbit/s Passive Optical Networks, 2009.

14 IEEE P802.3ca 100G-EPON Task Force, Physical Layer Specifications and Management Parameters for 25 Gb/s, 50 Gb/s, and 100 Gb/s Passive Optical Networks, http://www.ieee802.org/3/ca, accessed in March 2018.

15 Broadband Forum TR-059, DSL Evolution – Architecture Requirements for the Support of QoS-Enabled IP Services, 2003.

16 Society of Cable Telecommunications Engineers (SCTE) Standard 159-01 2017, Multimedia Application and Service Part 1: IP Cablecom Multimedia, 2017.

17 CableLabs, *DOCSIS 3.1: MAC and Upper Layer Protocols Interface Specification*, 2015.

18 CableLabs, PacketCable 2.0: Quality of Service Specification, 2014.

19 Broadband Forum TR-101, Migration to Ethernet-Based DSL Aggregation, 2006.

20 Abbas, H.S. and Gregory, M.A. (2016). The next generation of passive optical networks: a review. *Journal of Network and Computer Applications* 67: 53–74.

21 Broadband Forum TR-156, Using GPON Access in the context of TR-101, 2008.

22 Broadband Forum TR-178, Multi-service Broadband Network Architecture and Nodal Requirements, 2014.

23 Broadband Forum TR-280, ITU-T PON in the Context of TR-178, 2016.

24 Janevski, T. (2014). *NGN Architectures, Protocols and Services*. Chichester: Wiley.

25 R. Sanchez, L. Raptis, K. Vaxevanakis, Ethernet as a carrier grade technology: developments and innovations, *IEEE Communications Magazine*, September 2008.

26 MEF Technical Specification, MEF 6.1, Ethernet Services Definitions – Phase 2, 2008.

27 IEEE 802.1Q - 2005, Virtual Bridged Local Area Networks, 2006.

28 IEEE 802.1ad - 2005, Virtual Bridged Local Area Networks – Amendment 4: Provider Bridges, May 2006.

29 IEEE 802.1ah - 2008, Virtual Bridged Local 14 Area Networks – Amendment 6: Provider Backbone Bridges, August 2008.

30 MEF Technical Specification, MEF 22.1, Mobile Backhaul Phase 2, January 2012.

31 MEF Technical Specification, MEF 23.2, Implementation Agreement MEF 23.2: Carrier Ethernet Class of Service – Phase 3, August 2016.

32 ITU-T Recommendation Y.1541, Network Performance Objectives for IP-based Services, December 2011.

33 ITU-T Recommendation Y.1540, Internet Protocol Data Communication Service – IP Packet Transfer and Availability Performance Parameters, July 2016.

34 ITU-T Technical Paper, How to Increase QoS/QoE of IP-based platform(s) to regionally agreed standards, March 2013.

35 ITU-Development sector, Impact of Broadband on the Economy, Broadband Series, 2012.

36 ITU, Working Together to Connect the World by 2020 – Reinforcing Connectivity Initiatives for Universal and Affordable Access, a Discussion Paper to Partners working to Connect the World, 2016.

37 ITU Broadband Commission, The State of Broadband: Broadband Catalyzing Sustainable Development, 2017.

38 ITU, Working Together to Connect the World by 2020 – Reinforcing Connectivity Initiatives for Universal and Affordable Access, 2016.

5

QoS for Mobile Ultra-Broadband

The mobile and wireless network evolution has sped up over the past several decades, since the appearance of digital mobile systems in the 1990s. However, if one wants to see the complete mobile and wireless evolution with the main events in its history, then the timeline will be the following:

- Invention of electro-magnetic theory in 1861 by J.M.C. Maxwell, who established the theoretical basis for further development of wireless communications.
- Start of the era of wireless communications with the invention of the wireless telegraph by Guglielmo Marconi in 1895.
- Appearance of NMT (Nordic Mobile Telephony) in 1981 as the first widely deployed mobile system (such analogue mobile systems were different in different countries, and later referred to as 1G – the first generation).
- In 1991 GSM (Global System for Mobile communications) appeared and started the second generation (2G) mobile networks.
- In 1997 came IEEE 802.11 (a.k.a. WiFi), which later became a unified wireless local area network WLAN.
- Around the year 2000 appeared 2.5G, which introduced IP connectivity via mobile networks (e.g. General Packet Radio Service (GPRS) was added to GSM for IP connectivity, with bitrates in the range of tens of kbit/s at that time).
- In 2002 3G (the third generation) started, which in fact introduced mobile broadband access (with up to Mbit/s speeds in the radio access networks (RANs)), with the main representative being the UMTS (Universal Mobile Telecommunication System) standardized by 3GPP.
- The 2000s saw the integration of WLAN and cellular mobile networks for additional bitrates in WiFi hotspots (many telecom operators have implemented hotspot deployments at cafeterias, airports, and other public places).
- In 2005 the first global mobile standard from IEEE appeared, called Mobile WiMAX, which joined the 3G family of standards in 2007 (around half a decade later than other 3G technologies).
- In 2008 3GPP's LTE (long term evolution) standard was standardized, which marked the start of the fourth generation mobile systems, 4G.
- Around 2010 appeared the only two 4G standards, LTE-Advanced from 3GPP and Mobile WiMAX 2.0 from IEEE, of which the first has proven to be more successful on the global mobile scene. At the end of the 2010s the global 4G mobile broadband is LTE/LTE-Advanced.

QoS for Fixed and Mobile Ultra-Broadband, First Edition. Toni Janevski.
© 2019 John Wiley & Sons Ltd. Published 2019 by John Wiley & Sons Ltd.

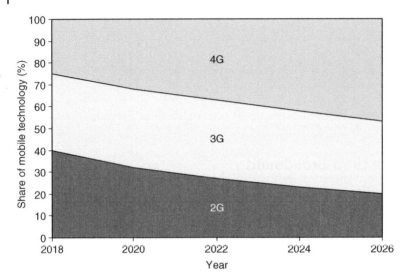

Figure 5.1 Global shares of 2G, 3G, and 4G mobile networks.

- By the end of 2018 the first 5G standard by 3GPP had been completed, which will mark the 2020s.

So, mobile systems have enjoyed a continuous evolution since the beginning of the 1980s, with global penetration since the 1990s.

Figure 5.1 shows the global shares of different mobile generations on a global scale, which coexist in parallel. This includes 2G, 3G, and 4G mobile systems and networks. As one may expect, the share of older generations (e.g. 2G) is decreasing due to increasing shares of the newer mobile generations (3G and 4G). However, the mobile broadband technologies include both 3G and 4G, so in the second decade of the twenty-first century they are the dominant way for mobile Internet access (regarding bitrates and volume of IP traffic). Going toward 2020 4G, which is mainly represented by the LTE standard from 3GPP (and its advanced or professional versions, such as LTE-Advanced and LTE-Advanced-Pro), is continuously increasing its share. One may expect a similar trend by the end of the 2020s with 5G, and further in the 2030s with 6G (assuming that the one mobile generation per decade trend will continue).

5.1 Mobile Ultra-Broadband Network Architectures

In order to discuss the mobile ultra-broadband architectures, let's summarize the evolution of mobile networks through the generations [1]:

- *1G.* This refers to the analogue mobile systems used in the 1980s, which used FDMA (frequency division multiple access) in radio access and circuit-switching as networking technology.
- *2G.* This generation brought the first all-digital mobile systems, starting from the 1990s, based primarily on TDMA (time division multiple access) and FDMA (e.g. GSM from 3GPP). 2G was initially circuit-switched-based (CS), with the novelty of

global roaming. The main services in 2G were mobile telephony (i.e. mobile voice) and short message service (SMS), while later 2.5G introduced the packet-switched (PS) network (e.g. GPRS from 3GPP) in parallel with the CS part (i.e. GSM).

- *3G.* This was based on a 2.5G "hybrid" network approach, which included both the PS part (e.g. for Internet access and multimedia messaging service (MMS)) and the CS part (for voice and SMS), while the radio part in the most known 3G representative (UMTS) was based on WCDMA (wideband code division multiple access) in combination with TDMA/FDMA.
- *4G.* Appearing around the 2010s, 4G is noted as the first mobile generation which is all-IP by default in all network parts. The RAN is based on OFDMA (orthogonal frequency division multiple access), again combined with TDMA (which is required in mobile networks for QoS support).
- *5G.* This is the mobile generation for the 2020s, which is all-IP by default (like 4G). It introduces higher data rates than its predecessor (which is standard with each new mobile generation) and additionally provides the standardized possibility for network virtualization (i.e. different network slices for different types of services). The radio part again is new, called New Radio (NR), which has been established practice (to have a novel radio access) in each of the mobile generations.

The development of the mobile network architectures has direct influence on the QoS in the given mobile network. This is especially important in all-IP mobile networks such as 4G. Why? Because in CS mobile networks the end-to-end delay is typically limited with the propagation delay of signals which travel approximately at the speed of light (including radio, copper, and fiber transmission links), that is $3 \times 10^8 \, \mathrm{m \, s^{-1}}$, which provides a delay of up to 100 ms between any two points on Earth connected with terrestrial connection (i.e. not through satellite links), bearing in mind that the equator is around 40 000 km long and its half-length gives a delay of $20\,000\,\mathrm{km}/3 \times 10^8 \, \mathrm{m \, s^{-1}} = 67$ ms, while the full length of the equator requires 2×67 ms $= 134$ ms. Therefore, the requirement for end-to-end delay for voice below 150 ms (specified in ITU-T recommendation G.114) can be satisfied in CS mobile (and fixed) networks. However, PS networking principles result in increased total delay and delay variation due to packetization, coding/decoding, buffering at networks nodes (switches, routers), etc. Therefore mobile network architectures have also evolved to accommodate specific characteristics of IP transfer end-to-end, which is always higher than in CS networks for the same distance (in CS networks the path is established before any call/session by using signaling, such as SS7) [2].

5.1.1 3G Network Architecture

Initially, the 3GPP mobile IP network architectures were introduced in GPRS, with two main gateway nodes, the central node SGSN (serving general packet radio service support node) and the network edge node GGSN (gateway general packet radio service support node). The same two nodes defined the main core network (CN) architectures in UMTS, which was 3GPP's 3G standard (note that there were different 3G standards in the 2000s, such as cdma2000 in Americas, Mobile WiMAX 1.0, and several others [3]). The bitrates of UMTS increased in both downlink and uplink, with better modulation and coding schemes in later 3GPP releases during the 2000s, which resulted in standardization of HSPA (high speed packet access), also noted globally as 3.5G.

The 3GPP mobile network evolution started during the 3G era, which transformed the hierarchical mobile architecture (with a radio network controller (RNC) between the base stations, i.e. eNodeBs, and SGSN) into a flat architecture, with UTRAN (Universal Mobile Telecommunication System Radio Access Network) which consisted only of NodeBs (base stations in UMTS) and centralized core network nodes (3G-SGSN and 3G-GGSN, together with subscribers database, the Home Subscriber Server (HSS)). This was referred to as system architecture evolution (SAE).

Why did such 3GPP evolution result in a flat architecture? The main reason was the QoS in the mobile networks based on IP, including UTRAN (in 3G) and the mobile core network, i.e. for provision of lower delay in the mobile network which is needed for real-time services over all-IP mobile networks (e.g. carrier-grade VoIP as the primary service in this regard) as well as emerging critical services over a mobile network (e.g. control of driverless vehicles or drones, remote surgery in real-time, and similar examples in the future).

5.1.2 4G Network Architecture

The 4G mobile network from 3GPP is LTE/LTE-Advanced. LTE was standardized initially with 3GPP Release 8, together with the common IMS (Internet Protocol multimedia subsystem) for signaling based on SIP and Diameter (for communication with the users' database HSS) as well as with evolved packet core (EPC). Evolution of 3GPP mobile standards is shown in Figure 5.2. Overall, there are three segments that are standardized by 3GPP:

- High-speed access, which refers to radio access technology.
- IP core network, which includes all controllers/gateways and databases in the mobile networks as well as their mutual interfaces.
- Services, which include service overlay network implemented over a given mobile network, with IMS as the central node in the services part.

LTE-Advanced was initially standardized as 4G with 3GPP Release 10 and further continued with the following Releases 11–14 (Figure 5.2). In fact, the requirements

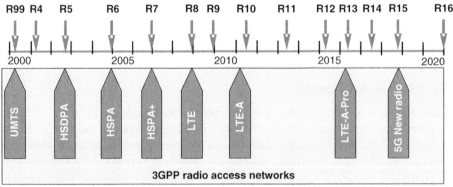

HSDPA – High speed downlink packet access
HSPA – High speed packet access
LTE – Long term evolution

LTE-A – LTE-advanced
UMTS – Universal mobile telecommunication system

Figure 5.2 Timeline of 3GPP mobile networks.

for 4G were specified by the ITU-R in International Mobile Telecommunications (IMT)-Advanced umbrella, and in 2010 the assessment of all candidates for 4G mobile broadband, i.e. IMT-Advanced, resulted in the acceptance of two technologies as 4G:

- LTE-Advanced (from the 3GPP);
- Mobile WiMAX 2.0 (i.e. Wireless MAN-Advanced), with IEEE 802.16m standard in the radio access interface.

The development in the 2010s showed market dominance by the LTE/LTE-Advanced mobile standards, driven by the evolution of the predecessor 3G/3.5G technologies (UMTS/HSPA). Although LTE was initially referred to as 3.9G because it did not satisfy the required 1 Gbit/s downlink speeds (initially required for 4G mobile networks), in practice LTE is marketed as 4G technology because it is almost the same as LTE-Advanced, which provides higher bitrates with carrier aggregation (i.e. simultaneous use of single IP layer over multiple underlying LTE frequency carriers in adjacent and non-adjacent spectrum bands).

3GPP Release 8 also introduced evolved packet system (EPS), which consists of a flat-IP core network EPC and Evolved Universal Mobile Telecommunication System Radio Access Network (E-UTRAN) as LTE-based RAN. However, the transition from hierarchical to a flat architecture started with 3GPP Release 7 with direct tunnel, and it was completed in 3GPP Release 8, then continued in further 3GPP releases (Figure 5.3).

Base stations in LTE/LTE-Advanced mobile networks, eNodeBs (i.e. eNBs), which are the main nodes in E-UTRAN, are connected directly to the EPC core network gateway via the "S1 interface" (Figure 5.4). Additionally, adjacent eNodeBs are directly connected through an "X2 interface," which provides fast handovers by rerouting the IP packets from the old eNodeB toward the new eNodeB after the handover execution (which is initially managed by the adjacent eNodeBs over the X2 interface). This approach provides lower handover delay because otherwise it would take longer to establish communication with centralized nodes in EPC and then to instigate rerouting (also, certain IP packet losses may occur in the absence of the X2 interface between the old and the new eNodeBs).

The EPC consists of several network nodes, where the control and user data plane are further separated (considering the architecture in UMTS/HSPA). The mobility-related signaling is targeted to the MME (mobility management entity), while the user data traffic from E-UTRAN toward the Internet and vice versa traverses through the serving gateway (S-GW) and the packet data network gateway (PDN-GW, or shortly P-GW), much like SGSN and GGSN in GPRS and UMTS in previous mobile generations.

The main database in EPC is HSS, which contains user-related information, such as the subscriber's profile (e.g. services available to that user), as well as user authentication information. However, the main node in the EPC regarding QoS in the mobile network is the policy and charging rules function (PCRF). It was created as a software-based network node responsible for policy and charging control (PCC). The PCRF's goal is to detect the service flow and to determine policy rules for it in real-time, while enforcing appropriate charging policy for such service in the given mobile network. In the 4G architecture the PCRF provides PCC information that is required for bearer setup to the policy enforcement functions located in P-GW.

3GPP approved a new LTE marker for the LTE-Advanced specification in Releases 13 and 14, called LTE-Advanced-Pro. As the name says, it is fully based on LTE/LTE-

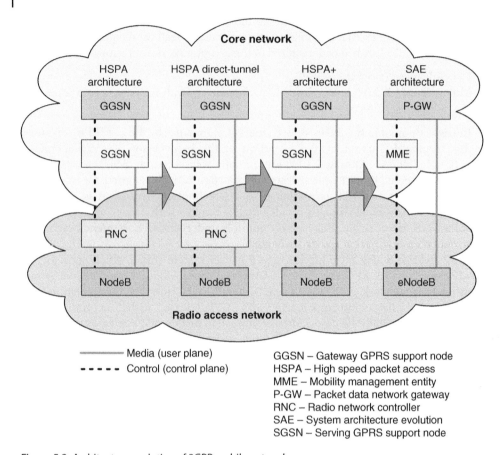

Figure 5.3 Architecture evolution of 3GPP mobile networks.

Advanced but with additional features. With such a professional version the LTE mobile networks pointed toward addressing new markets as well as new functionalities to improve efficiency. Major novelties in LTE-Advanced-Pro include the following:

- *Cellular Internet of Things.* This refers to LTE enhancements for machine type communications (MTCs), which are targeted at definition of lower device category (Cat-0) in 3GPP Releases 12 and 13, with 1 Mbit/s in uplink and downlink.
- *LTE device-to-device (D2D) communication.* This is targeted at public safety as well as building a mobile network architecture for support of emergency services.
- *LTE for V2X (vehicle-to-X) communication.* This increases its importance starting from LTE-Advanced-Pro and then continues in 5G specifications.
- *Low latency LTE.* This is developed to enable new delay-critical services, which further continue their development in 5G.
- *LTE assisted access.* This is the approach to use the standardized LTE technology in unlicensed spectrum (on 5 GHz) for mobile data traffic offload or backhauling WiFi access points (APs).

LTE, LTE-Advanced, and LTE-Advanced-Pro differ in maximum possible bitrates. LTE can use frequency carriers with different widths, including 1.4, 3, 5, 10, 15, and

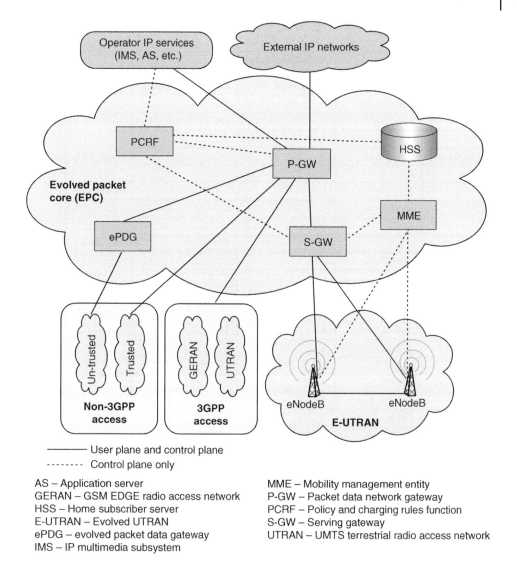

Figure 5.4 4G mobile network architecture for LTE/LTE-Advanced.

20 MHz carrier bands (wider carrier gives proportionally higher bitrate). For example, LTE-Advanced has higher bitrates due to the possibility for frequency carrier aggregation on layer 2 (i.e. below the IP layer) of up to five non-continuous frequency carriers (each with up to 20 MHz). LTE-Advanced-Pro provides further enhancement allowing carrier aggregation of up to 32 frequency carriers (each up to 20 MHz), with up to 640 MHz (= 32 × 20 MHz) carrier bandwidth in total. The drawback is that most of the mobile operators do not have such spectrum to utilize the maximum possible aggregation. Also, the mobile equipment categories typically support (as maximum) lower bitrates than theoretical maximums specified in the 3GPP standards. The theoretical maximum bitrates for user equipment (UE) for 3GPP release up to 3GPP Release 14

Table 5.1 Theoretical maximum bitrates for user equipment for 4G mobile broadband and ultra-broadband by 3GPP.

User equipment category	Maximum downlink bitrate (Mbit/s)	Maximum parallel streams (MIMO) in downlink	Maximum uplink throughput (Mbit/s)	3GPP release
NB1	0.68	1	1.0	Release 13
M1 (Rel. 13)	1.0	1	1.0	
NB2	2.5	1	2.5	
M1 (Rel. 14)	1.0	1	3.0	Release 14
M2	4.0	1	7.0	
0	1.0	1	1.0	Release 12
1	10.3	1	5.2	
2	51.0	2	25.5	
3	102.0	2	51.0	Release 8
4	150.8	2	51.0	
5	299.6	4	75.4	
6	301.5	4	51.0	
7	301.5	4	102.0	Release 10
8	2 998.6	4	1 497.8	
9	452.3	4	51.0	
10	452.3	4	102.0	Release 11
11	603.0	4	51.0	
12	603.0	4	102.0	
13	391.6	4	150.8	
14	3 916.6	8	9 585.7	Release 12
15	750.0–807.7	4	226.1	
16	979.0–1 051.4	4	105.5	
17	25 066.0	8	2 119.4	
18	1 174.8–1 211.6	8	211.0	Release 13
19	1 566.3–1 658.3	8	13 563.9	
20	1 948.0–2 019.4	8	316.6	Release 14
21	1 349.0–1 413.1	4	301.5	

(the last release before 5G) [4] is given in Table 5.1. The practical individual user bitrates depend upon the installed capacity in the cell(s), the mobility of the users, the number of users in the same cells(s), as well as UE and its category (i.e. which bitrates are supported on the end-user's side).

Considering Table 5.1 one may note that LTE-Advanced-Pro, besides higher bitrates, provides additional support for the emerging IoT and enhanced machine type communication (eMTC) via the LTE mobile networks. eMTC is also abbreviated as LTE-MTC or LTE-M (long-term evolution for machines). Also, NB-IoT (narrow band Internet of Things) is targeted at New Radio to the LTE standard optimized for the low end of the market. The IoT and MTC do not need very high bitrates, therefore Cat-M1 and

Cat-M2, NB1 and NB2, provide lower bitrates (i.e. NB, in contrast to broadband) which are needed and can be processed by various current and future IoT devices connected to the network via the mobile network.

Overall, LTE-Advanced-Pro is the last version of 4G mobile networks before the appearance of the first 5G standard, which tends to provide a smooth transition of 3GPP network architectures from 4G to 5G.

5.1.3   5G Network Architecture

The 5G mobile network architectures start with 3GPP Release 15, which is the first standard to enter into the IMT-2020 family (the 5G family). It is very likely that the number of standards within the given IMT family reduces from one mobile generation to the next, so we had seven standards under the IMT-2000 (the 3G umbrella), two standards under the IMT-Advanced (the 4G umbrella), and one standard under the IMT-2020 (the 5G umbrella). It is very likely that similar to WiFi in the WLAN environment, the 5G standard from 3GPP will become the unified mobile standard, which, however, must coexist with other IP-based mobile technologies (e.g. UMTS, LTE).

The 5G mobile architecture is ultra-broadband architecture which aims to provide Gbit/s and tens of Gbit/s aggregate bitrates and over 100 Mbit/s individual bitrates; however, this is based on the design of the mobile network (that is, installed capacity versus average number of active users at peak hours), size of the cell, users' location in the cell and their mobility, etc. For example, users at the cell edge have worse signal-to-noise ratio, which results in use of a modulation and coding scheme that carries fewer bits per symbol in the radio interface (e.g. 1024 quadrature amplitude modulation [QAM] can carry 10 bits on a single symbol because $2^{10} = 1024$, while 16 QAM can carry 4 bits on a single symbol on the same location and with the same frequency band in use, which gives $10/4 = 2.5$ time higher bitrates by using the same radio resources). To have higher capacity, 5G mobile networks require more bits/s/Hz, which can be provided (as one approach) with better modulation and coding schemes.

Another approach to increase bit/s/Hz is using multiple antennas on the sender's or receiver's side, i.e. multiple input multiple output (MIMO), which adds more capacity on the same frequency band in use between the base station and the mobile users. For example, if the bitrate in the mobile network with single antennas on both ends (mobile terminal and base station) is B bit/s, then $N \times N$ MIMO (where n is an integer) will result in $N \times B$ bitrate. For example, if $B = 10$ Mbit/s with SISO (single input antenna and single output antenna), then 1024×1024 MIMO will provide 1024×10 Mbit/s $= 10$ Gbit/s. Of course, to have the advantage of MIMO use, the same MIMO should be supported on both ends, the mobile terminal and the base station. Hence, the 5G mobile network targets constant improvement in MIMO (from one release to the next), developing toward massive MIMO with the aim of maximizing the bits/s/Hz.

However, to increase capacity as needed or expected in 5G requires additional spectrum. But the "good" spectrum for macro mobile coverage is below 6 GHz, and it is becoming more crowded with each mobile generation. Therefore, 5G also enters the spectrum above 6 GHz (e.g. different bands between 24.25 and 86 GHz, etc.) which can be used on smaller areas in hotspot locations, besides allocating new bands below 6 GHz to 5G (e.g. 600 MHz band, 1.5 GHz band, 3.5 GHz band, 5–6 GHz band). Whatever the spectrum in use in a mobile network, there is always a demand for more capacity (e.g.

in hotspot locations in cities). To provide capacity requires reusing the spectrum on smaller distances (because the spectrum that can be used is limited in a given location), which results in scenarios with many small cells deployments for 5G mobile networks. Also, the small cells by default improve the QoS because they are smaller, so the users at the cell edges experience better quality than in macro cellular environments in the same scenario. The small cells deployments can be used by nomadic users (low-mobility users) or stationary devices (e.g. things connected via the mobile network), while the coverage of high mobility users is provided by macro cellular coverage (e.g. with cells ranging from several kilometers in radius in urban areas to 20–30 km in rural areas).

Another important aspect in the 5G mobile networks is their target to massive IoT, such as support for at least 1 million devices per km². IoT devices require ultra-broadband speeds, but only tens or hundreds of kbit/s, referred to as NB access. Hence, such NB access is standardized in 5G architectures in parallel with the gigaspeed access. With the massive IoT, the 5G mobile networks' target is to enter different verticals, such as intelligent transportation service (e.g. vehicle-to-X communication), automation of the industry, etc., which adds the demand for delay-critical services that are going to be provided via the 5G mobile network. Such critical service (e.g. direct automation, intelligent public transport) has strict QoS requirements regarding guaranteed bitrates, end-to-end delay (which has to be very low, lower than voice delay), and delay variation requirements (i.e. jitter requirements).

So, the 5G mobile architectures in general have three main targets: to provide in parallel ultra-broadband speeds for Internet access and legacy telecommunication services (e.g. voice, video), massive IoT support (to expand the IoT services on a global scale, in a standardized manner), and support for new critical services (over the 5G mobile network), which have very strict QoS requirements.

The 5G network architecture consists of two main parts (Figure 5.5): the CN and RAN domains. The 5G mobile network also allows placement of network functions in radio network sites such as access and aggregation sites. 5G RAN, called NG-RAN (next generation radio access network), uses New Radio. The 5G base station is also being referred to as gNodeB (or shortly gNB). In 5G the gNB functions can be split into separate parts such as centralized unit and distributed unit. Additionally, there is a split of functions into control plane functions (CPFs) and user plane functions (UPFs), and with that decoupling of the control and user planes for higher flexibility in provisioning of new emerging services. In NG-RAN there are gateways for each of the planes, such as the CPF gateway and the UPF gateway, connected with 5G base stations via control plane and user plane interfaces, respectively (Figure 5.5). CPF includes access and mobility management functions (AMFs), which has a similar set of functions as MME in the LTE/LTE-Advanced networks. CPF also includes a session management function (SMF), the authentication server function (AUSF), and the Policy Control Function (PCF), which may be integrated into a single node (e.g. the CPF gateway).

Because 5G will certainly coexist with the 4G mobile network during the 2020s and beyond, 3GPP has defined the architecture for interworking of its 5G and 4G mobile networks, which provides seamless handovers (due to the mobility of users) between LTE/LTE-Advanced (as 4G) and New Radio (as 5G). The interworking in the core part can be provided by interfaces between the EPC nodes (on the 4G side) and CN nodes (on the 5G side), such as Nx interface between CPF (5G) and MME (4G), and Xn interface between 4G base stations (eNodeBs) and 5G base stations (gNodeBs). Regarding the

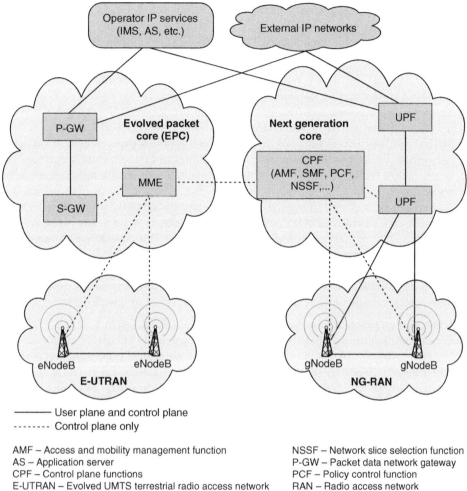

Figure 5.5 5G mobile network architecture.

AMF – Access and mobility management function
AS – Application server
CPF – Control plane functions
E-UTRAN – Evolved UMTS terrestrial radio access network
IMS – IP Multimedia subsystem
MME – Mobility management entity
NG-RAN – Next generation radio access network

NSSF – Network slice selection function
P-GW – Packet data network gateway
PCF – Policy control function
RAN – Radio access network
S-GW – Serving gateway
SMF – Session management function
UPF – User plane functions

user plane, the EPC user plane nodes S-GW and P-GW are replaced in 5G with UPF nodes (different UPF nodes are interconnected via the N9 interface). In the longer term the EPC can be removed and eNodeBs can be connected with direct interfaces to CPF and UPF nodes in the 5G CN.

5.2 QoS in 3G Broadband Mobile Networks

Mobile broadband access in fact started with 3G mobile networks. However, these networks were based on a hybrid approach, that is circuit-switching used for voice and packet-switching (IP connectivity) used for Internet access service. However, the main representative of 3G mobile networks, the UMTS, had well-defined QoS support from the beginning.

For the packet-switching traffic (i.e. IP traffic) in UMTS, four different traffic (i.e. QoS) classes are defined:

- *Conversational class.* This is targeted at services with requirements for low delay and preserved time variation between different IP packets, i.e. very low delay variation. This scheme is targeted at carrier-grade real-time services over IP, such as VoIP, video telephony (based on VoIP), and conferencing tools (based on VoIP).
- *Streaming class.* This is targeted at traffic which requires preserved time variation between IP packets within a given flow, but it does not have certain strict requirements on low transfer delay. Hence, this class is suitable for streaming services (audio and video).
- *Interactive class.* This is targeted for non-real-time services which have certain requirements on transfer delay (which is however not as strict as for real-time services) based on request-response use by the end-user. This class refers to interactive client-server services/applications, such as Web, database retrieval, etc. The key QoS parameters are low round trip delay time and low bit error rate.
- *Background class.* This is targeted at services which are flexible to delay or delay variation, such as email, file downloading, etc. This class is serviced with the least priority.

Additionally, QoS classes in UMTS can be further differentiated by parameters allocation and retention priority (ARP) and traffic handling priority (THP) [1]. In practice, in 3G mobile networks not all QoS classes are used because typically telephony is provided via the CS part, while only Internet access service over the mobile network is based on IP connectivity to which the QoS classes refer. However, they exist as means for QoS differentiation when different carrier-grade services are provided by telecom operator, such as carrier-grade VoIP (with QoS guarantees) or video streaming services with QoS guarantees.

The QoS support in mobile networks standardized by 3GPP is accomplished by the definition of bearers. What is a bearer in UMTS (the 3G network from 3GPP)? The bearer service is an enabler for any data transmission between two defined endpoints in the mobile network, including the mobile RAN and mobile core network. In UMTS the bearer may refer to control (i.e. signaling) or data information. For example, there is the SMS bearer and USSD (unstructured supplementary service data) bearer for SMS and USSD messaging, respectively, both provided via GSM signaling. Then there is a voice CS bearer for voice services and a PS bearer in UMTS for Internet access service. In all cases, the bearer specifies the configuration on layer 2 and layer 1 (i.e. the physical layer) of the protocol stack, with the aim to have defined QoS attributes and traffic flow characteristics. The concept of bearer services continues in the same manner in the 4G standards. The UMTS bearer architecture is shown in Figure 5.6. It consists of three main bearers:

- Radio bearer service (RAB) between the mobile terminals and NodeB (base stations in UMTS), i.e. UTRAN.
- Iu bearer service between UTRAN and core network gateway, which is SGSN in UMTS.
- Backbone bearer services between the core network main gateways, i.e. SGSN and GGSN.

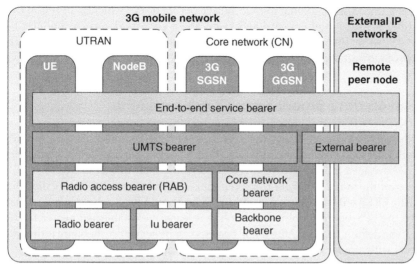

GGSN – Gateway GPRS support node
SGSN – Serving GPRS support node
UMTS – Universal mobile telecommunication system

Figure 5.6 Bearer architecture for QoS support in UMTS.

All three bearers in UMTS form the UMTS bearer service (Figure 5.6), which should be mapped on the external bearer (between the mobile core network and external networks with which the core network is interconnected). To have end-to-end QoS requires appropriate QoS mapping between each adjacent pair of bearers. For example, in the backbone network they may use Carrier (i.e. Metro) Ethernet, which requires mapping of UMTS bearers to available CoS. MPLS, meanwhile, requires different bearers to be mapped with different labels. The main bearer services in UTRAN and in the core network are established by using tunneling protocols, which gives the possibility of aggregation of the same type or the same class of traffic, thus providing the possibility for QoS class differentiation (not differentiation per flow). The standard for tunneling in 3GPP mobile networks (GPRS, UMTS, LTE/LTE-Advanced) is GTP (General Packet Radio Service Tunneling Protocol), which is standardized by 3GPP [5] and not by the IETF (although there are also IP tunneling standards established by the IETF, such as IP-in-IP tunneling [6]). GTP is used in PS parts of GSM/GPRS (2.5G) and UMTS (3G), as well as in LTE/LTE-Advanced/ LTE-Advanced-Pro mobile networks. There are three main GTP types: GTP-C (General Packet Radio Service Tunneling Protocol-Control plane) for transfer of signaling traffic, such as bearer activation, modification; GTP-U (General Packet Radio Service Tunneling Protocol-User plane) for carrying user data traffic; and GTP (GTP for charging data). GTP-C and GTP-U are used in all 3GPP mobile systems with IP connectivity. With that approach 3GPP has full control over the standards used for delivery of QoS in its mobile systems, although all such mobile systems are all-IP starting from 4G and onward.

How is QoS bound with the user connection? Well, in 2.5G and 3G mobile networks from 3GPP that is accomplished by PDP (packet data protocol), which contains session information when mobile user has an active IP connection. Each PDP context includes the subscriber's IMSI (international mobile subscriber identity), which is stored in the SIM (subscriber identity module) card and given an IP address (by using the DHCP) which may be IPv4 or IPv6, as well as a logical connection with a QoS profile (which includes a set of QoS attributes) through the UMTS mobile network.

5.3 QoS in 4G Ultra-Broadband: LTE-Advanced-Pro

In 4G mobile networks from 3GPP, the QoS approach has similarities to the QoS approach in 3G networks, being directly related to bearers. The number of bearers in the LTE/LTE-Advanced mobile networks (Figure 5.7) is similar to that in UMTS.

The bearer over the mobile network (between user equipment and P-GW) is called the EPS bearer. The EPS itself has two main parts: E-UTRAN and EPC. Each mobile terminal connected to the EPS-based mobile network has at least one established bearer for IP connectivity (the default bearer). Additional bearers can be set up for the same terminal. Different flows with similar QoS requirements can be mapped on the same bearer, with the same QoS treatment (e.g. different flows over the public Internet accessed through the mobile network). The EPS bearer consists of the following three main bearers:

- radio bearer, which covers the LTE/LTE-Advanced radio interface;
- S1 bearer for the S1 interface, which is between eNodeB and S-GW;
- S5/S8 bearer on the S5/S8 interface, between S-GW and P-GW.

The two main gateways in the EPC for the user data traffic, S-GW and P-GW, perform QoS enforcement as policy enforcement points (PEPs). On one side, the S-GW performs QoS control at the IP packet level, while on the other side P-GW provides QoS and policy enforcement at the service level. S-GW serves also as anchor node for connection

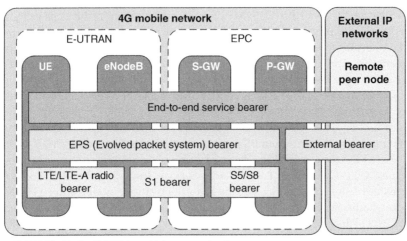

E-UTRAN – Evolved UMTS terrestrial access networks
EPC – Evolved packet core
P-GW – Packet data network gateway
S-GW – Serving gateway
UE – User equipment

Figure 5.7 QoS support in LTE/LTE-Advanced mobile networks.

of previous 3GPP RANs to the EPC, which include UTRAN (for UMTS) and GERAN (Global System for Mobile communication EDGE Radio Access Network, that is, 2.5G from 3GPP). However, for non-3GPP IP networks (e.g. WiFi, WiMAX), the mobility anchor point is the P-GW.

The bearers in 3GPP mobile networks are based on IP traffic tunneling in the EPS with GTP, which in fact started with GPRS (as 2.5G) and further was well established with UMTS (as the 3G technology) and then continued in 4G mobile networks from 3GPP. Regarding 4G bearers, GTP-U tunnels are established for carrying user data traffic between the eNodeBs and S-GW as well as between S-GW and P-GW. In fact, S-GW acts as the GTP-U relay between the eNodes and the P-GW. On the other side, GTP-C tunnels are established between S-GW and MME (for the signaling related to handovers, due to mobility of the mobile users) and between S-GW and P-GW. Also, for fast handovers there is a GTP-U tunnel established between adjacent eNodeBs to transfer the IP packets from the old eNodeB (before the handover) and the new eNodeB (after the handover), i.e. for fast path switching (between S-GW and the mobile terminal) through the EPS at the handover events.

One should note that in LTE/LTE-Advanced mobile networks besides the bearers there is an option to use Proxy Mobile IPv6 (PMIPv6) for interconnection with non-3GPP networks, such as IEEE standardized wireless networks like WiFi and WiMAX [3].

The PCC architecture in LTE/LTE-Advanced mobile networks works on a service data flow (SDF) level [7]. For the purpose of QoS-related messages, S-GW and P-GW are connected with the PCRF via control interfaces by using the Diameter protocol. These interfaces control the policy and charging enforcement function (PCEF), which is located in the S-GW and P-GW. For access to non-3GPP networks, both EPC gateways (S-GW and P-GW) must also implement Diameter interfaces toward external AAA network nodes.

When a given application server requests an EPS bearer, it is established by using signaling for lower layer bearers, which include LTE/LTE-Advanced radio bearer, S1 bearer, and S5/S8 bearer. The main goal of bearer setup and existence is minimization of QoS knowledge and configuration in mobile terminals [8].

An EPS bearer is characterized by the following main parameters:

- *Quality of service class identifier (QCI)*. This is a scalar value which defines a set of QoS parameters such as delay and packet loss/error rate and sets the priority of the IP packets depending on their type and the policy decision of the mobile operator (the targeted applications are not always the ones that are used for certain QCI because each network is autonomous in definition of its QoS rules).
- *ARP*. This parameter additionally defines the priority used for the allocation and retention mechanisms for the same QCI. ARP was also used in UMTS mobile networks for QoS differentiation within the same traffic class.
- *Guaranteed bitrate (GBR)*. This refers to guaranteed (minimum) bitrate for certain real-time services provided by mobile operators such as carrier-grade (i.e. QoS-enabled) voice and live streaming.
- *Maximum bitrate (MBR)*. This refers to setting a limit on the bitrate which is expected for the given service.

The GBR parameter is used to differentiate between resource types in LTE/LTE-Advanced mobile networks. In general, there are two resource types, one with

Table 5.2 QoS class identifiers (QCI) for LTE/LTE-Advanced (3GPP Release 14).

QCI	Resource type	Priority	Delay budget (ms)	Packet loss rate	Targeted services
1	GBR	2	100	10^{-2}	Voice over IP (VoIP)
2		4	150	10^{-3}	Video call and live streaming
3		3	50	10^{-3}	Real-time gaming, V2X messages
4		5	300	10^{-6}	Buffered video streaming (not conversational)
65		0.7	75	10^{-2}	Mission critical push to talk (MCPTT) voice
66		2	100	10^{-2}	Non-mission critical push to talk voice
75		2.5	50	10^{-2}	Vehicle-to-X (V2X) messages
5	Non-GBR	1	100	10^{-6}	IMS signaling
6		6	100	10^{-3}	Buffered video streaming, interactive TCP-based applications (Web, email, file sharing, etc.)
7		7	100	10^{-3}	Voice, live video streaming, interactive gaming
8		8	300	10^{-6}	Buffered video streaming, interactive TCP-based applications (Web, email, file sharing, etc.)
9		9	300	10^{-6}	
69		0.5	60	10^{-6}	Mission critical signaling (e.g. MCPTT signaling)
70		5.5	200	10^{-6}	Mission critical data from buffered video streaming, interactive TCP-based applications (Web, email, file sharing, etc.)
79		6.5	50	10^{-2}	V2X messages

guaranteed bitrates (i.e. GBR) and the other one with non-guaranteed bitrates (i.e. non-GBR). However, each of the resource types has several QCIs which are targeted to be used for QoS differentiation in LTE/LTE-Advanced mobile networks. Initially, LTE/LTE-Advanced had 9 QCI; however, with later releases (e.g. 3GPP Release 14, which is the last release before the appearance of the 5G, which appears with 3GPP Release 15) further granulation that introduced several new QCIs was required, having priorities in between some of the existing (initial) QCIs, as shown in Table 5.2. From all types of traffic, highest priority is given to IMS signaling, although it is served by non-GBR resource type. Signaling has no predefined bitrates because it depends on the load (e.g. volume of voice call setup/release signaling), but it must have the highest priority because it is a prerequisite for establishing call/sessions (e.g. voice calls over the mobile network). Further, priority two is given to voice services over LTE (e.g. VoIP over LTE –VoLTE); however, the delay budget for voice is more flexible than, for example, the required delay for real-time gaming (which is an emerging service) or V2X (vehicle to X) messages.

For mission critical services, such as mission critical push to talk (MCPTT) voice, and non-mission critical services there are defined QCIs 65 and 66, respectively, which were later added to the initial 9 QCIs. In a similar manner, QCI = 75 is also defined with GBR resource types, with a very low delay requirement for the V2X communication. QCI value 2 is targeted at video calls (and live streaming) which are sensitive to delay (e.g. in video calls the video has the same delay requirements as voice, due to the necessary synchronization between the audio and the accompanied video), while QCI = 4 is targeted at transfer of buffered video (i.e. VoD) provided by the mobile operator. All services served with GBR are in fact carrier-grade services because for Internet access services i is not possible to guarantee bitrates due to the network neutrality principle which is applied also for mobile Internet access. Therefore, Internet-based applications and services typically use QCIs values 8 and 9, which are served with the lowest priority, targeted for use by TCP-based applications (Web, email, Web-based services such as social networking, picture, and files sharing, etc.).

There are defined non-GBR QCIs for voice, video calls, live video, critical and non-mission critical services. There are also defined QCI = 69 for mission critical signaling (which has the highest priority of all, noted as 0.5, which is higher priority than IMS signaling), QCI = 70 for mission critical data services and other TCP-based services (such as Web-based, email, etc.) with priority between GBR and non-GBR buffered video streaming, and QCI = 79 for V2X messages (with priority set in the range for non-GBR real-time and interactive services such as video streaming, voice, gaming, browsing, etc.).

While in the past the most demanding service regarding delay was the voice, currently it is being "overruled" by emerging mission critical services and others such as online interactive gaming (which cannot be provided as GBR when network neutrality is in use, only as carrier-grade service). There is less QoS differentiation for non-real-time traffic due to the current network-neutral approach for such traffic, and in such cases all services are served in a "best-effort" manner, where the network does its best effort to carry the IP packets from the source to the destination. This is due to the already noted network neutrality principle for the services and applications provided through public (i.e. open) Internet access, where all such services are also referred to as OTT services. The GBR and services are provided via managed IP networks on the side of the mobile operator, and hence they are not part of the public Internet (e.g. such services are not accessible through mobile Internet access).

Considering Table 5.2 as an example, a mobile operator running a LTE/LTE-Advanced mobile network can use QCI = 5 for IMS signaling, QCI = 1 for carrier grade voice, QCI = 2 for video telephony, and QCI = 9 for mobile data (i.e. Internet access through the mobile network). With emerging IoT services over the mobile network, the newer QCIs may be used (as an example) for V2X services or mission critical services.

However, to provide the required QoS the mobile network needs binding of PCC and QoS rules with SDFs. The binding mechanism has three steps:

- *Session binding*. It is performed by the PCRF node and includes an IPv4 or IPv6 network prefix (for the mobile terminal's IP address) as well as user equipment identity.
- *PCC rule authorization and QoS rule generation (where applicable)*. It refers to the selection of the QoS parameters (including QCI, ARP, GBR, MBR, etc.) for the PCC rules.
- *Bearer binding*. It is an association of PCC rule and QoS to a given bearer. When GTP is used, the bearer binding function is located at the PCEF, otherwise it is located at the bearer binding and event reporting function (BBERF) [7], which does event reporting to the PCRF as well as exchange of (sending and receiving) Internet Protocol connectivity access network (IP-CAN) specific parameters with the PCRF.

The QoS rules in the mobile network are derived from the PCC rules. The PCRF is the node that can activate, modify, and deactivate QoS rules over a given reference (i.e.

Table 5.3 QoS rules in 3GPP mobile networks.

	Information	Rule category	Modification by PCRF allowed?
Rule ID	Rule identifier (for active QoS rule for UE)	Mandatory	No
Service data flow detection	Precedence	Mandatory	Yes
	Service data flow template	Mandatory	Yes
QoS control	QCI	Mandatory	Yes
	Uplink maximum bitrate	Conditional (if used)	Yes
	Downlink maximum bitrate	Conditional (if used)	Yes
	Uplink sharing indication		No
	Downlink sharing indication		No
	ARP	Conditional (if used)	Yes
	PS to CS session continuity	Conditional	No

network interface) point. Active QoS rule means that the SDF template should be used for the detection of the given SDF, and further mapping of downlink traffic to downlink bearer determined by bearer binding as well as mapping of uplink IP packets to the uplink bearer determined by the bearer binding. Table 5.3 provides QoS rule information and its category (mandatory or conditional). At bearer deactivation, all active QoS rules on that bearer are being deactivated.

5.4   QoS and Giga Speed WiFi

Since the standardization of the IEEE 802.11 standard in 1997, later known as WiFi, it has become the unified WLAN technology on a global scale. Together with Ethernet (IEEE 802.3 standard) as the unified fixed LAN standard, WiFi is implemented in every home and office, and in many public places. Besides use as WLAN technology as an extension of fixed or mobile access networks (toward home, office, or public place), WiFi is also used for traffic offloading in mobile IP-based networks as well as for interconnection of different devices.

However, WiFi does not use TDMA on the physical layer, so there is no possibility for explicit (i.e. strict) QoS provisioning due to absence of concrete radio resource allocation (e.g. with time slots reservation, which is possible in mobile networks that use TDMA in combination with other multiple access technologies). Additionally, WiFi operates in unlicensed frequency bands (2.4 GHz, 5 GHz, and lately 60 GHz), which are not reserved (as spectrum bands for mobile networks), so they can become congested when many users are using WiFi networks that are close to each other and overlap in the coverage area (e.g. in dense residential areas or in enterprises).

Similar to Ethernet, IEEE defines the WiFi standard on the lowest two layers of the protocol stack, that is, the physical layer (i.e. layer 1) and the data-link layer (i.e. layer 2). However, after being frozen as the main standard IEEE 802.11, additional developments of WiFi are given as amendments, noted as additional letters added to the standard. There are different amendments on protocol layer 1 and layer 2. The initial IEEE 802.11 standard provided bitrates up to 2 Mbit/s; it was followed by IEEE 802.11b with up to

Table 5.4 WiFi maximum and bitrates.

WiFi standard	Theoretical maximum bitrate	Throughput (assuming 70% efficient layer 2)	Frequency bands
IEEE 802.11	2 Mbit/s	1.4 Mbit/s	2.4 GHz
IEEE 802.11b	11 Mbit/s	7.7 Mbit/s	2.4 GHz
IEEE 802.11a	54 Mbit/s	37.8 Mbit/s	5 GHz
IEEE 802.11g	54 Mbit/s	37.8 Mbit/s	2.4 GHz
IEEE 802.11n	600 Mbit/s	420 Mbit/s	2.4 and 5 GHz
IEEE 802.11ac	6.9 Gbit/s	4.8 Gbit/s	5 GHz
IEEE 802.11ad	6.9 Gbit/s	4.8 Gbit/s	60 GHz
IEEE 802.11ay	40 Gbit/s	28 Gbit/s	60 GHz

11 Mbit/s (in 2.4 GHz), and further continued in the 2000s with IEEE 802.11g (2.4 GHz) and IEEE 802.11a (5 GHz), which provided bitrates up to 54 Mbit/s (Table 5.4). Regarding the physical layer amendments, the development of WiFi continued with IEEE 802.11n, which provides up to 600 Mbit/s (in both 2.4 and 5 GHz), and IEEE 802.11ac with up to 6.9 Gbit/s (operating in 5 GHz band). WiFi further extends into the 60 GHz unlicensed spectrum with IEEE 802.11ad, and further with IEEE 802.11ay, which is expected to be completed by the end of 2019 [1]. However, 60 GHz band is completely different than 2.4 GHz and 5 GHz bands regarding the propagation of radio signals.

IEEE 802.11ad is defined at 60 GHz bands (microwave WiFi), low power, but has very high performance regarding maximum bitrates (up to almost 7 Gbit/s). But such bitrates are available only in very short ranges, such as 1–10 m, within a single room or office. Although IEEE 802.11ac and 802.11ad both provide much higher data throughputs than their predecessors, such as IEEE 802.11n/a/g/b, they have much different potential uses. On the one hand, IEEE 802.11ac is a kind of an evolution of previous WLAN standards toward higher bitrates, offering the possibility of the "wireless office" to compete directly with gigabit wired systems while having wireless connection flexibility. On the other hand, IEEE 802.11ad provides ad-hoc short-range connectivity in support of extremely high data rates. However, in all WiFi standards the theoretical maximum bitrates that can be achieved on MAC layer are not possible due to the WiFi radio access scheme, and hence the expected throughput (achievable aggregate bitrate) is around 70% of the theoretical maximum, as shown in Table 5.4. In practice, all given bitrates can be much lower at longer distances between AP and wireless stations (e.g. lap-top, smartphones with WiFi) due to lower signal to noise ratio which results in a different modulation and coding scheme that carries fewer bits per symbol in the radio interface, thus resulting in lower bitrates than the maximum possible throughput with a given WiFi standard.

Besides the standardization of different WiFi amendments on the physical layer, there are many amendments on protocol layer 2, which are created for different purposes, such as application support (e.g. 802.11p/z/aa), network convergence (e.g. 802.11u), network management (e.g. 802.11k/v), QoS (e.g. 802.11e), security (e.g. 802.11i), etc.

In general, WiFi network architecture is based on APs, which work on layer 2. However, as any other LAN, a WiFi network is connected to the Internet via a router, which

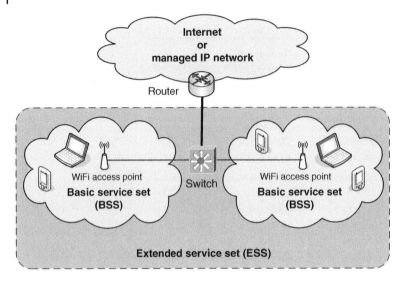

Figure 5.8 WiFi architectures.

can be integrated within a given WiFi AP. Overall, there are two main types of WiFi architectures:

- *Infrastructure mode.* In this case each WiFi AP is connected on an Ethernet switched network, which is further connected to the Internet (or certain intranets) via a router. As illustrated in Figure 5.8, each AP and its coverage area are referred to as a basic service set (BSS), while several APs in the same WiFi local network (on the same side of the router through which they are connected to the Internet) define the extended service set (ESS).
- *Ad-hoc mode.* In this case WiFi is used for direct ad-hoc communication between wireless hosts (e.g. smartphones, laptops, and other user devices with WiFi connectivity).

The main WiFi standard for QoS provisioning is IEEE 802.11e. To explain the WiFi QoS one needs to understand the way in which wireless stations communicate with the AP. The main access to wireless interface in almost all deployed WiFi networks is through the DCF (distributed coordination function), which is based on avoidance of collision between competing wireless stations by using CSMA/CA (carrier-sense multiple access with collision avoidance) and a binary backoff algorithm (it backs off WiFi stations which may collide in accessing the wireless link, by using random backoff within a given contention window (CW)). Besides DCF, WiFi has as option to use the PCF (polling coordination function) in which AP polls wireless stations to transmit data on PCF intervals PIFS (PCF interframe space) shorter than DCF intervals DIFS (distributed coordination function interframe space).

QoS support in WiFi is provided with the IEEE 802.11e standard. For the purposes of QoS support, the IEEE 802.11 standard defines the hybrid coordination function (HCF), which has two working modes:

- *Enhanced distributed coordination function channel access* (*EDCA*). Similar to DCF, but provides different priority levels for different services.
- Hybrid coordination function controlled channel access (HCCA) is CSMA/CA – compatible polling technique (like PCF, but with improvements).

The QoS support with EDCA is based on traffic differentiation by using traffic prioritization, so higher priority traffic has a greater probability of being sent over the wireless interface than lower priority traffic. The access to the wireless channel in ECDA is controlled with four parameters:

- *CWmin.* It is the minimal size of the CW.
- *CWmax.* It is the maximal size of the CW.
- *Arbitration interframe space (AIFS).* It is variable DIFS (shorter AIFS is used for higher priority packets and vice versa).
- *Transmission opportunity (TXOP).* It specifies the maximum time interval during which wireless stations can transmit WiFi layer-2 frames back-to-back (contention-free access).

The WiFi QoS support typically is provided with EDCA, which offers service differentiation of traffic into four (ACs and eight priority levels with the aim of providing traffic differentiation, as shown in Table 5.5.

The defined ACs in IEEE 802.11e EDCA can be used over different physical layers for WiFi (e.g. 802.11a/g, 802.11n, 802.11ac) because the QoS support is implemented on the MAC layer (i.e. layer 2). WiFi multimedia (WMM) APs have enabled EDCA and also TXOP. In general, QoS support for WiFi is optional. Default values for EDCA parameters per access/category are given in Table 5.6.

The QoS support in WiFi is relative, not absolute. For example, considering the given values for parameters in Table 5.6, voice traffic has CW for backoff in the range (CWmin, CWmax) = (3, 7), which gives voice traffic the possibility to access the wireless channel before video traffic which has (CWmin, CWmax) = (7, 15) and best effort and background traffic with (CWmin, CWmax) = (15, 1023) because voice traffic will have a backoff interval (to try to access the medium again to avoid collision at the initial moment) after a random number of time slots (this is not the TDMA approach, but each slot is a predefined time interval), which in this case are a minimum of three and a maximum of seven. The same priority is given to video traffic with additionally set TXOP parameters, which provides the possibility for back-to-back transmission of video packets (needed due to the bursty nature of video) for the given time period (e.g. 3.008 ms according to Table 5.6 values) before it releases the wireless channel. TXOP is convenient for variable bitrate (i.e. bursty) traffic such as video or Internet traffic (e.g. Web-based). The packets

Table 5.5 WiFi access categories and priorities for QoS provisioning.

Priority	Access category (AC)	Designation (informative)
7	3	Voice
6	3	Voice
5	2	Video
4	2	Video
3	1	Video probe
2	0	Best effort
1	0	Best effort
0	0	Best effort

Table 5.6 Default EDCA parameters for WiFi QoS provisioning.

Access category	CWmin	CWmax	AIFSN	Max TXOP
Background	15	1023	7	0
Best effort	15	1023	3	0
Video	7	15	2	3.008 ms
Voice	3	7	2	1.504 ms

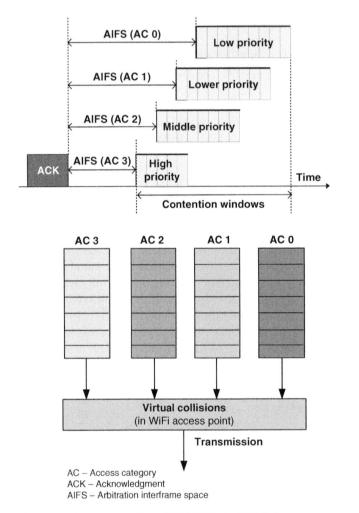

AC – Access category
ACK – Acknowledgment
AIFS – Arbitration interframe space

Figure 5.9 Access categories in IEEE 802.11e for WiFi QoS support.

belonging to different ACs are separately queued at the given AP (Figure 5.9), which implements virtual collisions to avoid physical collisions in the wireless channel.

Another approach to provide QoS support in WiFi networks is by using HCCA, which extends the EDCA access rules. It is based on a hybrid approach which has a contention period and a contention-free period.

In the contention period, TXOP is given and used after AIFS + Backoff period. QoS polling is done after PIFS (based on the PCF access scheme for WiFi).

In the contention-free period there is no contention between wireless stations connected to the same WiFi AP. In this period, TXOP is also given; however, its start time and duration are specified by a hybrid coordinator (HC), which uses QoS polling of wireless stations.

Overall, similar to the dominant use of DCF (and not PCF) in WiFi networks, HCCA is rarely used, while the QoS provisioning in WiFi is typically implemented via EDCA. However, to have end-to-end QoS requires mapping of WiFi ACs and priorities onto network classes of wired IP networks, such as IETF DiffServ classes (Table 5.7) [9].

Using the QoS in WiFi and mapping the WiFi priorities onto DSCP values in the IP packer header also has potential security concerns. For example, it is possible for a device connected to a QoS-enabled WiFi network (which does QoS mapping from layer 2 to DSCP on layer 3) to mark packets in a way that degrades existing QoS policies in that network. For example, if gaming applications are set to WiFi user priority 6 (according to Table 5.7) and if such traffic traverses through an enterprise, it may interfere with the business's QoS policies which (for example) intend to provide that priority to business voice service. To avoid such cases requires setting a traffic conditioning policy which reflects business objectives and policy, so that traffic from authorized users and applications only will be accepted by the network, while in all other cases (e.g. unauthorized devices) IP packets will be "bleached" (i.e. re-marked for DSCP, e.g. to 0). Such an approach for the QoS policy in WiFi networks is especially relevant for IoT deployments because in such case many devices (globally it is tens of billions devices) are being connected to IP networks with little or no security capabilities, so such IoT devices are vulnerable (e.g. they can be used for distributed denial of service [DDoS] attacks).

Table 5.7 IETF DiffServ classes mapping to WiFi access categories and priorities.

IETF DiffServ class	WiFi user priority	WiFi access category
Network control	7	Voice (VO)
Network control	0	Best effort (BE)
Telephony	6	Voice (VO)
Voice-admit	6	Voice (VO)
Signaling	5	Video (VI)
Multimedia conferencing	4	Video (VI)
Real-time interactive	4	Video (VI)
Multimedia streaming	4	Video (VI)
Broadcast video	4	Video (VI)
Low-latency data	3	Best effort (BE)
Operation and maintenance	0	Best effort (BE)
High-throughput data	0	Best effort (BE)
Standard	0	Best effort (BE)
Low-priority data	1	Background (BK)

5.5 WiFi vs. LTE/LTE-Advanced in Unlicensed Bands: The QoS Viewpoint

The LTE mobile networks in the 4G era have also entered the unlicensed frequency bands where WiFi has been the dominant technology since the end of the 1990s. From 3GPP Release 12, each next release further specifies the use of LTE in unlicensed bands with different purposes, so we have the following technologies (Figure 5.10):

- *LTE in unlicensed spectrum (LTE-U).* This is first version of LTE in unlicensed bands which has been standardized in 3GPP Release 12, and it is backward compatible with all 4G releases from 3GPP (i.e. Release 10 and beyond). It uses 5 GHz unlicensed band, and it is targeted to add capacity to the LTE network by using it on a 5 GHz unlicensed spectrum besides the licensed one. Of course, it coexists in this spectrum with WiFi, which operates on 5 GHz.
- *Licensed assisted access (LAA).* This was introduced with 3GPP Release 13 in 2015, as part of LTE-Advanced-Pro. It provides aggregate simultaneous use of LTE in licensed and unlicensed spectrum bands in the downlink direction to provide the most efficient use of both available bands. This is targeted to give higher bitrate in the range of Gbit/s with a single LTE carrier of 20 MHz and unlicensed 5 GHz band, providing a better user experience. Enhanced LAA (eLAA) from 3GPP Release 14 allows licensed and unlicensed spectrum aggregation in the uplink, in addition to the downlink aggregation by the LAA.
- Long term evolution WiFi link aggregation (LWA). This enables aggregation of LTE in a licensed band and WiFi carrier, which is easy for implementation with legacy LTE and WiFi equipment. However, LWA lacks use of LTE in unlicensed bands [10].
- *MulteFire.* This is fully operational use of LTE in an unlicensed spectrum on 5 GHz, specified by the MulteFire Alliance [11], which is different than LAA and LWA, which aggregate LTE carriers in a licensed spectrum with LTE or WiFi carrier in an unlicensed spectrum, respectively. MulteFire adds more capacity in crowded areas, as well as providing enhanced local broadband access even without a SIM card (similar to WiFi) and open possibilities for IoT verticals for LTE value added services.

Regarding the use of LTE and WiFi in unlicensed bands (at 5 GHz), they will both continue to be utilized. However, there are differences regarding the utilization of spectrum

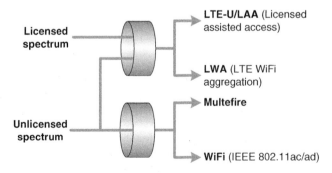

Figure 5.10 LTE-WiFi aggregation scenarios.

as well as QoS possibilities. For example, WiFi cannot guarantee strict QoS due to the DCF mechanism for access to the wireless channel (it can provide certain traffic differentiation by using the EDCA, as discussed in the previous section). Therefore, in the case of LWA, Internet traffic should be carried over the WiFi carrier while real-time traffic is preferably carried over the LTE carrier due to available strict QoS support based on bearers and associated QCIs. What is the QoS effect of transmitting mobile LTE signals over an unlicensed band? It provides better QoS and higher efficiency in radio resource utilization than WiFi in the same spectrum.

There are two types of interworking between LTE and WiFi for serving the same users:

- offloading mobile data traffic to WiFi, which is a traditional approach used in previous mobile generations (e.g. GPRS, UMTS);
- offloading WiFi traffic to LTE (opposite to traditional offloading between mobile networks and WiFi), which is targeted at provision of better QoS to the users transferred from WiFi to LTE in the unlicensed band in a local area (e.g. home, office, public place).

However, there are several challenges regarding the use of unlicensed spectrum between multiple technologies such as LTE and WiFi. One challenge is to have fair and efficient unlicensed spectrum sharing, e.g. to avoid possible interference. Another is to ensure QoS for the LTE traffic.

The traffic steering for LTE in unlicensed bands is driven by several factors [10]:

- *QoS and QoE enhancements.* By aggregation of different bands including licensed and unlicensed bands, the user throughput can be significantly improved, which has higher influence on the perceived quality of services (because higher throughput also provides lower latency).
- *Better radio resource utilization.* The traffic steering from congested licensed LTE bands to uncongested unlicensed bands and vice versa can improve the utilization of resources and provide higher average throughputs per user, which gives better QoS/QoE (also on average).
- *Reduction of operational cost.* The use of unlicensed bands for LTE traffic steering reduces operational costs due to no license fees for that spectrum, which on the other hand adds capacity, has lower delay in the radio part due to smaller cells, and hence can result in better QoS. Also, lower frequency LTE cells (e.g. below 1 GHz) can be used for providing coverage while LTE in unlicensed band (together with WiFi) can add more capacity in the network.

Overall, the use of unlicensed bands in combination with licensed bands is not novel in mobile networks since the introduction of Internet connectivity (e.g. with GRPS by 3GPP) and standardization of WiFi (1997 onwards). However, with carrier aggregation available with 4G, there is the possibility to aggregate different frequency bands in a given direction (e.g. downlink), including licensed bands (e.g. LTE 700 MHz, 800 MHz, and other bands) and unlicensed bands (e.g. 5 GHz for LTE and WiFi, 2.4 GHz for WiFi carriers), which provides several benefits regarding QoS in mobile and wireless networks, where the most important QoS parameter that is improved is the available throughput as well as mobile/wireless network availability.

5.6   The ITU's IMT-2020

Each new generation of mobile systems and networks has requirements specified by the ITU-R in an umbrella specification. In that manner, 3G mobile systems requirements were defined in IMT-2000 around the year 2000, which further resulted in seven technologies being accepted as 3G (by 2007). The 4G requirements were specified in IMT-Advanced around 2010, and it has resulted in two technologies that have fulfilled the requirements set in it – they are LTE-Advanced (3GPP Release 10) and Mobile WiMAX 2.0 (with IEEE 802.16m standard on radio interface). The following generation of mobile networks is noted as 5G and its requirements are specified in an umbrella called IMT-2020, again pointing to the year around which it is expected that the recommendation will be used for selection of the first 5G standard.

The IMT-2020 umbrella has been triggered by ITU-R M.2083 [12], which initiated creation of the IMT-2020 regarding its framework and overall objectives. As usual, each next generation (IMT-2020 sets the requirements for 5G mobile networks and systems) targets higher bitrates than predecessors (e.g. 4G); however, IMT-2020 mobile systems (5G) are expected to have even more diverse services requirements by entering of ICTs into different verticals, (e.g. public sector, industry sectors) and at the same time to provide better QoE at lower costs (per bit or byte) as well as better affordability. Regarding QoS, its improvement is considered by use of small cells, which provide better QoS experience at the cell edges (where the signals are much attenuated in the macro cellular environment). Table 5.8 provides comparison of selected ITU objectives in IMT-Advanced (for 4G) and IMT-2020 (for 5G).

There are three considered usage scenarios for IMT-2020:

- *Enhanced mobile broadband.* This is targeted at human centric access to the Internet with higher bitrates than 4G, including different mobility of users. Typically, this scenario is used for access to Internet OTT services and applications, including Internet video on mobile devices, which may be expected as the dominant traffic type in the 5G era (the 2020s).

Table 5.8 Comparison of ITU objectives in IMT-Advanced (4G) and IMT-2020 (5G).

	IMT-Advanced	IMT-2020
Minimum peak bitrate	Downlink: 1 Gbit/s Uplink: 0.05 Gbit/s	Downlink: 20 Gbit/s Uplink: 10 Gbit/s
Bitrate experienced by individual mobile device	10 Mbit/s	100 Mbit/s
Peak spectral efficiency	Downlink: 15 bit/s/Hz Uplink: 6.75 bit/s/Hz	Downlink: 30 bit/s/Hz Uplink: 15 bit/s/Hz
Mobility	350 km h^{-1}	500 km h^{-1}
User plane latency	10 ms	1 ms
Connection density	100 thousand devices per km^2	1 million devices per km^2
Traffic capacity	0.1 Mbit/s/m^2	10 Mbit/s/m^2 in hot spots
Frequency bandwidth	Up to 20 MHz/carrier (up to 100 MHz aggregated)	Up to 1 GHz (single or multiple frequency carriers)

- *Ultra-reliable and low-latency communications.* This scenario is targeted to services that have strict requirements on performances such as throughput, latency, and availability (e.g. smart city transportation, driverless vehicles and drones, industrial processes, remote surgery).
- *Massive machine type communications* (*massive MTCs*). This scenario refers to the very large number of devices connected via the mobile networks. Regarding QoS, this type of communication requires relatively low bitrates (e.g. hundreds of kbit/s or couple of Mbit/s) and in most cases is not too sensitive on delay. Such devices (e.g. sensors) are intended to have low cost and very long (lifelong) battery duration.

Considering the general principles of IMT-2020 mobile systems, they should support services with different end-to-end QoS requirements [13]. Also, the IMT-2020 network should be designed to avoid congestion for massive numbers of different MTC devices (with concurrent connections) at a given location, with aim of providing the required end-to-end QoS. For third parties to be able to guarantee a certain level of QoS, the mobile operator (operating IMT-2020 system) should be able to provide information about QoS capabilities (as well as other capabilities, such as connectivity, mobility, etc.) via certain application interfaces. However, to have a common core for different radio access technologies, the QoS mechanisms in the core should be able to work independently from the QoS mechanisms in the access parts, of course with a given mapping of the QoS parameters between them. The end-to-end QoS control in IMT-2020 requires QoS mapping between the QCIs and 5QIs (in the access part, standardized by 3GPP), and DSCP on the network layer (standardized by the IETF). Furthermore, for certain services, IMT-2020 mobile systems require finer granularity regarding QoS, such as end-to-end QoS provisioning on a per flow basis. Also, IMT-2020 is required to provide the possibility for user specified QoS parameters, such as minimum or guaranteed bitrate for a given service (smart home, gaming, etc.).

So, IMT-2020 is intended to satisfy unified QoS control mechanisms which are independent of the access technology, end-to-end QoS mechanism (because application/service QoS is always end-to-end), implicit and explicit QoS policy enforcement mechanisms, user-initiated QoS control mechanisms, as well as interworking with legacy IMT networks (e.g. 3G) and NGNs [13].

The IMT-2020 networks will coexist with existing UMTS, LTE, and/or WiFi; however, they provide new technologies that are designed to provide low latency, high reliability, or massive types of communications (considering the three main scenarios listed in this subsection). Such new capabilities of IMT-2020 mobile networks require bigger expenditure and operational expenditures (CAPEX (capital expenditure) and OPEX (operating expenses)), so mobile operators require different ways of optimization of networks and equipment, which lead further to network virtualization. Similar to the transition from separate networks for separate services to a single (IP) network for all services, the following step is further flexibility in allocation and control of network resources by means of network virtualization and softwarization for the purpose of provision of new services with desired cost and quality. However, for support of the existing as well as emerging new services IMT-2020 requires management and orchestration for IMT 2020 mobile networks. The term orchestration refers to automated arrangement, coordination, instantiation, and use of network functions and resources for both physical and virtual infrastructures by optimization criteria [14].

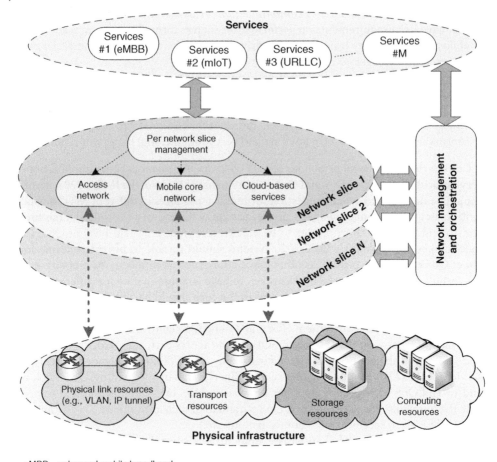

eMBB – enhanced mobile broadband
mIoT – massive Internet of Things
URLLC – Ultra-reliable low latency communications

Figure 5.11 Network slicing for IMT-2020.

Generally, network virtualization based on software defined networking (SDN), network function virtualization (NFV), and cloud computing technologies contributes to the IMT-2020 framework [15] by use of software for mobile network design, implementation, deployment, management, and maintenance. With such virtualization, the underlying physical infrastructure is abstracted as network, storage, or computing resources. The main logical building blocks in IMT-2020 are network slices. By one definition [14], a network slice in IMT-2020 is a logical network that provides certain specific capabilities and characteristics for a given mobile services market scenario (Figure 5.11).

There are different possible uses of network softwarization in IMT-2020. For example, video traffic is becoming the dominant type of traffic which shows a continuously increasing trend [15]. However, in certain regions disaster resilience is very important (e.g. due to hurricane disasters, floods, earthquakes), where IMT-2020 offers the possibility to be used as the first option for communication. Softwarization may contribute to the creation of a separate network slice (or slices) for massive IoT deployments. Due to very low latency in IMT-2020 mobile networks, a separate network slice (or slices) can be allocated to ultra-reliable low-latency services. One may expect SDN and NFV to be used for realization of network slicing and even further extend the possibilities for offering customized services with different requirements on QoS (and other network functions) between slices and within a given network slice. So, different slices can provide different QoS for the same type of services (ToSs) (managed VoIP and OTT VoIP as one explicit example).

Considering the requirements in IMT-2020 to have access-agnostic network architecture [13], with a unified core network, there is a need to decouple control mechanisms from the radio access technologies. IMT-2020 leads the fixed-mobile convergence (FMC) one step further, allowing complementary network performance by using both fixed and mobile connections. However, service continuity at handover events between fixed and mobile networks should support the guaranteed QoS for the given service [16]. Guaranteed QoS is required in all cases, including single and multiple access network connections. The following two scenarios outline the FMC application in IMT-2020:

- *Fixed access via both fixed and mobile access networks.* In this case the user terminal (desktop, lap-top, etc.) at a fixed user location (e.g. home, office) can access the broadband service controlled by the IMT-2020 FMC network by simultaneously using fixed and mobile access, or only one access type at given time. The traffic policies and QoS should be supported simultaneously through both access network types, aiming to provide high user experience.
- *Mobile access via both fixed and mobile access networks.* In this case the mobile terminal (e.g. a smartphone) connects through fixed access (via WiFi as wireless LAN for the fixed access network) and/or mobile access network, again by using both access network types simultaneously or one access network at a time for a given service (e.g. a voice call can be transferred via the mobile access network, while an open Internet connection for file download/upload can be active at the same time via WiFi through the fixed access network).

For FMC purposes IMT-2020 requires unified authentication and authorization for both fixed and mobile access technologies. Also, FMC means that IMT-2020 QoS control needs to support unified measurement of network performance, unified QoS policy management, and unified QoS decision making in the control plane for both networks. However, one should note that FMC is already fostered in NGNs, which include 4G mobile networks, where the standardized service stratum nodes are access independent (e.g. IMS) [2].

5.7 QoS in 5G Mobile Ultra-Broadband

The QoS support in 5G mobile networks from 3GPP, which started with 3GPP Release 15, is based on further evolution of QoS support defined for LTE/LTE-Advanced mobile networks (i.e. 4G networks from the 3GPP). Similar to LTE, the 5G QoS supports two resource types, i.e. QoS flows that require GBR and QoS flows that do not require GBR (i.e. non-GBR) [17].

To identify a QoS flow in 5G systems, a quality of service flow identifier (QFI) is used. So, the user plane traffic which has the same QFI receives the same traffic management approach (e.g. the same admission control or scheduling). Such QFI can be assigned dynamically or can be equal to standardized 5QIs (5G QoS indicators). The list of 5QI is given in Table 5.9. This list includes all QCI from LTE/LTE-Advanced/LTE-Advanced-Pro (i.e. up to 3GPP Release 14) and adds new QoS indicators which refer to emerging services with 5G deployments such as remote control, intelligent transport systems (ITSs), discrete automation, and augmented reality. The new resource type in 5G is delay critical GBR, which has a requirement for very low delays (up to 20 ms) as well as low jitter (up to 20 ms), targeted to critical services such as control of automatic processes or transportation vehicles via the mobile network, thus excluding the need for building a separate network for such delay critical services.

For each QoS flow in 5G there is an associated 5G QoS profile, which must include the following two parameters:

- *5QI parameter*. It is defined for 5G by 3GPP, given in Table 5.9.
- *ARP parameter*. Similar to 4G and 3G mobile networks from 3GPP, this parameter defines the priority of the flows that have the same 5QI. The range of ARP is 1–15, where ARP = 1 has the highest priority. ARP values 1–8 can be assigned only for services that are authorized for prioritized treatment within the mobile operator's domain.

For the GBR only flows, the QoS profile also must include the bitrate parameters:

- guaranteed flow bitrate (GFBR) for both uplink and downlink;
- maximum flow bitrate (MFBR) for both uplink and downlink.

For the GBR flow the QoS parameters list may be completed by two additional parameters:

- *Notification control*. That is availability of notifications from the RAN when GBR for the flow can no longer be fulfilled.
- *Maximum packet loss rate in uplink and downlink*. These parameters refer to maximum error rate that can be tolerated for the QoS flow.

For non-GBR flow the QoS profile may also include the parameter RQA (reflective quality of service attribute). The reflective QoS in 5G refers to the possibility to map user plane traffic to QoS flows in the uplink without having QoS rules from the SMF by deriving the QoS rules from the downlink traffic (hence, there is reflection of the QoS rules from downlink into the uplink). This applies to both the IP PDU (protocol data unit) session and the Ethernet PDU session. The derived QoS rule contains several parameters which include one packet filter, QFI, and precedence value.

The packet filter set is used to identify one or more IP or Ethernet flows. The IP packet filter set is based on a combination of the following parameters from IPv4 or IPv6 packets:

- source and destination IP address (IPv4 or IPv6);
- source and destination port numbers (e.g. TCP port, UDP port);
- Protocol Identifier (ID) (in IPv4) or Next Header (in IPv6) giving the protocol above the IP layer in the stack (e.g. TCP, UDP);

Table 5.9 5G QoS indicator (5QI) list.

5QI	Resource type	Priority	Delay budget (ms)	Packet loss rate	Default maximum data burst volume	Targeted services
10	Delay Critical GBR	11	5	10^{-5}	160 bytes	Remote control
11		12	10	10^{-5}	320 bytes	ITS (intelligent transport systems)
12		13	20	10^{-5}	640 bytes	Other delay critical services
16		18	10	10^{-4}	255 bytes	Discrete automation
17		19	10	10^{-4}	1358 bytes	Discrete automation
1	GBR	20	100	10^{-2}	—	Voice over IP (VoIP)
2		40	150	10^{-3}	—	Video call and live streaming
3		30	50	10^{-3}	—	Real-time gaming, V2X messages
4		50	300	10^{-6}	—	Buffered video streaming (not conversational)
65		7	75	10^{-2}	—	Mission critical push to talk (MCPTT) voice
66		20	100	10^{-2}	—	Non-mission critical push to talk voice
75		25	50	10^{-2}	—	Vehicle-to-X (V2X) messages
E		18	10	10^{-4}	255 bytes	Discrete automation
F		19	10	10^{-4}	1358 bytes	Discrete automation
5	Non-GBR	10	100	10^{-6}	—	IMS signaling
6		60	100	10^{-3}	—	Buffered video streaming, interactive TCP-based applications (Web, email, file sharing, etc.)
7		70	100	10^{-3}	—	Voice, live video streaming, interactive gaming
8		80	300	10^{-6}	—	Buffered video streaming, interactive TCP-based applications (Web, email, file sharing, etc.)
9		90	300	10^{-6}	—	Buffered video streaming, interactive TCP-based applications (Web, email, file sharing, etc.)
69		5	60	10^{-6}	—	Mission Critical signaling (e.g. MCPTT signaling)
70		55	200	10^{-6}	—	Mission Critical Data from buffered video streaming, interactive TCP-based applications (Web, email, file sharing, etc.)
79		65	50	10^{-2}	—	V2X messages
80		66	10	10^{-6}	—	Low latency eMBB augmented reality

- ToS field in IPv4 header or Traffic class (i.e. DSCP field) in IPv6 header, for traffic differentiation on the IP layer based on defined classes (e.g. with DiffServ);
- flow label, which is available only in IPv6 headers (identification of flows in IPv4 uses the five-tuple, consisting of source and destination IP addresses, source and destination ports, and the transport protocol above the IP);
- security parameter index, which is used in case of IPsec protected packets;
- direction of the filter, which can be either downlink or uplink.

The QoS rule precedence value determines the order for evaluation of the QoS rule. Lower precedence value gives higher priority in evaluation of the QoS rules, and vice versa.

5.7.1 5G QoS Control and Rules

Regarding QoS control in 5G, one should note that QFI can be equal to 5QI in some cases, but also these two parameters can have different values. Why? The parameter QFI is defined (in 3GPP Release 15) with 6 bits, which gives possible values in the range between 1 and 64 (because $2^6 = 64$). So, 5QI values that do not belong to this range cannot become values for QFI. The approach is the following: for non-GBR flows when the 5QI parameters are used, the value of the 5QI can be used (when it is in the range 1–64) as the value of the QFI for the given flow. When a non-GBR flow goes over a non-3GPP network (e.g. fixed network, WiFi), a default ARP value should be preconfigured in the access network, while in the case of 5G access networks the ARP and QFI are sent to RAN via PDU session establishment/modification mechanisms. In all other cases (including both GBR and non-GBR flows) dynamically assigned QFI values are used, while 5QI values can be the standardized ones, preconfigured (except for roaming scenarios) or dynamically assigned to the flow. However, the QoS profile and the corresponding QFI for the given QoS flow are also provided via 5G RAN using the PDU session establishment/modification mechanisms.

5.7.2 5G QoS Flow Mapping

The classification and marking of the user data traffic in the uplink direction (that is association of uplink traffic to QoS flows) is done by the user equipment (i.e. mobile terminal), based on QoS rules which may be explicitly provided to the user terminal (via PDU session establishment or modification), preconfigured, or implicitly derived by using the reflective QoS (i.e. deriving the QoS rules from the downlink direction and applying them in the uplink).

The QoS flow mapping in 5G is shown in Figure 5.12. In 5G networks from 3GPP the function responsible for service flow detection, authorized QoS, charging, traffic reporting, etc. is SMF. The SMF performs also the binding of SDFs to QoS flows by interacting with the mobile terminal (i.e. user equipment), 5G RAN, and UPFs. SMF should be able to determine the authorized QoS of a given QoS flow based on PCC rules that are associated with the flow, as well as to notify PCF when the QoS targets cannot be fulfilled [18].

For QoS control at the PDU session level the PCF provides a 5QI/ARP combination for the PDU session. Association of a PCC rule to a QoS flow within a PDU session is called

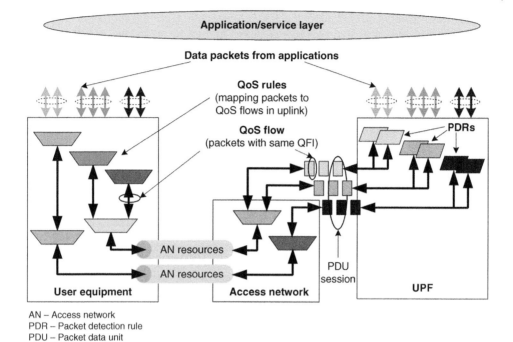

AN – Access network
PDR – Packet detection rule
PDU – Packet data unit
UPF – User plane functions

Figure 5.12 QoS flow mapping in 5G.

QoS flow binding. Besides the 5QI/ARP pair of parameters, if available in the PCC rule, the QoF flow binding can use the following parameters: quality of service notification control (QNC), priority level, averaging window, and maximum data burst volume.

The binding of the PCC rule and the QoS flow causes the downlink transfer of the SDF to be directed toward the QoS flow. For the uplink direction the QoS rule associated with the QoS flow instructs the mobile terminal (i.e. user equipment) to direct the SDF in the uplink direction to the QoS flow in the established association. If authorized QoS from the given PCC rule is changed, the bindings are reevaluated and then the given SDF can be bound to another QoS flow.

So, we have three levels of QoS control in 5G:

- QoS control at the SDF level;
- QoS control at the QoS flow level; and
- QoS control at the PDU session level.

Figure 5.12 illustrates the mappings between the three QoS levels in 5G by 3GPP. Note that one or more SDFs can be mapped on a single QoS flow. Also, one or several QoS flows can be mapped on a given PDU session. In the downlink direction data packets (in the user plane) are classified by the UPF based on the packet filter sets of the downlink packet detection rules (PDRs), applying them in order of their precedence. The UPF passes on the classification of user data traffic which belongs to a QoS flow through the interface between the UPF and 5G RAN (radio access network, i.e. radio AN) based on user plane marking with QFI. Further, the AN does binding of QoS flows to AN resources, which are the data radio bearers (DRBs) in the RAN. However, the relationship between QoS flows and AN resources is not strictly one-to-one. Further, AN establishes and releases the AN resources and in the case of QoS flows indicates that to the SMF.

In summary, the main novelty (regarding the 3GPP mobile networks) in the 5G QoS approach is the possibility for differentiation of data flows, which provides means for differentiation of traffic from different applications/services with diverse QoS requirements while at the same maximizing the resource utilization in the AN. Also, 5G (starting with 3GPP Release 15) is designed to support different types of AN, including fixed ANs. QoS provisioning is symmetric because it is provided in the same manner in both directions, downlink and uplink, which is possible with the reflective QoS approach (copying the downlink QoS rules for the uplink). Finally, while in UMTS and LTE/LTE-Advanced the QoS support is based on bearers, the 5G mobile core is designed to have flow-based QoS rather than bearer-based one.

5.8 Mobile Broadband Spectrum Management and QoS

The spectrum management in mobile broadband networks is crucial for QoS support. Why? Because the most important parameter for the mobile broadband (e.g. 3G and 4G) and mobile ultra-broadband (5G) is to have more bits per second. And that can be accomplished by using more frequency spectrum bands for a given RAN, either 3G, 4G, or 5G, and by using wider frequency carriers (e.g. 20 MHz was the maximum width of a frequency carrier in LTE, 5 MHz in UMTS, and 0.2 MHz in GSM/GPRS). To provide higher aggregate bitrates in a given cell it is also required to have bigger width of frequency carriers. If the technology development at a given moment in time is limited by the maximum possible frequency carrier width (which increases with the time), then the tool to increase it on the radio interface level is to aggregate different carriers below the network (i.e. IP) layer and then the aggregated spectrum (consisting of different carriers, which may be located in different spectrum bands) can appear as a single pipe in a given cell from the networking and application point of view. So, the capacity in a given cell is directly related to the available spectrum and the possibility for aggregation of frequency carriers, and that directly influences the speeds available to end-users. Of course, the support of higher speeds is also dependent upon the capabilities of user equipment (smartphone, lap-top computer, etc.), which is ranged into categories. For example, in LTE-Advanced mobile networks, smartphone Category 6 supports up to 300 Mbit/s, Category 9 supports up to 450 Mbit/s, Category 12 supports up to 1024 Mbit/s, i.e. 1 Gbit/s, and Category 18 supports up to 1200 Mbit/s. Each new release of the leading smartphone models on the market introduces a new category regarding achievable bitrates. So, even though the network will support Gbit/s speeds, a smartphone of Category 6 cannot achieve a speed above 300 Mbit/s. Additionally, mobile operators may limit (by certain configurations of the network nodes) the maximum speed that can be achieved by a certain user, which is a subject of network design and planning as well as product offers from operators to customers (via SLAs). Also, the spectrum in use at the base stations is required to be supported by the mobile terminals. Older mobile terminals have no support for recently added new frequency bands or new modulation schemes or new MIMO supported by the mobile operator's base stations. Therefore, every new mobile generation is gradually introduced into the market, first in the hotspot

urban areas and then into the suburban and rural areas, while the older mobile generations (e.g. 3G, 4G, and even GPRS, in respect to 5G) are used to provide macrocellular
coverage.

However, spectrum management is carried out by governments and regulators to
ensure that there is sufficient frequency spectrum for those services most in demand
among their citizens, and in such a way that provides the highest possible socio-
economic benefits [19]. The spectrum for terrestrial mobile and wireless networks is
typically divided into two main types:

- *Licensed spectrum*. This is given to telecom operators on the basis of individual
  licenses, which specify the bands for exclusive use by that operator. Such spectrum
  provides the best possible QoS controls because it is dedicated to the operator for
  defined radio access technology (or technologies). The recent trend is to provide
  bands for the IMT spectrum, which includes 2G, 3G, 4G, and 5G. The operators that
  have licenses for certain spectrum bands can use them according to their network
  design and planning, through the given requirements on the spectrum use (antenna's
  output power, interference, etc.)
- *Unlicensed spectrum*. This can be used by everyone at their own location (e.g. home,
  office, public place) without a need for licenses to be issued by the government or
  regulator. Such bands are license-free and are located on 2.4 GHz (the most used),
  5 GHz, and lately 60 GHz. However, unlicensed bands can become crowded with
  higher interference ratio when many wireless/mobile networks are overlapping (e.g.
  in neighboring homes, offices). Therefore, QoS cannot be strictly guaranteed as
  with licensed bands. Typically, WiFi is the technology used in all three unlicensed
  bands, although the 3GPP mobile network (e.g. LTE-Advanced-Pro, 5G) has defined
  standards for use in these bands, besides their use in traditional licensed bands.
  However, the 3GPP technologies in unlicensed bands offer the possibility of better
  QoS because they are based on the TDMA scheme in the RAN (unlike WiFi, which
  is not), offering the possibility to reserve certain radio resources (time slots) for a
  given mobile user terminal.

Overall, the licensed spectrum on average gives higher QoS in mobile networks
because the mobile operator can autonomously design, deploy, and optimize the mobile
network. However, the very high price of the licensed spectrum (e.g. the new spectrum
for 5G) may limit the amount which mobile operators are ready to invest in the network,
which can result in less capacity for the given number of users and hence lower QoS,
considering that the bitrates are one of the most important parameters for ultra-
broadband Internet access via the mobile network.

To provide device mobility and user mobility from one network and/or country to
another requires harmonization of frequency bands around the globe. That is done
by the ITU-R through its almost periodical World Radiocommunication Conferences
(WRCs). For example, WRC 2012 set the initial bands for 4G and WRC 2015 completed
the list of IMT bands to be used for 4G (as well as for previous IMT networks, such
as 3G). In fact, IMT includes IMT-2000 (3G), IMT-Advanced (4G), and IMT-2020
(5G), hence IMT spectrum bands refer to all existing generations under the IMT
umbrella. WRC 2019 will set the initial bands for 5G mobile networks. Using the same
bands around the globe is crucial to ensure mobility through countries without the
need to change end-user equipment, which increases QoE for mobile services.

5.9   Very Small Cell Deployments and Impact on QoS

The 4G and 5G mobile networks have included small cells in their standards. But let's define what these are. Small cells are deployed by using low powered radio communication equipment controlled by the mobile operator and typically have coverage in the range of tens or hundreds of meters (similar to WiFi coverage). However, there are different classifications for small cells, in which some definitions include all cells smaller than macro cells (e.g. micro cells, metro cells, pico cells, femto cells) into the group of small cells [20], where micro cells typically are used outside the urban areas. Other classifications state that small cells are mainly hotspot and pico cells [12].

Small cells have multiple benefits. They provide better signal to noise ratio for users in all cell locations, including cell edges. They require lower transmitting power of mobile devices due to lower distances to the base stations, which saves battery power and hence gives better QoE for use of the mobile services. Also, small cells decrease the interference due to lower transmitting power in both directions (downlink and uplink) and hence improve network performances and with that QoS. However, the number of small cells and their sizes depends on the projected number of active users (e.g. it is different when small cells are used in enterprise environments on one side and rural areas on the other). With the frequency reuse on smaller distances, small cells add capacity with the same spectrum. Additionally, small cells provide the possibility to use millimeter wave spectrum (above 6 GHz, such as the 24–86 GHz spectrum in different bands), due to smaller distances and the possibility of line-of-sight communication between the mobile network equipment (e.g. radio units) and the mobile devices.

Regarding spectrum type, small cells can be deployed in both the licensed and unlicensed spectrums, unlike macro cells which can utilize only the licensed (i.e. reserved) spectrum. The mobile networks have entered the unlicensed spectrum with the LTE standards for such a case. If one compares LTE vs. WiFi in the unlicensed spectrum [21], LTE gives better QoS due to its radio interface, which is well designed and standardized to guarantee QoS to substantial numbers of users in different traffic scenarios.

However, to have QoS in mobile networks with small cells, the backhauling for small cells is also very important, i.e. their connection toward the operator's core network. The following requirements exist for small cells backhauling from the QoS point of view:

- *Delay in the backhaul*. Different QCI in 4G and 5QI in 5G mobile networks have delay requirements that should be met. For example, for defined QCI in LTE-based mobile networks (up to 3GPP release 14) there is required backhaul delay of up to 20 ms for 98% of packets for high priority CoS or in uncongested conditions [20]. However, for 5QI defined for 5G mobile networks, there are even stricter delay requirements for delay critical services, such as below 5 ms for remote control services (Table 5.9), or below 10 ms for ITS, discrete automation or ITSs.
- *Capacity of the backhaul*. Small cells backhauling should carry traffic during both busy time intervals and quiet time peak traffic. So, the backhaul networking for N cells should satisfy the following:
  Backhaul capacity = Max {peak bitrate, N × busy time mean bitrate}
- *Availability of the backhaul*. This refers to the proportion of time that a backhaul connection is available (i.e. fully functional), while the outage time is the time when it is

not available. Recovery from outage is referred to as resilience time. There is no strict value recommended for availability; however, the range of 99.9–99.99% availability is expected to cover most implementations [20], which is normally below the availability requested by the transport network (that is 99.999%, i.e. five nines) [2] because the transport networks connect different backhaul and core networks.

- *Frequency and time synchronization.* The backhaul needs to provide synchronization for small cells with accuracy of ±0.1 ppm (parts per million) as well as phase accuracy for the TDD (time division duplex) systems of 1.5 μs.

Typically, backhaul network for small cells is Carrier Ethernet, which has built-in QoS support. For the purpose of synchronization of Carrier Ethernet (required also in macro mobile network backhaul, not only for small cells backhaul), the ITU-T SyncE standard, IEEE 1588-2008 standard (the Precision Time Protocol), or Network Time Protocol (NTP) from the IETF can be used.

Regardless of the mobile generation (e.g. 4G, 5G), there are three main deployment scenarios of small cells (as shown in Figure 5.13):

- *Different frequency carriers for macro cellular network and for small cells.* This scenario consumes more spectrum bands; however, it provides the best QoS due to absence of interference between the carriers.

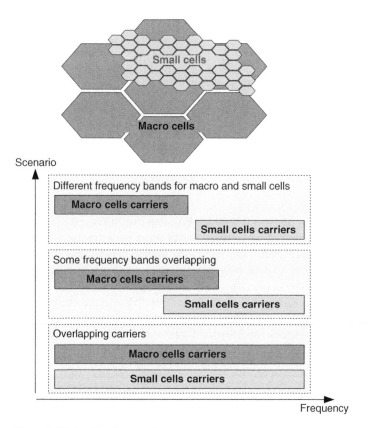

Figure 5.13 Small cells scenarios.

- *Same frequency band used for overlapping macro and small cells.* This is the most typical scenario because the spectrum is scarce and limited, as well as the licensed spectrum having a cost. However, regarding QoS, there is higher interference in this scenario and hence lower bitrates on average.
- *Overlapping carriers.* In this scenario the frequency carriers in macro and small cells are partly overlapping, so this approach provides a hybrid solution by combining the other two scenarios, thus having QoS level in the middle of the other two deployment scenarios.

Considering the standardization aspects of small cells, LTE (as 4G technology) has offered deployment on a large scale with LTE-U, LAA, LWA, and MulteFire. However, the expansion of small cells is coming with 5G deployments. Using small cells, 5G can utilize higher frequency spectrum (e.g. 24–86 GHz), which can cover only shorter distances from the base stations. With the capacity of small cells, the 5G mobile networks can compete in capacity with fixed networks. Typical 5G architecture uses macro cells for control, all around coverage and fallback (e.g. when users leave the small cells area), while small cells are used for capacity for ultra-broadband Internet access or for communications with IoT devices deployed in the given area. So, different 5G network slices can be overlaid over small cells in a given area, which may have different QoS requirements (e.g. higher bitrates are required for ultra-broadband Internet access while lower bitrates are needed for the most of the IoT services).

5.10 Business and Regulation Aspects for Mobile Ultra-Broadband

Mobile broadband access has become globally the most used access for all telecommunication/ICT services. Mobile ultra-broadband, with individual user bitrates exceeding 100 Mbit/s, will further improve the QoE for services in mobile networks, becoming comparable with fixed access.

5.10.1 Business Aspects

There is a continuous need for revenue growth, which began in the 2000s and continued in the 2010s. That can be accomplished while there is still room for attracting users to mobile broadband and ultra-broadband. However, with higher penetration of users, growth decreases because there are fewer new users to join up. With mobile ultra-broadband, operators continue to be network providers offering Internet access service, which is mainly used for access to various OTT services. Overall, the mobile broadband business is saturating horizontally (i.e. regarding new individual users), so new sources of revenue are needed. These sources for growth of the mobile sector are directed toward digitalization of all segments of public life, public administration, and industry (Figure 5.14). Such expansion of the mobile markets is referred to as entering vertical markets [22].

So, mobile operators may achieve larger growth by focusing on several steps in the value chains. 5G can be used for digitalization of the manufacturing industry (one may

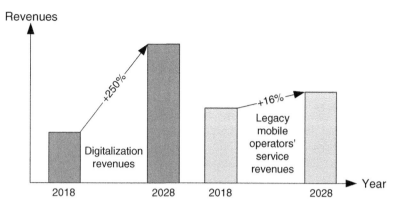

Figure 5.14 Revenue forecast of mobile operators in the 5G era.

refer to 5QI, which are defined for direct automation as critical services in 5G). The digitalization of previously non-digital industries provides new business models and new possibilities for mobile operators. Such opportunities include B2B2X (broadband to business to X), which is a newer segment, besides B2C (broadband to customer), B2B (broadband to business), and wholesale provisioning of ICT services. The B2B2X is trying to capture various digital solutions, and unlike conventional B2B services, the final service customers, noted as "X," are providing non-ICT solutions or services in which the telecom (i.e. mobile) operator's solution is in fact embedded into the value chain of the service by the X party. However, such non-ICT industries do not have the necessary capacity or experience to provide the given services by themselves. So, the digitalization of initially non-digital industries is giving telecom operators the possibility to include themselves in the new business models and value chains.

5G is the main technology driver for the industry digitalization. According to forecasts [22], there are eight main sources of digital industrialization revenues: financial services, automotive industry, public transport, media and entertainment (which is already using the online provision of services massively), emerging healthcare, energy utilities, manufacturing, and public safety. Regarding the capabilities of 5G mobile networks and the digitalization of non-ICT sectors there are three main use case business scenarios:

- *Massive machine-type communications.* This scenario is intended for use of massive IoT via the 5G mobile network (also available in the previous generation of mobile networks).
- *Critical machine-type communication.* This scenario is introduced in the mobile networks with 5G because it requires very low delays (e.g. below 10 ms), which can be provided by 5G mobile networks (of course, such delays cannot be provided on the global scale due to limitations of transfer speed of signals by the speed of light, which is up to 3×10^8 m s^{-1}).
- *Enhanced mobile broadband.* This scenario targets customers who are using mobile broadband before 5G (e.g. in 3G and 4G mobile networks) for use of various OTT services (WWW, email, video, social networks, etc.). This scenario is also the initial driver for 5G.

The mobile operator can take one or more of the three main roles for 5G industry digitalization: network developer, service enabler, and service creator. However, one may expect the largest increase in 5G revenues to come from digitalization of manufacturing and energy utilities.

5.10.2 Regulation Aspects

The development of 5G mobile networks raises certain regulatory challenges. They may be grouped into several groups, which tackle different aspects of the new business models in 5G mobile networks. The following aspects are some of the most important:

- Planning regulation for small cells sites, which should remove the barriers (where they exist) for mobile masts, which should be placed on different commercial buildings.
- The mobile network must take into account the location of the deployed "things" to provide enough coverage and capacity, as well as appropriate QoS. That is particularly important in the case of higher frequency bands (e.g. 24 GHz, 60 GHz, or generally frequency bands above 6 GHz) because different physical factors (trees, buildings, etc.) influence the propagation of radio waves at high frequencies.
- Spectrum regulation for 5G mobile systems should provide enough capacity and coverage as well as flexibility regarding previous mobile generations with which 5G will coexist during the 2020s (e.g. 4G mobile networks). Also, the harmonization of the 5G spectrum on a large scale, which is done at ITU's WRCs, is important for mobile ultra-broadband broadband (individual bitrates in the mobile network above 100 Mbit/s).
- Safety regulations for 5G mobile services are needed for critical services, which will digitalize different sectors such as previously non-ICT industry or the public transportation sector, as emerging examples. In general, customers need to have confidence to use existing and new services over 5G mobile networks.

The regulation regarding mobile ultra-broadband access which is coming with 5G and beyond should be targeted to support the innovation of services and application in parallel with existing enhanced services (e.g. ultra-broadband access to the Internet for use of OTT services via mobile network with very high bitrates). Network neutrality also influences mobile ultra-broadband, including the mobile operators, service providers, and customers. However, network neutrality may have a certain negative effect on innovation capabilities of mobile operators, which have to invest in mobile ultra-broadband access and core networks, but it has a positive effect on OTT services and applications for which mobile broadband is used in practice. Mobile operators also enjoy increased revenues due to higher usage statistics of mobile users via ultra-broadband access of OTT services. In that manner network neutrality has a positive effect on mobile operators' revenues from broadband access, which brings higher revenues than mobile voice in the second part of the 2010s, and such a trend will continue.

However, when there are stricter network neutrality rules, it may limit the possibility for quality differentiation, thus reducing the motivation of mobile network operators to invest in new technologies (e.g. 5G in the 2020s, or 6G, expected in the 2030s) and to improve the return on their investment in new technologies. Nevertheless, strict network neutrality fosters innovations from OTT service providers.

Generally, it will be nearly impossible for regulation to determine the point of optimal performance for network neutrality regulation which gives the highest investment-innovation incentives, which may be referred to as a zone of acceptable performance or "workable" zone between lower and upper bounds of regulatory intensity [23]. However, it is almost impossible to find the workable zone in advance, especially considering the new emerging services which have stricter QoS requirements.

Besides the network neutrality aspects, mobile broadband networks may have experienced QoS degradation for carrier-grade services such as voice (which is provided with QoS-enabled VoIP, as a replacement for digital PLMN). The QoS regulation of voice and other carrier-grade services is required to achieve satisfactory performance on the end-users' side. Of course, one should note that QoS is coupled with pricing in the mobile network because higher QoS requires more investment in network equipment in access and core parts, and vice versa. Additionally, QoS for critical services is raising a need for its regulation, especially when such services will provide control and automation of different industries in different verticals. That requires new types of performance monitoring for critical services and QoS enforcement when QoS degradation will be detected (e.g. high delay, traffic congestion or unavailability for mission critical GBR services in 5G mobile networks). Thus, QoS regulation may need to transform itself from naming and shaming strategies (e.g. publishing the results from QoS monitoring campaigns, to allow customers to make informed decisions) to stricter QoS regulation for mission critical services.

References

1 Janevski, T. (2015). *Internet Technologies for Fixed and Mobile Networks*. Norwood, MA: Artech House.
2 Janevski, T. (2014). *NGN Architectures, Protocols and Services*. Chichester: Wiley.
3 Janevski, T. (2003). *Traffic Analysis and Design of Wireless IP Networks*. Norwood, MA: Artech House.
4 3GPP TS 36.306, Evolved Universal Terrestrial Radio Access (E-UTRA); User Equipment (UE) Radio Access Capabilities (Release 14), March 2018.
5 3GPP TS 29.060, GPRS Tunnelling Protocol (GTP) Across the Gn and Gp Interface (Release 13), March 2015.
6 IETF RFC 2003, IP Encapsulation Within IP, October 1996.
7 3GPP TS 23.203 version 14.5.0 Release 14, Digital Cellular Telecommunications System (Phase 2+) (GSM); Universal Mobile Telecommunications System (UMTS); LTE; Policy and Charging Control Architecture, October 2017.
8 Holma, H. and Toskala, A. (2011). *LTE for UMTS: Evolution to LTE-Advanced*, 2e. Wiley.
9 IETF RFC 8325, Mapping DiffServ to IEEE 802.11 February 2018.
10 Zhang, N., Zhang, S., Wu, S. et al. (2016). Beyond Coexistence: Traffic Steering in LTE Networks With Unlicensed Bands. *IEEE Wireless Communications* 23 (6,): 40–46.
11 MulteFire Alliance, https://www.multefire.org, accessed in May 2018.
12 ITU-R Recommendation M.2083.0, IMT Vision – Framework and Overall Objectives of the Future Development of IMT for 2020 and Beyond, September 2015.

13 ITU-T Recommendation Y.3101, Requirements of the IMT-2020 Network, January 2018.

14 ITU Recommendation Y.3100, Terms and Definitions for IMT-2020 Network, September 2017.

15 ITU Recommendation Y.3150, High-level Technical Characteristics of Network Softwarization for IMT-2020, January 2018.

16 ITU Recommendation Y.3130, Requirements of IMT-2020 Fixed Mobile Convergence, January 2018.

17 3GPP TS 23.501 V15.1.0, System Architecture for the 5G System; Stage 2 (Release 15), March 2018.

18 3GPP TS 23.503 V15.1.0, Policy and Charging Control Framework for the 5G System; Stage 2 (Release 15), March 2018.

19 GSMA, *Introducing Spectrum Management*, Spectrum Management Series, GSMA, 2017.

20 ETSI TR 103 230 V1.1.1, Fixed Radio Systems; Small cells Microwave Backhauling, January 2018.

21 Huawei, LTE Small Cell v.s. WiFi User Experience, 2013.

22 Ericsson: The 5G Business Potential Industry Digitalization and the Untapped Opportunities for Operators 2017.

23 International Telecommunications Policy Review, 22, 2, June 2015.

6

Services in Fixed and Mobile Ultra-Broadband

The main legacy telecommunication services from the twentieth century are telephony and television. In the beginning they were provided via separate networks such as telephone networks for telephony (e.g. PSTN/ISDN, PLMN) and broadcast TV networks (for television). However, the convergence of all telecommunication networks onto IP networks has triggered the transition of legacy telephony services to VoIP and the transition of television toward IPTV (although this lags behind VoIP because TV requires far more bits per second to be transmitted with desired QoS). Besides legacy telecommunication services, the twenty-first century provides a plethora of new services enabled by IP networks and Internet technologies, which has paved the way for deployment of new services without a need for change of the underlying network infrastructure. However, some services are provided with QoS support, while others (e.g. Internet-based services) are provided in a best-effort manner. With the deployment of ultra-broadband access, the difference between the two types of services (with and without QoS guarantees) becomes smaller, especially when broadband speeds are sustainable between end-users and other end-peers (e.g. servers) with which users communicate (also, the user can be a human or a machine on both ends, which depends upon the service type).

6.1   QoS-enabled VoIP Services

By definition, VoIP refers to provision of voice communication service (i.e. legacy telephony as a conversational type of service between two or more humans as the end-users) in both directions (between the end-users) over all-IP networks by using Internet technologies end-to-end. With the transition of all telecommunication networks to IP networks, all services have become IP-based, implemented in the application space. So, all telephony becomes VoIP, including fixed and mobile access. In the case of fixed access networks, there can be analogue voice in the last meters in cases where legacy analogue telephone devices are used (as dumb devices, without any software implemented regarding voice service), connected via twisted-pair copper wires on the user's side. So, every telephony service becomes VoIP, a transition that started in fixed networks in the early 2010s and then continued in the late 2010s

QoS for Fixed and Mobile Ultra-Broadband, First Edition. Toni Janevski.
© 2019 John Wiley & Sons Ltd. Published 2019 by John Wiley & Sons Ltd.

in mobile networks. In the 2020s all voice services will in fact be VoIP-based. But regarding QoS, there are two main types of VoIP services:

- QoS-enabled VoIP services, which are in fact a replacement for legacy digital (or analogue) telephony by telecom operators (including fixed and mobile ones) with VoIP with guaranteed QoS end-to-end (in the same manner as legacy telephony).
- OTT VoIP services, such as Skype, Viber, WhatsApp, which are provided as OTT services over network neutral Internet access, without any QoS guarantees, with equal treatment to the other Internet traffic.

For QoS provision to voice services in IP-environments the fundamental part is signaling. As was practice on a global scale for legacy telephony in the twentieth century, the signaling needs to be standardized globally since the voice services are global (aiming to be vendor independent). How is standardized signaling provided for VoIP in telecom operators' environments? That is accomplished by standardization of NGNs by the ITU [1], which defines end-to-end QoS support for voice by using standardized architectures (by ITU-T) and protocols (by the IETF) [2].

6.1.1 NGN Provision of VoIP Services

The NGN was initially developed to provide backward compatibility for operator's voice services over IP networks to PSTN/ISDN [3], and that was accomplished using two main approaches [4]:

- *Emulation approach for operator's VoIP provision.* It provides most of the PSTN/ISDN service capabilities by using their adaptation (via emulation) to an all-IP environment in the NGN.
- *Simulation approach for operator's VoIP provision.* This refers to the same service provision in the NGN as the emulation approach but without guarantees for all PSTN/ISDN features that were available before the transition to VoIP telephony.

NGN QoS-enabled voice provision is used for both mobile and fixed access networks. While in the fixed access network the point of end-user network attachment does not change over time, in mobile access networks the terminal can change its point of attachment during an ongoing voice call (that is called a handover), which requires additional mobility management support regarding QoS (e.g. to have the same QoS in the new cell in the mobile network, after the handover, as was in the old cell, before the handover).

QoS-enabled mobile VoIP in the NGN uses service control functions, mobility management and control functions, network attachment control functions, and resource and admission control functions, as shown in Figure 6.1.

The SCFs perform mobile terminal registration in the network, service establishment/ service release, by using the CSCFs of the IMS [1]. The VoIP service is always established by using signaling messages, unlike the legacy Internet client-server communication. Why? Because in the case of voice, a caller user (the originating user terminal, i.e. user A) requires to establish a connection with the called party (the terminating user terminal, regarding the given call, typically denoted as user B) by using a certain addressing scheme (e.g. ITU-T E.164 numbering schemes are used since the PSTN era, recently updated with ITU-T E.212, and continue to be used for operator's VoIP, while OTT VoIP service providers typically use different proprietary addressing schemes).

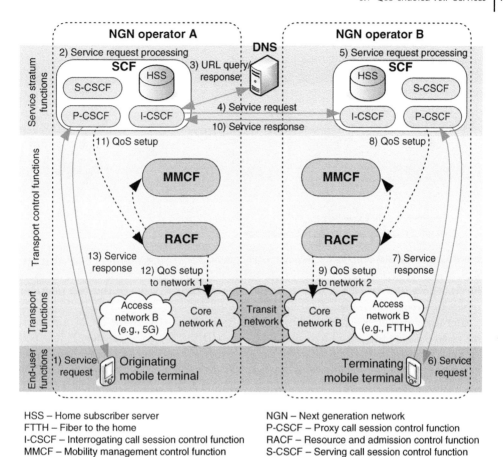

HSS – Home subscriber server
FTTH – Fiber to the home
I-CSCF – Interrogating call session control function
MMCF – Mobility management control function
NACF – Network access control function

NGN – Next generation network
P-CSCF – Proxy call session control function
RACF – Resource and admission control function
S-CSCF – Serving call session control function
SCF – Service control function

Figure 6.1 Voice service in the next generation network (NGN) environment.

For voice call establishment in the NGN, the originating user terminal (i.e. terminal A) sends a request for establishing a voice call to the P-CSCF, which then forwards it to the S-CSCF, and it finally reaches the I-CSCF of network operator A (which is needed in cases when the called user, B party, belongs to another operator). To obtain the IP address of the I-CSCF in the target operator B, the I-CSCF node of operator A performs DNS resolving of the domain name of operator B. Let's say the called party has telephone number +38912345678@operatorB.example.com (which is a typical addressing scheme for SIP signaling, used in IMS and in NGN). The I-CSCF node of operator B receives the call request from operator A and forwards it to the S-CSCF entity in its IMS, which further obtains the P-CSCF to which the called user B is registered in the operator B network. Finally, the service request from user A is forwarded to called user B (to terminating user terminal) and its response is carried back to the operator A network by using the reverse path back. Each P-CSCF in both networks (operator A and operator B) communicates with the RACF for QoS provision, including admission control (for the given voice call) and transfer of QoS setup information (via QoS intra-network signaling) in

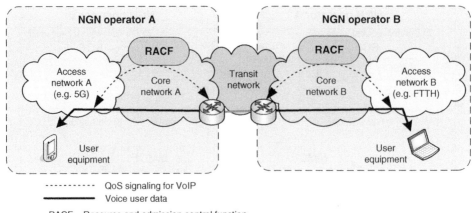

-------------- QoS signaling for VoIP
———————— Voice user data

RACF – Resource and admission control function

Figure 6.2 VoIP transfer over NGN.

the transport IP networks (which is done independently in each of the two operator networks, i.e. on both ends).

After the VoIP call is established by using SIP signaling between the NGN operators, the voice data is transferred by using RTP over UDP/IP for data transfer (Figure 6.2) [2]. For that purpose RACF (in each of the operators' networks) sets up the path (based on previously completed admission of voice calls) to maintain the required QoS for the given voice call.

6.1.2 Discussion on Telecom Operator vs. OTT Voice Service Quality

As already noted, VoIP is used in two main types, operator VoIP and OTT VoIP. However, all VoIP types have differences compared with PSTN, which may influence the QoS. In general, the differences between VoIP and PSTN may be summarized as follows:

- *Bit rates.* PSTN provides digital telephony in each direction with a bitrate of 64 Kbit/s by using ITU-T G.711 codec. VoIP can support various other codecs (G.723, G.729, etc.) with different bitrates, which may provide higher utilization of resources due to statistical multiplexing (i.e. there is no more channel with a given bitrate dedicated to a single voice connection in each direction between caller A and called party B).
- *Packet delay (i.e. latency).* PSTN has constant packet delay end-to-end which is below 150 ms in each direction (due to reservation of a channel end-to-end, and due to no need for buffering in telephone exchanges, i.e. network nodes in PSTN), while packet delay is higher in IP-based networks due to different causes that increase the end-to-end delay budget, including (but not limited to) voice packetization (creating IP packets from digitalized voice on the end-user's side), statistical multiplexing (mixing different IP packets with the same priority over the same IP network paths), and higher processing time of IP packets in network nodes (e.g. buffering of packets, traffic shaping by delaying packets at peak traffic periods, scheduling at different network nodes on the path of IP packets).
- *Jitter (i.e. packet-delay variation).* PSTN has very low jitter (again, due to the reservation of channels with a given bitrate of 64 Kbit/s in each of the two directions between

the end-users), while VoIP packets accumulate more jitter due to statistical multiplexing, traffic shaping, packet scheduling, etc., which introduces variable delay to VoIP packets. Therefore, voice playout buffers are used in receiving terminals, which means that voice IP packets are replayed a certain time after they are received in order to provide smooth reproduction of the digitalized voice into its analogue form that can be heard by humans.

- *Network equipment.* For provision of voice services, PSTN uses the telephone network with exchanges as network nodes and overlay signaling network (for call management and QoS), while VoIP in general can utilize more intelligent terminals (e.g. smartphones, computers) and generic IP network nodes (e.g. routers) that are used for all IP traffic. However, again in this case QoS requires an overlay signaling network. In the case of the operator's VoIP, signaling is also used to provide QoS-enabled VoIP provision, in the same (or a similar) manner as voice service previously found in PSTN. However, the protocols used for signaling are different in IP networks (e.g. SIP is used for VoIP, while SS7 is used in PSTN [1]).

Besides the different aspects of data transfer and signaling in the case of PSTN (digital telephony networks) and operator's VoIP services (with QoS support), the voice service is the same toward end-users (i.e. they are not likely to note any difference between PSTN and NGN-based VoIP services).

Overall, voice has been the primary form of communication between humans since the birth of civilization, and voice services will continue to be important in the future. However, because of the single-dimension nature of the voice-only service (unlike video, as an example), it requires lower bitrates (e.g. compared with video services and most data services). Bitrates are decreasing on one hand due to the higher processing power of user terminals (to use different voice source coding codecs, for compression of voice bits), while on the other hand voice bitrates may increase with the extension of the used frequency band for voice itself (the human ear can hear on frequencies between 20 Hz and 20 kHz), such as using not only 300–3400 kHz frequency band for voice as in PSTN, but gradually up to 20 kHz. Additionally, there more than one byte per sample of voice can be used, which is the standard used in G.711 codec in PSTN (at the analogue/digital conversion of voice at the user terminal in mobile networks or home gateways/IP phones in fixed networks), which should provide higher QoE for voice services than PSTN (of course, this requires higher bitrates per voice connection end-to-end, which gives a lower number of connections being served with the same bandwidth). This calculation refers to both OTT VoIP and operator's VoIP; however, in the case of operator's VoIP the changes in codecs are less likely to be frequent while for OTT voice services (where the same application is present on both ends of the connection) such change can be more easily provided than with operator's VoIP. So, in cases where Internet bandwidth (end-to-end) is high, OTT VoIP may provide even higher QoE than operator's VoIP (e.g. for fixed access or for nomadic users in the case of mobile access).

6.2 QoS-enabled Video and IPTV Services

Video is the most demanding type of traffic on the Internet and managed IP networks. Around two-thirds of all IP traffic (around 2020), including Internet and managed

Table 6.1 Main video resolutions and respective bitrates (for display aspect ratio 16:9).

Video type	Video picture resolution (pixels × lines)	Bitrate per TV stream (user's side) (Mbit/s)
SD (standard definition)	720 × 576 (for PAL)[a] 720 × 480 (for NTSC)	1.5–5
HD (high definition)	1280 × 720	5–10
Full HD	1920 × 1080	8–20
4K-UHD (4K ultra HD)	3840 × 2160	15–40
8K-UHD (8K ultra HD)	7680 × 4320	20–100

a) PAL/SECAM is the analogue TV standard in Europe, while NTSC is the analogue TV standard in the Americas.

IP networks, comes from video traffic, including OTT video and QoS-enabled video and IPTV. So, the development of broadband and ultra-broadband speeds in the access networks (fixed and mobile) contributes to better user experience with all different services and applications, but particularly with video services. Why? Because possible video improvements regarding picture resolution, frame rate, and quality of accompanied audio (synchronized with video) are almost endless. For example, the traditional quality of TV and VHS video since the last decades of the twentieth century (e.g. since the 1970s/1980s) is constantly increasing from below standard definition (SD) digital picture resolution to higher resolutions (HD video, full HD, 4K, 8K) [5]. Higher resolution requires more bit/s due to larger volume of individual video frames (pictures); however, that is also dependent upon the video codecs being used for the video on both sides (video encoding on the transmitter's side and decoding on the receiver's side). Another parameter which influences video bitrate is number of frames per second (fps). While the initial minimum fps for video is 25 (for the European PAL video standard) and 30 (for the American NTSC standard), the number of fps is also increasing (gradually) to provide better QoE for video users. Further, picture resolution and fps are also influenced by the processor and memory capabilities of end-user devices (setup box for TV, or mobile devices in mobile network, or personal computers, etc.) to process incoming packets from the video and to provide synchronization (audio and video) and continuous reproduction of the video content at the receiving side (with certain playout buffering which is required for smooth video reproduction by the video player). At the present time (around 2020), the required bitrates for different video resolutions and fps are given in Table 6.1. Such bitrates are dependent upon the type of video codec and can have different values for different codecs at the same video picture resolutions. With the increasing processing power of devices (it doubles almost every two years, according to Moore's law), the required bitrates for a given resolution and fps normally decrease over time.

6.2.1 IPTV and QoS

TV is the second service type (besides telephony, i.e. voice services) to be transferred to the IP environment. TV over IP with QoS guarantees end-to-end is called IPTV.

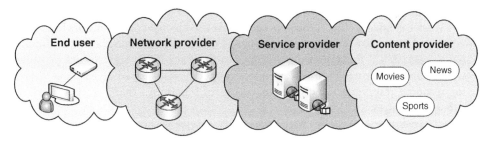

Figure 6.3 IPTV domains.

However, one should note that due to much higher demand for bandwidth of TV compared with voice services, the transition of analogue to digital TV happened later than the transition from analogue to digital telephony (analogue to digital TV transition was completed globally by the middle of the 2010s, while the transition for voice from analogue to digital telecommunication networks was completed by the end of the twentieth century). TV refers to a real-time service where the same content (e.g. live content) is broadcast to all recipients tuned to watch a given TV channel (on a given frequency, either via terrestrial broadcast or cable broadcast of the channels). In the case of IP delivery of TV content, the TV is not more the broadcast, because Internet network do not use broadcast for delivery of user contents, but its delivery is based on unicast (point to point) and multicast (point to multipoint) transport.

Considering IPTV definition, the ITU-T defines IPTV as multimedia services such as TV, video, audio, pictures/graphics, and data, delivered over IP-based networks that are designed to support the necessary level of QoS and QoE [6], interactivity, reliability, and security. So, a TV stream provided over the public Internet network cannot be considered as IPTV (according to the given definition) because the public Internet network does not provide QoS guarantees to individual streams of data due to the network neutrality principle.

For the provision of IPTV services four IPTV domains can be differentiated: (i) content provider; (ii) service provider; (iii) network provider; and (iv) end-user (Figure 6.3). However, one provider entity can manage more than one IPTV domain. That is the typical case when telecom operators provide IPTV service. In fact, with IPTV the telecom operators (first fixed ones) have entered the TV delivery business by using the convergence of all services onto IP networks (which telecom operators have and operate), while in the twentieth century broadcast TV was typically delivered via separate networks (rather than telecom operator's telephone networks) and hence it was a separate business.

There are three main types of IPTV functional architecture:

- *Non-NGN IPTV architecture.* In this approach IPTV services are provided by using proprietary platforms in combination with standardized IP networking technologies and Internet application level technologies (e.g. HTTP can be used for presentation purposes). In this approach the IPTV streams are created in the operator's network, and also set-top boxes (on the end-user's side) are proprietary, developed and provided to end-users by the telecom operator (which is the IPTV service provider). From the four IPTV domains, only the content provider is outside the telecom operator's network. QoS is provided by bandwidth reservation per IPTV channel combined with traffic shaping in the network and playout buffering on the end-user's side (at the IPTV set-top boxes). This is the approach most used by telecom operators in the late 2000s and in the 2010s, and it will not change drastically in a short period of time.

- *NGN-based IPTV functional architecture.* In this case IPTV services are provided with NGN functional entities. For QoS support, NGN-based IPTV uses RACF for resource reservation and admission control functions. There are two possible sub-architectures:
  - *IPTV over non-IMS NGN.* It provides IPTV service without deployment or usage of an IMS.
  - *IPTV over IMS-based NGN.* It uses IMS functional entities for IPTV service provisioning. In the longer term the IMS-based NGN should be a typical approach for IPTV provisioning by using also inter-operator transfer of IPTV streams (which will be possible due to the standardized nature of the IMS-based NGN architecture, similar to its use for operator's VoIP services).

IPTV functional architecture consists of different functional entities, which include end-user functions, application functions, SCFs, CDFs, network functions, management functions, and content provider functions. Such IPTV functional architecture can be used for different delivery mechanisms, including multicast delivery and unicast delivery of IP packets carrying the IPTV content. Regarding the design of IPTV transfer over IP networks, the most efficient way for linear IPTV channels (linear TV refers to TV content which is transferred to all users that have chosen the given TV channel at the same time) is the use of multicast routing. In fact, multicast routing is intended for use in real-time services where many recipients listen and/or watch the same content.

Why is multicast routing for IPTV better for the telecom operator's network than unicast? To answer this question, let's assume that the operator provides 100 IPTV channels to its customers. Each channel with SDTV requires 3 Mbit/s per channel and 6 Mbit/s per HDTV channel. Let's have 100 000 active IPTV users (watching one IPTV channel per user at the same time) in a given region being served by the given IPTV provider (of course, this is an example; in practice it is dependent upon the size of the country and the telecom operator). If only unicast is used in the network, then on a network link between nodes A and B which (in this example) carries all traffic to all 100 000 users, capacity of $100\,000 \times 3\,\text{Mbit/s} = 300\,\text{Gbit/s}$ will be needed in the case of SDTV channels, or $100\,000 \times 6\,\text{Mbit/s} = 600\,\text{Gbit/s}$ capacity in the case of HDTV channels. Although such capacities in the transport networks in the 2010s and the 2020s can be provided, multicast saves the capacity needed in the network, so IPTV can be provided at a lower price or with a higher profit margin (since less capacity is used in the network), or the remaining capacity can be used for other IP traffic (e.g. Internet traffic, IP managed traffic), which can bring additional revenues from such services using the given network capacity. Let's make the simple calculation for the required capacity between network nodes A and B, which are multicast capable routers in this second case (in the first case here they were only unicast capable routers). Multicast needs only one copy of each IP packet to be delivered (from each of the 100 TV channels), which gives required capacity of $100 \times 3\,\text{Mbit/s} = 3\,\text{Gbit/s}$ for all SDTV channels, and $100 \times 6\,\text{Mbit/s} = 600\,\text{Mbit/s}$ for all HDTV channels (Figure 6.4). So, with multicast, the required network capacity is significantly lower (for a large number of IPTV customers, which is a typical case for having a sustainable IPTV business) and it is dependent upon the number of TV

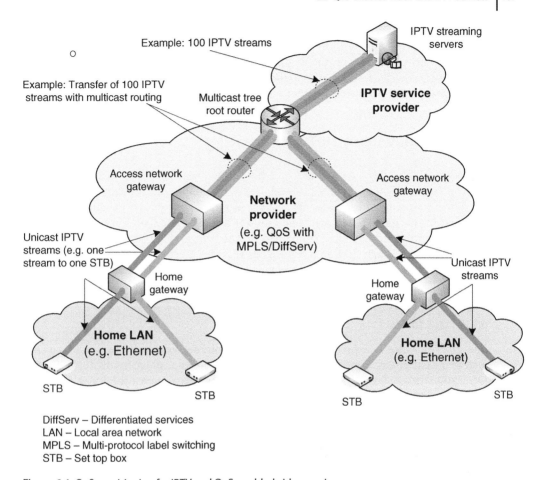

O

Figure 6.4 QoS provisioning for IPTV and QoS-enabled video services.

channels being offered to the end customers (which is a fixed number in a given period of time, and hence provides easier network dimensioning and planning) and it is not dependent upon the number of customers. So, with multicast usage in the transport and core IP networks for IPTV, QoS can be guaranteed for end-users.

There are several different multicast protocols standardized by the IETF [2]; however, not all of them are being used equally. Multicast is based on the creation of multicast groups and usage of multicast routing protocols. For this purpose the creation of the multicast groups in IPv4 networks is done with IGMP (Internet Group Management Protocol), which is standardized in three versions (IGMPv1, IGMPv2, and IGMPv3) by the IETF [2]. In the case of IPv6 networks, multicast listener discovery (MLD) is used, which has two versions (MLDv1 and MLDv2). Each of the routers engaged in multicast delivery of IPTV traffic joins a multicast group by using a multicast type of IP address assigned to the network interface (that belongs to that router) which joins the given multicast group. Although there are different standardized multicast routing protocols, the most widespread and typically used for delivery of IPTV traffic is PIM-SM (protocol independent multicast – sparse mode) [7]. However, in the last leg of the IPTV transmission (between network nodes and set-top boxes on the end-users' side), unicast can be used, while in the IPTV provider network, multicast is used. For QoS in the transport network, multi-protocol label switching (MPLS) and the DiffServ approach (Figure 6.4), as well as Ethernet QoS mechanisms (e.g. IEEE 802.1p), are used and can be mapped to MPLS and DiffServ QoS traffic types (and vice versa).

6.3   QoE for VoIP and IPTV

QoE is directly related to humans and how they judge QoS by using their senses, such as ears for voice services (i.e. telephony) and audio (accompanying video) and eyes (for watching video content). QoE cannot be related to machines, such as machine-to-machine (M2M) type of communications, while QoS can be associated to all types of services, including human-to-machine (e.g. client-server communication over the Internet, such as Web browsing, or real-time services such as IPTV or video streaming), human-to-human (e.g. telephony), and M2M services (e.g. various IoT services). Therefore, this section covers QoE for VoIP and IPTV.

6.3.1   QoE for VoIP

The QoE for voice is expressed with the mean opinion score (MOS), which is a subjective measurement done by human listeners placed in a "quiet room" (ITU-T P.800 [8], P.800.1 [9], P.800.2 [10]). Such an approach has existed since the PSTN era, so it was not invented for VoIP, but it continues to be used for voice services regardless of whether they are analogue telephony, digital telephony, or VoIP telephony. The QoE of voice is rated with a five-point MOS for voice quality assessment (where 1 = bad, 2 = poor, 3 = fair, 4 = good, 5 = excellent). MOS is defined as "the mean of opinion scores" [9], where the opinion score is initially defined (in ITU-T P10/G100 [11]) as "the value on a predefined scale that a subject assigns to his opinion of the performance of the telephone transmission system used either for conversation or for listening to spoken material." There are definitions for different types of speech codecs, so there is wideband MOS (for voice in the range 50–7000 Hz), super-wideband MOS (20–14 000 Hz), and full-band MOS (10–20 000 Hz) [9]. Also, for voice communication, MOS can be related to listening only situations or to conversational situations. However, one should note that further MOS is not exclusively defined for voice services (e.g. there is MOS for video and audiovisual services).

To avoid the need for real people evaluation of the voice quality, there is standardized automated assessment of the speech quality called PESQ (perceptual evaluation of speech quality), defined in ITU-T P.862 [12] and P.863 [13]. This way, apart from subjective testing (which includes people as quality evaluators), PESQ uses MOS for objective assessment in network planning and/or optimization. PESQ is defined as a one-way listening model. It estimates user-perceived MOS by comparing the transmitted reference speech signal and the received degraded signal. It includes impairment effects due to voice compression and IP network parameters (e.g. packet delay, jitter, packet loss), in addition to conventional circuit-switched network impairments such as noise and echo (ITU-T P.862 [12]). The QoE assessment defined in ITU-T P.863 [13] is also known as POLQA (perceptual objective listening quality assessment). POLQA incorporates

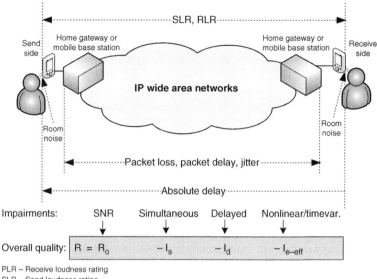

PLR – Receive loudness rating
SLR – Send loudness rating
SNR – Signal to noise ratio

Figure 6.5 E-model for voice quality assessment.

assessment of super-wideband speech, as well as networks and codecs that introduce time warping. This method includes various test factors as well as different voice coding technologies such as voice codecs standardized by the ITU-T G-series (e.g. G.711, G.723, G.729, etc.), voice codecs standardized by 3GPP (e.g. global system for mobile communication (GSM) codec, adaptive multi rate – AMR codec), etc.

ITU-T P.564 [14] provides conformance testing for VoIP transmission quality assessment models where VoIP uses the standard RTP/UDP/IP protocol stack on both ends which carry 3.1 kHz (i.e. 300–3400 Hz) narrowband voice, as well as extension for wideband (up to 7 kHz) voice services. In this approach the payload of the RTP stream carries generic voice payload (e.g. it can be artificially generated).

The E-model for use in voice transmission planning [15] is shown in Figure 6.5. It is split into two sides, the send side and the receiver side. The model is created to estimate the voice conversational quality from mouth to ear. All transmission parameters in the E-model are combined into a single transmission rating factor, R. Calculation of the R factor is based on the following equation [15]:

$$R = R_o - I_s - I_d - I_{e-eff} + A \tag{6.1}$$

In (6.1) the parameters are the following:

- R_o is the signal to noise ratio including background noise and transmission noise.
- I_s represents the combination of all impairments which occur simultaneously with the voice signals.
- I_d represents impairment caused by delay.
- I_{e-eff} is the equipment impairment factor representing impairments from low-bit codecs and packet loss.
- A is the advantage factor, which is a compensator to offset the other parameters.

Table 6.2 Categories for voice transmission quality.

Range of E-model rating R	Voice transmission quality category	User satisfaction
$90 < R < 100$	Best	Very satisfied
$80 < R < 90$	High	Satisfied
$70 < R < 80$	Medium	Some users dissatisfied
$60 < R < 70$	Low	Many users dissatisfied
$50 < R < 60$	Poor	Nearly all users dissatisfied

The idea of R factor is to be used to determine voice transmission quality. Table 6.2 maps *R* factor values to five voice transmission quality categories (similar to MOS grades), which can be used for planning purposes by network and service planners.

6.3.2 QoE for IPTV

IPTV services provided over the broadband and ultra-broadband access network require a certain level of QoS to support adequate QoE. QoE depends upon two main technical elements:

- *IPTV application (transmit and receive side).* Quality of the video source, resolution (HDTV, full HD, etc.), required bitrate, constant or variable bitrate encoder output, group of pictures (GoP) size, and structure in Moving Pictures Experts Group (MPEG) (e.g. more B frames in the GoP reduces the required bitrate, but video is more sensitive to bit errors and packet loss), etc.
- *Network transmission parameters.* The IPTV QoE is highly dependent on the type of data loss (especially on I and P frames losses from the MPEG streams). Further, it is dependent upon the codec being used, MPEG transport stream packetization, packet losses, outage (e.g. due to multicast router table recovery), etc.

Packet loss for IPTV is defined with loss period and loss distance. The loss period is the time duration of consecutive packet losses and the loss distance is the time between packet loss intervals. According to the Broadband Forum [16], the loss distance should be limited to at most one error event per 60 minutes for SD and one error event per 4 hours for HD. In practice, the error event is a loss or corruption of a group of a small number of IP packets where each of them contains, for example, up to 7 MPEG packets of 188 bytes in length (corresponding to MPEG-2 packet length). On the end-user's side, set-top box decoders typically employ error concealment techniques to minimize the impact of lost or corrupted IPTV video packets.

The errors which may appear in a sequence can cause noticeable quality degradation effects ranging from no noticeable audio or video impact from the losses to complete loss of the video and/or audio signal.

An example calculation for IPTV losses (considering use of MPEG-2 packets) is the following:

MPEG packets per second = 3 Mbit/s/8 bits per byte/188 bytes per MPEG packet = 1994.7 MPEG packets per second
IP packets per second = 1994.7/7 MPEG packets per IP packet = 285 IP packets per second
Loss of 16 ms corresponds to = 285 IP packets per second × 0.016 seconds = 4.56 IP packets lost (as an average value, while the number of lost IP packets in a concrete case normally is always an integer value).

To combat losses, which can occur (due to non-zero loss probability of every transmission media, either copper, fiber, or radio), a common configuration for DSL is using interleaver depth (i.e. forward error correction (FEC) block duration) of 8 ms or 16 ms. Hence, the corresponding loss period is targeted to be 8 ms or 16 ms. However, considering different video bitrates (Table 6.3), this will correspond to a different number of lost video IP packets.

Recommended minimum transport layer parameters for satisfactory QoE for MPEG-2 and MPEG-4 AVC or VC-1 encoded SDTV services are given in Table 6.3. Also, it assumes MPEG-2 transport stream with seven 188-byte packets per IP packet; however, 7×188 bytes + RTP/UDP/IP headers should overall be less than 1500 bytes, which is the maximum Ethernet MA) frame payload (because Ethernet is used as a unified LAN everywhere between the home gateway node and the set-top box). I the case of optical (fiber) access networks (e.g. FTTH), the packet loss rate is much lower than with DSL access (e.g. the optical network loss rate is in the range of 10^{-8}–10^{-9}).

Tolerable loss rates may be higher depending on degree and quality of set-top box loss suppression. For QoS provisioning, IPTV streams are differentiated from IP traffic originating from other applications which have different performance requirements.

Table 6.3 Requirements for satisfactory QoE for IPTV over DSL.

IPTV stream bitrate (Mbit/s)	Latency (ms)	Jitter (ms)	Maximum duration of a single error (ms)	Corresponding loss period in IP packets	Loss distance	Average video stream packet loss rate
1.75	<200	<50	≤16	4 IP packets	1 error event per hour	<6.68E-06
2.0	<200	<50	≤16	5 IP packets	1 error event per hour	<7.31E-06
2.5	<200	<50	≤16	5 IP packets	1 error event per hour	<5.85E-06
3.0	<200	<50	≤16	6 IP packets	1 error event per hour	<5.85E-06
8	<200	<50	≤16	14 IP packets	1 error event per 4 h	<1.28E-06
10	<200	<50	≤16	17 IP packets	1 error event per 4 h	<1.24E-06
12	<200	<50	≤16	20 IP packets	1 error event per 4 h	<1.22E-06

The QoE of IPTV can also be expressed via MOS, similar to QoE for voice. There are three main types of issues that can impact IPTV QoE:

- *Degraded IPTV service.* It appears when some amount of the IPTV stream is unavailable to the set-top box at the playout time. This happens due to packet loss at some network segment, but also can be due to certain content encoding issues in the case of bigger jitter at the set-top box.
- *IPTV service unavailable.* This is the case when no IP packets arrive at the set-top box (no picture is shown on the screen), which may be caused by broadband IP access interruption, an IPTV delivery problem (network congestion or link failure), an IPTV platform problem (e.g. failure of streaming servers), or interruption of the input TV channel (e.g. digital TV from satellite or terrestrial broadcast systems) to the IPTV encoders (at the service provider's domain).
- *Insufficiently responsive service.* This could happen due to long channel change times for IPTV or delay in responsiveness of video on demand commands (play, stop, pause, etc.).

So, the major influence on the IPTV QoE has packet loss that cannot be recovered. In practice, IPTV typically uses RTP/UDP/IP, in which case retransmissions are not used. However, in the case of IPTV over TCP/IP (which is also possible), that will add delays and bitrate fluctuation due to TCP retransmissions of lost packets, which may result in long playout times at the set-top boxes (e.g. tens of seconds or even minutes). Therefore TCP versions used in legacy operating systems are not feasible for IPTV. However, proprietary transport protocols (with a combination of TCP and UDP features) may be developed for IPTV service delivery by telecom operators when they are at the same time service provider, network provider, and user set-top box provider.

Why does packet loss have the greatest influence on IPTV QoE? Well, it leads to video pixelization or blocking, frame freezing, and/or set-top box lockup. However, the impact of packet losses is dependent on the type of video frame that is affected with it. Because IPTV is based on MPEG standards (MPEG-4 or MPEG-2), the worst case is when the I-frames are affected by the packet loss because they serve as the reference for all frames in a GoP and the loss of an I-frame (or part of it) can affect the whole GoP. Similarly, losses that affect P frames influence B frames in the same GoP. Packet loss also can cause longer channel change times because the decoders at the set-top box wait for the next I reference frame before presentation of an image to the viewer.

6.4   QoS for Popular Internet Services

The popular Internet services are those which have appeared on the Internet and which are inseparably connected with Internet technologies and IP networks. One of the oldest and still most used in all types of communications is the email, in both personal and business environments. However, the most important native Internet service (or application) is the Web (i.e. the World Wide Web), based on the HTTP/TCP/IP protocol stack, which contributed to the global growth of the Internet with its appearance in the early 1990s.

Email has replaced fax communications (used in the late decades of the twentieth century) and it is the basis of personal and business communication (with electronic means) in the twenty-first century. As the name (electronic mail) says, it is delivered in the same manner as regular mail, so the sender creates the email and sends to a recipient, the sender's email server communicates with the recipient's email server, and finally the recipient chooses when to check the mailbox for new emails (with email use on mobile devices, it has become popular to use the push email functionality, which notifies the recipient about new email messages). Considering this, email is not a real-time service. Email protocols use TCP/IP, so the messages are delivered to the destination email server in a reliable manner, which is based on the TCP protocol reliability. Also, email as a non-real-time service typically is classified with lower priority than real-time services (from the traffic management point of view), and it is served via Internet access service based on the network neutrality principle. However, email functionality on both sides is important. For example, if the email server is down on the recipient's side then the email cannot be delivered (typically the sending email server will periodically try to deliver a given email message until it is several days old). Also, many email servers have a limit on the total size of an email message (that size, measured in bytes, is typically bigger than the size of the email message text and all attachments due to their encoding in the creation of the email message at the sender's side), so some messages may not be delivered due to their size. Certain email servers may become less responsive due to overload, either with regular messages or due to denial of service (DoS) attacks. Overall, there are a number of standardized QoS parameters for email services [17], which are given in Table 6.4. One should note that such QoS parameters refer to all email protocols in use, including SMTP (Simple Mail Transfer Protocol), POP3 (Post-Office Protocol 3), and IMAP4 (Internet Message Access Protocol 4) [2]. Email as a non-real-time messaging

Table 6.4 Standardized QoS parameters for email.

Email QoS parameter	Description
Email session failure ratio [%]	Proportion of unsuccessful sessions and sessions that were started successfully
Email mean data rate [Kbit/s]	Average data transfer rate measured throughout the entire connect time to the email service. The data transfer shall be successfully terminated
Email data transfer cut-off ratio [%]	Proportion of unsuccessful data transfers and data transfers that were started successfully
Email data transfer time [s]	Time period from the start to the end of the complete transfer of email content
Email login non-accessibility [%]	Probability that the email client is not able to get access to the email server
Email login access time [s]	Time period from starting the login procedure to the point of time when the client is authenticated
Email notification push failure ratio [%]	Probability that the notification announcement was not successfully conveyed to the B-party
Email notification push transfer time [s]	Time period from starting the notification push to the successful confirmation of the email server of the end of the idle period
Email end-to-end failure ratio [%]	Probability that the complete service usage from the start of the email upload at the A-party to the complete email download at the B-party with an email client cannot be completed successfully

Table 6.5 Trigger points for QoS parameter email mean data rate.

Event from abstract equation	Trigger point from user's point of view
Successfully started email transfer	Start: SMTP email upload starts.
Successful email transfer	Stop: SMTP email upload is successfully completed by the A-party.
Successfully started email transfer	Start: POP3 or IMAP4 header download starts.
Successful email transfer	Stop: POP3 or IMAP4 header download is successfully completed by the B-party.

service is more tolerant to delays (e.g. in the range of 10 seconds [18]) as users do not expect instant delivery.

The use of TCP on the transport protocol layer means email is delivered without losses (because TCP retransmits all lost packets/segments, providing 100% lossless transfer). As for any Internet service, the data rate (i.e. bitrate) for the email transfer can be calculated. The trigger points for the email mean data rate are given in Table 6.5. For example, QoS parameter email mean data rate [Kbit/s] [17] is defined with the following abstract equation:

$$Email\_mean\_datarate = \frac{Email\_data\_transferred[kbit]}{(t_{successful\_email\_transfer} - t_{start\_email\_transfer})[s]} \tag{6.2}$$

The WWW is one of the most important Internet native best-effort services, which is standardized by the IETF. Like email, it is a non-real-time service. However, unlike email, the Web is an interactive service between HTTP client and HTTP server, where the HTTP server is always a machine, while the HTTP client can be operated by a human (e.g. via a Web browser, such as Chrome or Firefox) or a machine. In both cases, the required responsive time for the Web is in the range of approximately two seconds [18]. Although the bitrates are increasing over time, the Web pages are becoming heavier by using higher resolution for pictures and embedded videos, which influence the delay. The Web is based on the HTTP/TCP/IP protocol stack for transmission purposes between the HTTP client and the HTTP server, but the presentation of the Web content is done by protocols implemented over HTTP on the application layer (e.g. HTML). With the use of TCP on the transport protocol layer, the Web is also lossless transmission because TCP (below the HTTP) handles retransmissions of all lost packets/segments. However, many losses in a given shorter period of time may case a low data rate for the Web connection, due to TCP behavior at multiple losses in a congestion window, which will cause reset of the bitrate to the initial value at the start of the TCP connection that uses the HTTP (which is typically in the range of Kbit/s due to TCP Slow Start and Congestion Avoidance mechanisms). Generally, Web browsing is used over public Internet access, based on the network neutrality principles. However, the Web has become a graphical user interface (GUI) for various real-time services, such as video streaming services and IPTV. In such cases, when the Web is used as GUI for QoS-enabled services, normally such Web traffic also is transferred over the managed IP network (not

Table 6.6 Standardized HTTP QoS parameters.

HTTP QoS parameter	Definition
HTTP service non-accessibility [%]	Probability that a subscriber cannot access the service successfully
HTTP setup time [s]	Time period needed to access the HTTP service successfully, from starting the connection (that is sending TCP segment with SYN bit set to one, from the HTTP client to HTTP server) to the point of time when the content is sent or received
HTTP IP-service access failure ratio [%]	Probability that a subscriber would not be able to establish a TCP/IP connection to the server of a service successfully
HTTP IP service setup time [s]	Time period needed to establish a TCP/IP connection to the server of the Web service, from sending the initial query to a server to the point of time when the content is sent or received
HTTP session failure ratio [%]	Proportion of uncompleted HTTP sessions and HTTP sessions that were started successfully
HTTP session time [s]	Time period needed to successfully complete HTTP data session
HTTP mean data rate [Kbit/s]	Data transfer rate measured throughout the entire connect time to the service. The HTTP data transfer shall be successfully terminated.
HTTP data transfer cut-off ratio [%]	Proportion of incomplete data transfers and data transfers that were started successfully

Table 6.7 Trigger points for QoS parameter HTTP service non-accessibility.

Event from abstract equation	Trigger point from user's point of view
Web service access attempt	Start: User initiates the Web service access.
Successful attempt	Stop: First Web content is received.
Unsuccessful attempt	Stop trigger point not reached.

the public Internet) with certain QoS support. So, although the Web has appeared in the public Internet (which is served in a best-effort manner), nowadays it is used in both the public Internet and managed IP networks (depending upon the ToS in which it is used).

As with email there are standardized QoS parameters for HTTP services (i.e. Web services, such as Web browsing) [17], which are listed in Table 6.6. For example, QoS parameter HTTP Service Non-Accessibility [%] is defined with the following abstract equation (the trigger points for this HTTP QoS parameter are given in Table 6.7):

$$HTTP\_service\_non-accessibility = \frac{unsuccessful\_HTTP\_attempts}{all\_HTTP\_attempts} 100\% \quad (6.3)$$

Besides email and HTTP, similar QoS parameters are defined for other Internet services, such as FTP (File Transfer Protocol). However, FTP is almost completely replaced by HTTP for uploads and downloads of files as well as for testing purposes such as QoS measurements.

6.5 QoS for Business Users (VPN Services)

Business services have been present in telecommunication networks since the second part of the twentieth century. Since digitalization in the 1970s and the 1980s, business services have been provided over TDM telecommunication networks (created primarily for digital telephony), typically to provide site-to-site connections with predefined bitrates and guaranteed QoS. This was referred to as "leased lines," which were in the range of N × 64 Kbit/s up to PDH (plesiochronous digital hierarchy) reserved bitrates (e.g. multiples of 2, 34, and 140 Mbit/s in Europe, or multiples of 1.5, 6, and 140 Mbit/s in the Americas) and SDH bitrates whose brut values are N × 155.52 Mbit/s. With the transition from digital to all-IP networks, the leased lines continued to be provided by telecom operators to business users by virtual private networks.

However, the VPN by itself does not contain the QoS mechanisms and it is not tied with predefined bitrates as was the case with the leased lines in the past. Regarding security aspects, VPN provides encryption of user data by using IPsec, which provides encryption of the payload of the IP packet using security protocols, cryptographic algorithms, and keys. However, the IP packet header cannot be encrypted across the Internet simply because IP network routing is based on IP addresses placed in the packet header (for both IPv4 and IPv6). To provide security [2], IPsec uses the following two protocols:

- *Authentication header (AH).* This does source authentication, data integrity, but without encryption of the IP packets. The AH header is inserted between the IP header and payload fields, so all intermediate routers process packets as usual.
- *Encapsulating security payload (ESP).* This provides confidentiality, host authentication, and data integrity. The IP packet data field and ESP trailer are encrypted.

IPsec protocols (autonomous system (AS) and ESP) are used for creation of VPNs, which in practice can be used in the following two modes:

- *Transport mode.* In this mode AH and ESP protocols provide protection for protocols above the IP (which is on the network layer, i.e. protocol layer 3).
- *Tunnel mode.* In this mode AH and ESP are used for tunneled packets, that is, IPsec encrypts the whole IP packet (including IP header and payload, where the data from higher layer protocols and user/application data are placed in the payload) and adds a new IP header to the encrypted original packet. The newly created packet is tunneled through the network. The tunnel starts at the node where the packet is encrypted and the new IP header is added. The tunnel ends at the network node where the packet will be decrypted and the original packet will be processed further (it is also possible to be placed in another tunnel for further transport over the IP network). The new header that is added to the IP packet has as the source address the IP address of a network interface (the one through which it is sent) of the node which is the start point of the tunnel, and as the destination IP address the IP address of a network interface belonging to a network node that ends the tunnel (that is, drops the added IP header for tunneling purposes, decrypts the original packet, and processes it further).

The primary use of IPsec in telecom operators' networks is for creation of secured VPNs, which are used to carry aggregated Internet traffic of a given type (e.g. VoIP, IPTV, traffic from business users, Internet traffic, signaling and control traffic). The VPNs can be used over the public Internet (e.g. for secure access to an office computer from any place) or over managed IP networks (this is typically used by telecom operators for carrying different types of traffic in different VPNs). In fact, it is very unlikely that an original IP packet created by the end-user host will be routed as such through the telecom operator's core and transport networks. Although in theory there are routing protocols which can route the packets by using source and/or destination addresses, in practice that is not enough to provide the required QoS. And the QoS is provided not only to traffic that originates and terminates over the managed IP networks (e.g. carrier-grade VoIP as PSTN/ISDN replacement, IPTV, VPN for business users as leased-lines replacement), it is also needed for aggregated IP traffic that originates and terminates in the public Internet network.

So, VPN solutions are used in telecom operators' IP transport networks. For QoS provision of VPNs (e.g. to provide sustainable or constant bitrate for the VPN end-to-end) there is the possibility to use various approaches for QoS provision, such as Integrated Services (IntServ), Differentiated Services (DiffServ), as well as MPLS [19]. However, the QoS in transport IP networks from the beginning of this century is mainly provided with MPLS or MPLS/Border Gateway Protocol (BGP) IP VPNs [20]. MPLS is used for QoS provisioning in core and transport networks within a given telecom operator network domain, while MPLS/BGP [21] can be used as QoS solution for traffic exchange between ASs.

With the transition of a telecom operator's network to NGN, the same QoS mechanisms in NGN (e.g. used for voice services or IPTV) can be used for VPN services for QoS provisioning, including all types of end-users (not only to business customers, i.e. companies) [22]:

- *Site-to-site VPN*. This is used for connecting branch offices of a given company (so virtually they appear as though they are connected to the same network) as well as individual users for the purposes of mutual interest (collaboration, online gaming, etc.).
- *Access VPN*. This type is an individual VPN which is used to access corporate network from other networks such as a home network (e.g. work at a distance).
- *Multiservice VPN*. This is used for provision of multimedia services (e.g. voice, video, and data) to business users, with QoS support.

A VPN framework architecture for fixed and mobile NGN environments is shown in Figure 6.6. VPN functions in the NGN service stratum provide registration, authentication, and authorization for VPN service users, support security solutions and QoS provision, as well as management functions (e.g. mobility management in mobile networks), multicast service control, and VPN membership management (e.g. joining and leaving VPNs). The NGN transport stratum provides functions for VPN admission and resource control (admission of a new VPN and resource reservation, policy management, etc.).

For business VPN services in NGN environments the enterprise network uses transport control functions with the NACFs, the RACFs, and the MMCFs; enterprise network attachment functions (e-NAFs), enterprise resource and admission control function (e-RACF), and enterprise management and control function (e-MCF). For the purposes of QoS provisioning and enforcement for VPN tunnels, the VPN SCF trigger RACF or e-RACF aiming to provide QoS to VPN tunnels as the end-user equipment (fixed network node or mobile equipment) registers itself to the VPN SCF [23]. After that, the

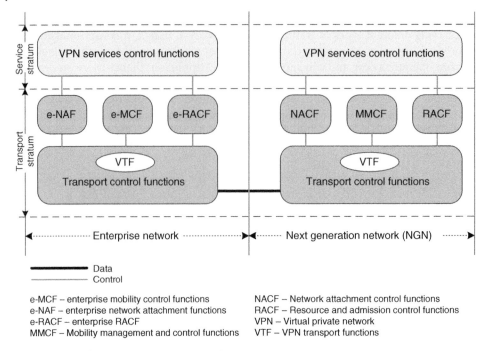

Figure 6.6 VPN framework architecture in NGN.

RACF/e-RACF enforces QoS for the VPN tunnels by using VPN transport functions in the NGN. When the user deregisters from the VPN service, the VPN SCF initiates to RACF/e-RACF release of the QoS for that VPN tunnel.

6.6 QoS for Internet Access Service and Over-the-Top Data Services

In the terminology used by telecom operators for their Internet access service, "data" includes all Internet-based services that are provided through the public Internet access, which is based on the network neutrality principle and served by using the best-effort approach (that is, the network does its best effort to deliver every IP packet, without particular QoS guarantees per packet of per flow). Other terminology which is widely used for services provided over public Internet access is OTT. OTT services (i.e. data services) include video, audio, and other multimedia data services (e.g. Web services, email) over the public Internet networks. Both "data" and "OTT" are used interchangeably in this book.

What belongs to the OTT (i.e. data) service space? It includes all services that telecom operators simply charge as data, which is in fact Internet traffic (not managed IP traffic). OTT services include but are not limited to FTP over the public Internet, email over the public Internet, Web services over the public Internet, proprietary (i.e. not standardized) services and applications over the public Internet (e.g. Skype, Viber, WhatsApp, BitTorrent), video streaming services over the public Internet (e.g. YouTube, Netflix, and many video sharing websites), social networking over the public Internet (e.g. Facebook, Twitter), cloud computing services over the public Internet (e.g. Google Docs, Amazon cloud services), and every existing and future service/application which will be provided over the public Internet. One may define OTT (i.e. data services) by specifying what they are not. OTT services are all IP-based services which are not provided over managed IP networks in the broadband access part (either fixed or mobile access).

Although served in a best-effort manner, OTT services also have QoS requirements, but they are indirectly specified. How? All OTT services which are used by a given user are multiplexed with all other IP traffic from other instances of the same OTT service or other OTT services from the same or from other users who use the same Internet access (e.g. the same fixed broadband access or the same cell/base station in mobile broadband access network). The QoS and QoE directly are dependent upon the bitrates provided via fixed or mobile broadband and ultra-broadband access, so the aggregate bitrate should be sustained. Therefore it is required for provided bitrates to be included in the SLA by the telecom operator (i.e. ISP), which is further monitored by a national regulator. For this purpose Internet traffic needs to be managed as well and all links that carry Internet traffic should be dimensioned to carry the aggregated IP traffic over given access, core, or transport IP network. So, traffic management, network planning, and dimensioning are well required for OTT services, i.e. for the Internet traffic to/from end-users.

Are there any issues regarding the traffic management on one side and network neutrality on the other? Well, network neutrality will be sustained if there is no differentiation of particular flows or toward concrete source or destination node connected to the Internet networks (e.g. traffic to/from one website should not have priority over traffic to/from another website). But network neutrality does not prevent justified traffic management, such as giving certain priority (via appropriate scheduling and buffering schemes in network nodes) to one traffic type over another. One such example is giving certain reasonable priority to conversational voice traffic over Web traffic or email traffic considering the required delay budget for voice, video, Web, email, etc. There is greater satisfaction for end-users from the Internet access services through which all such OTT services are provided. Overall, traffic management for the OTT services is influenced by:

- the type of Internet traffic, such as voice (requires lower bitrate) or video (requires higher bitrates and it is bursty by nature), or other data such as Web as an example (it has variable bitrates and is bursty by nature);
- the number of users attached to the broadband/ultra-broadband access network;
- the deployed network capacity and the possibility to provide the specified aggregate bitrates to end-users (according to the SLA), which is simpler to do in fixed networks and more complex in mobile networks (e.g. due to mobility of users, and dependence on the available bitrates from the distance between the mobile terminals and base stations, as well as the capabilities of mobile user equipment such as supported frequency bands or modulation and coding schemes);
- the capabilities for traffic management and link capacities (on network interfaces) of network nodes such as switches, routers, gateways, base stations (in mobile networks), etc.;
- existing business plans of the telecom operator (i.e. ISP) and national legislation and regulation of the Internet access service (e.g. is it network neutral or not, is differentiated charging for different OTT services allowed or not, etc.);

- human capacity of telecom operators and regulators regarding their knowledge for practical realization of broadband and ultra-broadband deployments and QoS/traffic management, and its regulation.

6.6.1 Traffic Management for OTT Services

The telecom operators, i.e. ISPs, may have different traffic management practices, but the main traffic management techniques are the same for all. When a traffic management decision is made, it is then implemented in the IP network as intervention. Such traffic management decision can take into account different inputs such as traffic type, the end-user's traffic, and the QoS profile, including any data caps or limits which are defined with the SLA. A data cap refers to maximum allowed data transfer (typically in MB, GB, or TB, depending upon the network access type, fixed or mobile). Primarily, for data services traffic management intervention is targeted to modification of the traffic priority or changing the allocated bandwidth such as guaranteed minimum bitrate or to impose a maximum bitrate.

There are in general two main types of traffic management [24]:

- *Implicit traffic management.* This is implemented with the network design where different traffic types are affected differentially from the beginning of network operation.
- *Explicit traffic management.* This is explicitly deployed through packet prioritization and bandwidth management mechanisms.

In addition to the types of traffic management (implicit and explicit), QoS of data services is influenced by the dimensioning of networks (access, core, and transport networks), the partitioning of access pipes (managed IP pipes, public Internet pipe), as well as the use of the content distribution networks. The use of CDNs is particularly important for OTT services, such as video streaming or sharing services, because it directly influences the RTT (round trip time). Also, CDNs are important for all Web-based services (e.g. including social networking) because TCP mechanisms (the Web is using the HTTP/TCP/IP protocol stack) directly depend upon the RTT. Additionally, shorter distances between clients and servers (which is provided globally by distributed CDNs) provides lower delay and jitter budget due to fewer network nodes and network links that IP packets traverse from the source to the destination (both ways, from client to server, and vice versa, since TCP-based Internet traffic is fully duplex).

6.6.2 Traffic Management Approaches

Although there are different scheduling and buffering mechanisms, in fact all traffic management is based on a decision and an intervention. For example, decision basis traffic management is control upon exceeding a weekly or monthly usage allowance for a given user. Then the intervention is the response of cutting the bitrate (lowering or stopping it) according to predefined policies (specified also in the SLA between the telecom operator and the end-user). Overall, one may distinguish among three common decision inputs:

- user QoS profile, which defines the QoS package that should be used for the given user;

- usage cap (e.g. daily, weekly, monthly, annual data allowance in MB, GB, or TB), which is usually set by the user's tariff (e.g. expensive tariffs have higher allowance, i.e. data caps, than cheaper ones);
- concrete traffic type (voice, video, Web, etc.).

For traffic management purposes, in telecom operators' networks traffic management interventions [25] typically are used, which belong to two main types:

- *IP packet prioritization.* The queues of IP packets normally occur in network nodes (e.g. switches, routers, gateways), and in such cases higher priority traffic will be served with priority over lower priority traffic (which can be delayed or may suffer packet loss in cases of packet dropping due to buffer overflow). Packet prioritization is applied today in the core and transport networks. However, it is likely in future to move closer to the ultra-broadband access networks (fixed and mobile) to provide better and optimized traffic management by maximization of the network utilization while having no or minimal effect on most end-users.
- *Bandwidth (bitrates) allocation.* The bandwidth (or data rate) offered to a given end-user or dedicated to a concrete type of traffic (e.g. audio, video, Web) may be actively controlled. End-users can be offered a minimum guaranteed rate or they can be limited (according to the bitrate cap) to a maximum bitrate. One should note that bandwidth allocation for data services does not cause packet loss by itself, unless it is below the required bitrate for a particular OTT service flow. Meanwhile, all OTT flows that use the TCP/IP protocol stack normally experience certain packet losses due to TCP congestion windows behavior, which in most TCP versions increases over time, either exponentially (in the Slow Start phase) or linearly (in the congestion avoidance phase), increasing the bitrate for the given OTT application flow until loss occurs (loss occurs when the aggregated bitrate from all OTT flows of that user's broadband access reaches the bandwidth allocation for that user's Internet access).

Table 6.8 provides the specifics of different types of traffic management interventions regarding the OTT services. However, the same also refers to real-time QoS-enabled

Table 6.8 Type of traffic management interventions.

Type of intervention	Possible actions	Intervention applied in	Protocol layers	Negative impact of the intervention		Best approach
				TCP/IP	UDP	
Packet prioritization	Prioritization or deprioritization	Core IP network	Layer 3 and Layer 4	Retransmission of packets	Data loss	TCP/IP interactive traffic is effectively managed by de-prioritizing this traffic type
Bandwidth allocation	Bitrate guarantee or bitrate cap	Broadband/ ultra-broadband access network	Layer 2 and Layer 3	Reduced throughput (service maintained, but at lower speed)	Reduced quality (e.g. video codec may drop to a lower resolution)	Video traffic is best managed by prioritizing in the access network

services such as carrier-grade VoIP (with QoS support), IPTV, and VPN business services. In such cases prioritization and bandwidth allocation are mandatory, with VoIP (as the PSTN/ISDN replacement) having the highest priority, IPTV and video on demand (with QoS) the second, while data services (i.e. all Internet traffic) has the lowest priority. Prioritization is also applied in the access network (fixed and mobile) besides core and transport network of telecom operators where prioritization is typically applied.

Regarding OTT traffic, to have traffic intervention, there is a need to classify the traffic according to the ToS. Although IP headers have a ToS field in IPv4 or a DSCP field in IPv6 headers, theses field are changed by network edge routers (i.e. network gateways) in each network domain (or in each telecom operator's network) because traffic management is autonomously applied in each network and it may be implemented by using different approaches, although the goal is the same.

Although the ToS and DSCP fields (in the IPv4 and IPv6 headers, respectively) can be used end-to-end, there is required inspection of IP headers and payloads (with deep packet inspection (DPI)), including higher layer protocol headers (transport layer headers and application headers and the type of user data in payload), with the aim of detecting to which application or service a given IP packet belongs and then to classify it to a given traffic type or class. General traffic management architecture for such purposes is shown in Figure 6.7. The packet inspections nodes implement IP packet manipulation based on the results obtained from DPI elements and then using the traffic policies set in the given network.

The DPI elements are typically deployed in the core network nodes (close to the network edges), which inspect the packets to determine their traffic types. With the DPI

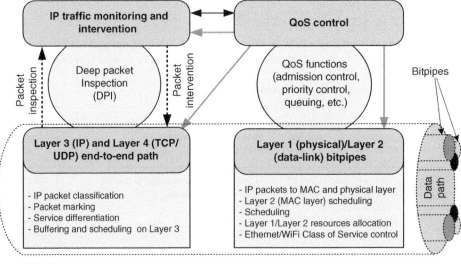

IP – Internet protocol
QoS – Quality of service
MAC – Medium access control
TCP – Transmission control protocol
UDP – User datagram protocol

Figure 6.7 General traffic management architecture.

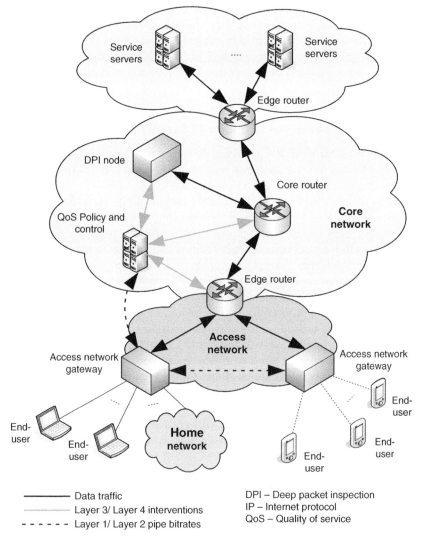

Figure 6.8 DPI-based OTT traffic management.

are inspected all headers and payload information in each IP packet. Association of the IP packets belonging to a particular flow can be provided by using the five-tuple (that includes five parameters: source and destination IP addresses, source and destination port numbers, and protocol being in use over the IP such as TCP, UDP, or IP in tunneling cases). The positioning of DPI and its relation to other network elements are shown in Figure 6.8.

The information collected from the DPI elements is further sent to the policy and control node in the network, which has traffic management policies implemented. Using the information obtained from the DPI elements, the policy and control nodes (e.g. RACF in the NGN) send control signals to the transport network nodes about how to treat the IP traffic.

The nodes at the core network edges remark the IP packets based on the priority decided by the traffic management policy, then the access nodes treat the packets accordingly. One of the uses of DPI could be monitoring of end-users' monthly (or daily or weekly) usage cap, so when the limit specified in the SLA is reached for the given end-user, that could trigger reduction of the bitrate (i.e. bitpipe) by allocation of fewer resources in the network.

6.6.3 Traffic Management Influence on QoE for OTT Services

The traffic management is targeted at controlling the network congestion being experienced by end-users. That is directly related to relation of the aggregated network traffic and the installed network capacity, as given in Figure 6.9.

One may assume that in the longer term, from the technical aspect, traffic management will continue to be a tool for network congestion control and QoS improvement. Also, to provide finer granularity in traffic management, it will be necessary for IP packet inspection (i.e. DPI) to move from core network to broadband access network and to be based on user specific needs. For example, if a given user wants to prioritize a certain OTT type of traffic or certain application over other OTT applications (e.g. gaming traffic or OTT VoIP traffic over Web, email, or torrent traffic), then it should be able to do it via the operator's management interface (which should be user friendly and easy to use). Also, that approach will not jeopardize network neutrality because one of its main goals is to provide choices to end-users (e.g. many OTT applications exist and many more will appear in the future, developed on the basis of Internet network neutrality), then the end-user can be given an additional choice for the QoS treatment of particular popular OTT services at a given time.

What is the relation between network capacity and traffic management? When Internet traffic demand is well below the available network capacity, then there is no need for traffic management intervention. When Internet traffic load approaches the installed

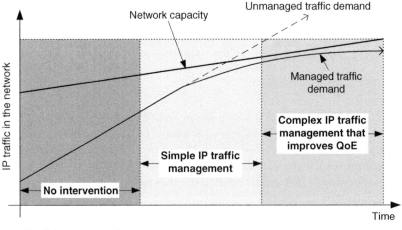

IP – Internet protocol
QoE – Quality of experience

Figure 6.9 Traffic management dependence upon network capacity.

network capacity, then simple traffic management can be applied (e.g. prioritization and bandwidth allocation in the core networks for given traffic types) [25]. When Internet traffic load is higher than the network capacity, traffic management is mandatory because it can help to improve the QoS and to increase the QoE.

6.7   Internet of Things (IoT) Services

The IoT started to appear as a paradigm with the spread of broadband access to the Internet around the mid-2000s. The ITU published its first book report on the IoT back in 2005 [26]. What else contributed to the growth of the IoT? It is driven by widespread use of the Internet and IP technologies and the availability of fixed and mobile access networks everywhere, continuous miniaturization of various devices and sensors, and development of big data analytics and cloud computing.

Let's define the IoT. According to the ITU [27], the IoT is a global infrastructure for the information society enabling advanced services by interconnection of different physical and virtual things, based on ICTs. The IoT defines a new dimension to the ICT world, called "any-thing communication" [28], in parallel with "any-time communication" and "any-where communication" (Figure 6.10).

The IoT refers to any communication between machines and objects over the Internet. Such technologies include (but are not limited to) Radio Frequency Identification (RFID) wireless sensor networks, and M2M communications.

The main challenge for IoT services is industry standardization. With the development of 4G and 5G mobile networks, particular attention is given to the IoT segment because it has spread due to standardized mobile and wireless networks, and the only prerequisite is to have end-to-end IP connectivity. In many countries, GPRS (General Packet Radio Service) and EDGE (Enhanced Data rates for Global Evolution), which appeared as the 2.5G mobile network in the GSM era (i.e. the 2G era), are the most used technologies for the IoT (GPRS modems can be found in different sectors, such as remote control of public utilities or measurement units, fleet tracking, credit/debit card mobile payment devices, and many more). Why? Because most of the IoT devices (e.g. sensors) do not require high bitrates for transfer of control information to the IoT device and

Figure 6.10 Three internet of things (IoT) dimensions.

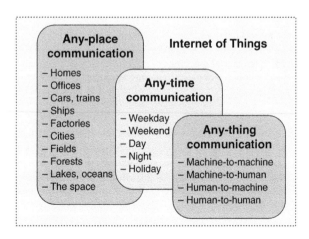

transfer the collected data from the IoT device to a given node (e.g. server, database) in the IP network. So, broadband is not required for the most of the IoT world, but narrowband speeds (which are in the range of hundreds of Kbit/s or couple of Mbit/s around 2020 – of course, one may note that such narrowband bitrates needed for the IoT today were in fact broadband bitrates one or two decades ago).

The 4G mobile networks LTE-Advanced and particularly LTE-Advanced-Pro, which comes from the last 3GPP releases before 5G (LTE-Advanced-Pro is standardized in 3GPP Releases 13 and 14), are focusing in a greater portion on the IoT. Since IoT devices are spread over a wide area, the typical way to connect them is through radio connections. To provide their interoperability there is increasing demand for standardization and unification of standards (e.g. few standards supported by most of the devices, similarly to what Ethernet and WiFi have become in fixed and wireless local area networks, respectively).

6.7.1 Mobile Cellular Internet of Things

The LTE-Advanced-Pro has focused on mobile IoT by introducing new narrowband technologies for that purpose. Table 6.9 illustrates how the LTE-Advanced-Pro is developed for IoT support, including further development of well proven (for IoT) GPRS technology.

The work on new IoT devices is also targeted at longer battery life, lower costs of devices and deployments, as well as extended coverage (which is successfully done with inclusion of IoT specific technologies within the 3GPP mobile standards). However, one may note that enhanced machine type communication (eMTC) and narrow band Internet of Things (NB-IoT) are designed for different types of IoT use cases. eMTC uses only 1.4 MHz (meanwhile, LTE carrier is up to 20 MHz, and LTE-Advanced/LTE-Advanced-Pro allows aggregation of multiple such carriers) to provide up to 1 Mbit/s of throughput (possible uses are asset trackers and wearables). High bitrates in mobile networks are reserved for more powerful devices (e.g. smartphones) which require at the same time high throughput (e.g. ultra-broadband access speeds) and low latency.

Table 6.9 3GPP mobile IoT.

LTE IoT technologies	Description	Bitrates
eMTC (enhanced machine type communication)	LTE/LTE-Advanced enhancement for machine type communications, which is also abbreviated as LTE-MTC or LTE-Ms (long term evolution for machines)	1 Mbit/s on smallest width LTE frequency carrier of 1.4 MHz
NB-IoT (narrow band Internet of Things)	New radio spectrum added to LTE, which is optimized for the low-end of the IoT market	10s–100s of Kbit/s (180 kHz narrowband frequency carrier)
EC-GSM-IoT (Extended Coverage GSM for Internet of Things)	Further enhancements on GSM/GPRS/EDGE technologies for the IoT segment	

The NB-IoT was developed with simplicity in mind, to be used in cheaper end-user equipment. It has low throughput and delay-tolerant applications (e.g. sensors as IoT devices). It supports bitrates of several tens of Kbit/s up to several hundreds of Kbit/s by using 180 kHz of bandwidth (which has size similar to GMS/GPRS/EDGE frequency carrier width). However, the NB-IoT was created to be deployed within existing LTE bands, by utilization of the spectrum between two "standard" adjacent LTE frequency carriers. Also, the NB-IoT can be used in its own standalone mode, thus providing easy migration path for the re-farmed GSM/GPRS/EDGE spectrum. So, due to its low price and affordability the NB-IoT is considered to be one of the "accelerators" for further explosion of the IoT services.

The Extended Coverage Global System for Mobile communication for Internet of Things (EC-GSM-IoT) has been developed to be backwards-compatible and to be deployed into existing GSM/GPRS/EDGE networks as a software upgrade. So, it can be supported on existing GSM/GPRS/EDGE deployments by using the common GSM spectrum on 900 MHz and 1800–1900 MHz bands.

However, the newer mobile IoT technologies, such as eMTC and NB-IoT, are further being developed in 5G standards, starting with 3GPP release 15 (completed in 2018) and further 3GPP releases.

What about the QoS for IoT services? Bearing in mind that most IoT devices are connected via radio connections, one may expect that in future almost all such IoT devices will be connected via standardized mobile networks (e.g. 2.5G–5G), which provide IP connectivity, or via WiFi local access networks (where WiFi networks are connected to IP networks via fixed or mobile access networks). Since mobile networks include QoS solutions in all mobile generations (unlike WiFi), for IoT services that demand QoS support the main deployment method will be over the deployed mobile networks (e.g. LTE-Advanced-Pro and 5G in the 2020s). One of the advantages of mobile networks is licensed spectrum because it provides more efficient traffic management. When the spectrum is not shared with other telecom operators (unlike WiFi which uses unlicensed spectrum), there is better reliability and QoS due to less interference. Also, mobile operators which use licensed spectrum (for all existing mobile generations, 2G–5G) can offer the same IoT services with the same QoS support and frequency spectrum on a global scale, thus extending the use of the same IoT devices and solutions in different countries around the world.

Regarding QoS, different IoT services may have different requirements (because they have wide areas of application), such as:

- *Power efficiency.* IoT devices for smart city services or personal wearables are required to have longer battery life, so that is an important key performance indicator (KPI).
- *Complexity.* IoT devices in smart homes and remote sensors require lower complexity.
- *Range and mobility.* These parameters are important for utility metering for various public services (e.g. electricity, water supply, heating, etc.) or object/fleet tracking services.
- *High reliability and low latency.* These parameters are important for IoT devices and objects that require significantly reduced end-to-end packet losses (e.g. 99.999% packet success rate) and reduced end-to-end delay (e.g. as low as 1 ms), which include autonomous vehicles, industrial automation, aviation, drones, and robotics.

- *Higher availability.* This parameter is required for IoT smart services which need to be available all the time, such as energy smart grids and medical IoT services. Higher availability in the telecommunication world is typically provided by having multiple links/path for failure tolerance.

So, different IoT services have different contexts and with that different QoS requirements. Overall, one may distinguish between two main types of IoT services regarding their QoS needs (Figure 6.11):

- *Massive IoT.* It includes IoT services based on low cost, low energy consumption, small data volume generated or transferred, and massive deployments. In this group belong capillary sensor networks, smart agriculture, smart metering, tracking and fleet management, smart buildings, etc. Massive IoT services do not have strict QoS requirements, i.e. they are more tolerant to packet losses, delays, and require lower bitrates. These services can be served over the best-effort Internet network, including fixed and mobile ones. For example, in 5G mobile networks that will be the eMBB (extended mobile broadband) network slice. Also, eMTC and NB-IoT can be used for this purpose.
- *Critical IoT.* This includes services which require ultra reliability, very low latency and very high availability. Such services include remote manufacturing, remote surgery, smart grids, traffic safety and control, and distance healthcare. These services require strict QoS, so they cannot be provided via the public broadband Internet network which is based on best-effort and network neutrality (these services belong to specialized services). Therefore, critical IoT deployments must be realized via managed IP networks, with QoS guarantees, such as the URLLC (ultra-reliable low latency communication) slice in 5G mobile networks [29].

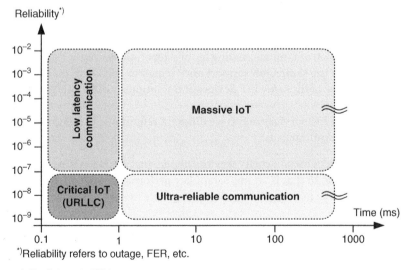

Reliability*)

10^{-2}
10^{-3}
10^{-4}
10^{-5}
10^{-6}
10^{-7}
10^{-8}
10^{-9}

Low latency communication

Massive IoT

Critical IoT (URLLC)

Ultra-reliable communication

Time (ms)

0.1   1   10   100   1000

*)Reliability refers to outage, FER, etc.

IoT – Internet of Things
FER – Forward error correction
URLLC – Ultra reliable low latency communication

Figure 6.11 QoS requirements for critical IoT vs. massive IoT.

The mobile IoT technologies are also referred to as low power wide area (LPWA) technologies [30]. Regarding KPIs, while the legacy mobile networks KPIs include, among others, radio coverage, interference (on the physical layer), or packet losses and delay (on the network layer), the IoT services may extend the set of KPIs to new ones such as endurance, efficiency, or reachability. Possible new IoT QoS parameters include meter report success rate (from IoT devices to network nodes), software/firmware upgrade (from network nodes to IoT devices), tracking report delay, surveillance video quality (e.g. are objects and people on the video identifiable?), etc. However, the set of QoS parameters which are important (i.e. KPIs) and their target values are strongly dependent upon the IoT service and its type (e.g. massive or critical).

6.7.2 IoT Big Data and Artificial Intelligence

The IoT provides a lot of data which increases over time from the existing devices and constantly increases with new deployed IoT devices (which are expected to grow significantly in numbers during the 2020s, to reach over 100 billions of such devices). Additionally, human users increase the volume of data because of ultra-broadband speeds and availability of end-user equipment (such as smartphones) with high capabilities for creation of content (e.g. pictures and videos, data download/upload in the clouds, etc.), which increases the amount of data that may be needed to be processed (for different purposes, which may span from marketing of end-users toward high quality provision of services). So, smart objects produce large amounts of data that needs to be transferred (e.g. to particular application or service), processed, and/or stored.

Big Data refers to a data set that is so large or complex that traditional computational analysis and processing cannot be used [31, 32]. However, Big Data is too complex to be processed with legacy computing approaches, therefore it is directly benefiting from the development of artificial intelligence and machine learning, which are gaining momentum in the telecom/ICT world [33]. One may expect wider use of AI in communication systems and networks in 5G mobile networks in the 2020s, for example for provision of better network efficiency.

AI can be used in a number of different cases for IoT services. For example, in smart city services AI can be used to provide human-like applications, which can make informed predictions and decisions (e.g. managing the energy-efficient city functions).

Figure 6.12 illustrates possible AI applications for measurement of QoE. In fact, bearing in mind that QoE is difficult to measure, unlike QoS (QoS can be measured on the basis of selected sets of KPIs, which are important to both telecom operators and customers), AI can be used to estimate the QoE for a given service by using complex analysis of network QoS for different services provided via broadband and ultra-broadband access, including non-technical data such as tariff plan, age of the customer, environment, end-user equipment capabilities, and other related contextual information that may influence the quality experienced by the end-users of the given service.

The uses of Big Data together with AI and ML can be targeted to automation of various network functions, including automation of the network design, operation and maintenance, network self-optimization, and so on. Such approaches may increase the QoE on the basis of better customization of services (e.g. use of digital assistants), or smarter use of the data (e.g. through machine learning) [34]. Further, Big Data and AI can be used for the creation of new business models, or for improvement of existing services – for

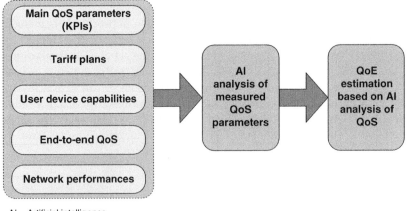

AI – Artificial intelligence
KPI – Key performance indicator
QoE – Quality of experience
QoS – Quality of service

Figure 6.12 QoE estimation based on artificial intelligence analysis of measured QoS and user contextual information.

example, important software updates for advanced smart metering, periodic reports from the devices, and network commands toward the devices. Another example could be important software update and event reports for smart trash cans. On the other side, security cameras require successful software updates, network commands, and continuously supported video streaming bitrates. This last example shows that not all IoT services are serviced by narrowband connections because good quality video surveillance requires higher bitrate than other IoT services (e.g. in video surveillance IoT services, important QoS parameters are video bitrates as well as video quality).

6.8 Cloud Computing Services

Cloud computing is used for most of the OTT services nowadays provided via broadband Internet access – for example, social networking, video or picture sharing, data storage, business documentations and applications, etc.

What is cloud computing? By definition, it is a paradigm for enabling network access to a scalable and elastic pool of shareable physical or virtual resources with self-service provisioning and administration on-demand [35].

The main roles in the cloud ecosystem include the following (Figure 6.13):

- *Cloud service provider (CSP).* This role provides the cloud computing services.
- *Cloud service customer (CSC).* It has a business relation with CSP (as well as with cloud service partner) for using the cloud services.
- *Cloud service partner.* It is a party that is engaged in support of either cloud service customer or provider.

The standardization of cloud computing is carried by ITU-T as a continuing development of ICT services after NGNs, which were standardized to provide easy transition of legacy telecommunication services to IP networks and Internet technologies.

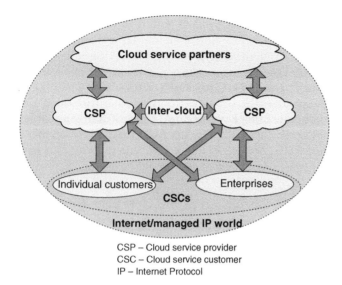

CSP – Cloud service provider
CSC – Cloud service customer
IP – Internet Protocol

Figure 6.13 Cloud ecosystem.

Cloud computing functional architecture includes four functional layers (from the user toward the cloud resources):

- *User layer.* This includes the user interface (e.g. Web-based interface).
- *Access layer.* This provides access mechanisms for presenting the cloud service capabilities. This layer is responsible for QoS enforcement for the cloud services.
- *Service layer.* This contains the implementation of the cloud services and controls the software components. This layer is dependent upon the resource layer and its capabilities, which directly influence the QoS (e.g. specified in the SLA).
- *Resource layer.* This contains the resources, such as servers (in data centers), switches, and routers for the networking, storage devices, and non-cloud software.

Most of today's Internet services are based on the cloud computing paradigm, particularly OTT services. All cloud service types are grouped into a limited number of cloud computing service categories. All such categories are based on the following:

- *Infrastructure as a service (IaaS).* It allows the CSCs to use the provided cloud infrastructure resources, which include processing, storage, or networking.
- *Platform as a service (PaaS).* CSP provides to CSC a platform that includes operating systems, execution environment for a given set of programming languages, databases, and web servers.
- *Software as a service (SaaS).* CSPs install and manage application software in the cloud, thus reducing the need to run the software on cloud users' own computers or devices.

All cloud computing services are using one or more cloud computing categories (IaaS, PaaS, and SaaS). Regarding the network impact on cloud services and their QoS requirements, there are two main types [2]:

- *Over the top (OTT) cloud services.* They are provided over the public Internet network. With network neutrality in force, there is no possibility for differentiation of cloud services from other Internet services regarding network performance. However, CSPs can provide QoS support by proper dimensioning of all four layers: resource layer, service layer, access layer, and user layer. In general, many OTT services utilize cloud services (e.g. SaaS is the most used cloud service category), such as web applications that offer storage, social networking services, video sharing sites, etc.

- *Telco cloud services.* Telecom operators (i.e. telcos) can provide enterprise cloud systems via managed IP networks because they have their own networks and can guarantee QoS end-to-end when they act as CSPs [36]. So, in this case QoS can be guaranteed end-to-end, between CSPs and CSCs. Also, telecom operators may provide QoS networking for the federators between different cloud entities, including public, private, community, and hybrid clouds [2].

6.8.1 QoS Metrics for Cloud Services

There are certain QoS metrics that can be specified for cloud services, as given in Table 6.10. Some can be further broken down into submetrics. For example, elasticity as a metric can be split into submetrics which include provisioning interval (that is, time needed to access or release the requested cloud resources), agility, and scaling (flexibility in service provisioning with increasing or decreasing number of CSCs) [37]. For higher agility, the CSP should have enough agile cloud system that can scale with the increased workload upon the needs of the CSCs. After elasticity as first basic metric for cloud QoS, the response time can be considered as the second basic cloud metric. Further, throughput can be considered as the third basic cloud metric, which is the most important one for the cloud services and can be measured also by cloud customers (on their side, i.e. client side, considering that the cloud services are based on the Internet/IP client-server principle).

Table 6.10 QoS metrics for cloud services.

QoS metric	Description
Elasticity	The ability of the CSP to accommodate the CSC's requirements
Agility	The ability of the cloud service to scale with the increased or decreased workload
Response time	It is defined as the time it takes for a request made by a client or workload generator to be received by the client
Throughput	Throughput should, be measured in the context of the CSC's workload, such as bit/s, packet/s, or tasks/second
Availability	The ability of a CSU to access the cloud resources (e.g. 99.9% uptime)
Durability	Probability of CSC data loss
Reliability	Ability of CSP to complete all required functions for a given period of time
Power	Total watt usage by the CSP system
Density	Refers to number of workload instances that can be run on the cloud system simultaneously before the QoS degradation occurs
Variability	This is the standard deviation of the other QoS metrics.
Latency	Round trip delay of packets between CSC and CSP

Important QoS metrics for cloud services are availability and reliability, which are related. While availability is the percentage of time when the cloud services are available to CSCs, reliability refers to the ability of the cloud system to perform any required functions under the stated conditions for a specified period of time.

For example, QoS metric reliability can be measured by the mean time between failures. If this is denoted with $T_{between\_failures}$, the number of disk drives is N_{Disk_drives}, and T_{hours} is number of hours run, then the reliability can be calculated by using the following equation:

$$Reliability = T_{between\_failures} = \frac{N_{Disk\_drives} T_{hours}}{N_{unit\_failures}} \qquad (6.4)$$

Density as a QoS metric is dependent upon the cloud service category. For example, when the cloud service is IaaS, density may refer to the number of virtual machines running. For PaaS, the density may be measured as the number of application servers that are running on the cloud system. For SaaS, the density refers to the number of users that the cloud system can service simultaneously.

Finally, latency (i.e. delay) is important for cloud services because most of the cloud services are interactive by nature (e.g. similar to Web services, which also can be based on cloud computing, for storage of user data for instance). Latency directly influences the amount of data that the CSC can back up, access, or restore [37]. Also, latency can limit the maximum amount of data that can be transmitted over the IP network (either public Internet or managed IP network) because cloud services are based on the TCP/IP protocol stack which ensures 100% lossless transfer, but it is directly dependent on the RTT due to TCP congestion control mechanisms (e.g. Slow Start, Congestion Avoidance). End-to-end delay is hard to measure for cloud services because the cloud data may traverse multiple network segments belonging to multiple telecom operators in different countries, so this metric is one of the least predictable since it does not depend only upon the access bitrates (and QoS being in force) on the side of the CSC or the CSP, but also is influenced by the transit networks. One may note that most of the QoS-supported cloud services are provided inside the telecom operator's network (e.g. the telecom operator is also the cloud provider), while in other cases when the CSP is a third party the cloud services are provided as OTT services (which can be charged by the CSP or given free of charge with certain data cap or/and time duration). However, with the development of ultra-broadband access networks, which typically are connected to core and transport networks that can support such ultra-broadband speed (based on proper traffic dimensioning), the low end-to-end latency can become reality for the cloud services. As is the case for most of the client-server interactive service, the delay is also influenced by the time required for the signals to travel between the client and the server with the speed of light (that is 3×10^8 m/s), which gives delays in the range of 100 ms when the client and servers are on the opposite sides of the globe (e.g. the perimeter of the Earth is around 40 000 km on the Equator, so half of the perimeter is 20 000 km, and the transmission path terrestrially will be significantly longer than 20 000 km, e.g. 30 000 km). So, to have lower RTT, most of the OTT service providers, and particularly OTT cloud services (e.g. video sharing sites such as YouTube, Amazon cloud services, social networking services) use CDNs which are located closer to the end-users in order to reduce the round trip

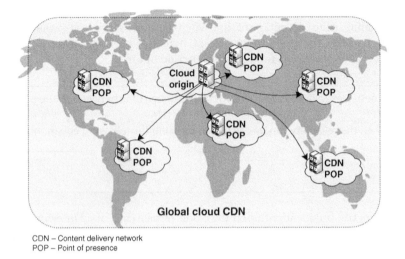

CDN – Content delivery network
POP – Point of presence

Figure 6.14 Content distribution networks (CDN) for cloud services.

time (or one way delay) as one of the most important QoS parameters in general for all services (Figure 6.14). There are many commercial cloud CDNs [38], which include Amazon elastic cloud, Microsoft Azure, Google clouds, etc. [2]. In all cases, one of the most important functions of a given CDN is to dynamically redirect cloud service clients to the most optimal cloud servers, based on the given set of QoS parameters which include (but are not limited to) cloud servers' load, end-to-end delay between the CSC and servers, network congestion on the path, type of access networks on the CSC's side, and proximity.

6.9 Business and Regulatory Challenges for Services Over Ultra-Broadband

Ultra-broadband access provides the basis for richer legacy services (voice with better audio quality, video with higher picture resolution, business services with higher bitrates, etc.) and lower RTT for OTT client-server services. The RTT is inversely proportional to the bitrates, so ultra-broadband bitrates may provide in the access and core networks delays in the range of 1 ms (for tactile Internet or tactile IP networks).

6.9.1 Business Aspects for Broadband Services

Each of the broadband and ultra-broadband services has its own challenges. Such services for residential users typically include the three main types of services: voice as VoIP with QoS guarantees, IPTV, and Internet access service. For business users there are VPN services as replacement for leased lines as well as telco cloud services with QoS guarantees. With the development of the IoT, particularly with deployments of 4G mobile networks (e.g. LTE-Advanced-Pro) and 5G mobile networks, it is expected that there will be an increase in telecom/ICT ecosystem revenues with the provision of new smart services based on either massive or critical IoT.

For example, VoIP is provided via broadband IP access (although voice as a service does not require high bit rates), but several different business models for VoIP can be distinguished:

- *Telco VoIP (as PSTN/ISDN replacement)*. This VoIP service is provided with QoS support end-to-end by using standardized network and service functions, typically NGN-based.
- *Bundling VoIP service with other IP-based services*. Telco VoIP is typically bundled with broadband Internet access services and/or IPTV as a resale product of the service provider, including fixed and mobile ones. Due to the global spread of mobile networks and their affordability in all countries (including developing ones), the bundled VoIP with QoS support via mobile network has significantly surpassed fixed telephony.
- *Separate provision of VoIP service*. VoIP is partly bundled with broadband access to the Internet, where the VoIP service can be offered to customers connected via other broadband access networks to Internet (i.e. VoIP provider and broadband access providers can be different business entities).
- *Best-effort VoIP service (i.e. OTT VoIP)*. In this case telecom operators do not guarantee delivery of VoIP packets because there is no relation between OTT VoIP service providers (e.g. Skype, Viber, WhatsApp) and telecom operators as broadband access providers. Although OTT VoIP is multiplexed with other Internet traffic on a best-effort basis, the telecom operators can apply traffic management techniques in their network to increase the QoE for OTT VoIP services. Some OTT VoIP services were offered free of charge (telecom operators do not charge the Internet traffic in such cases), aiming to attract new subscribers to Internet access service, but in certain cases such differentiated charging was seen as jeopardizing the Internet network neutrality principle [24].
- *Mobile VoIP service (e.g. voice over long term evolution – VoLTE in 4G networks, voice over New Radio in 5G networks)*. These services are provided in mobile environments with QoS as well as roaming support (i.e. use of VoIP through visited mobile or fixed networks). They have advantages over the OTT VoIP via mobile broadband access networks due to guaranteed QoS at different mobility of the end-users and roaming features. The OTT VoIP over mobile networks are attracting users due to their user-friendly interface and bundling different OTT services within the same application (VoIP, messaging, file sharing, etc.).

Another legacy service which is transferring to broadband/ultra-broadband IP networks is IPTV and TV streaming. Regarding the TV/VoD over IP one may differentiate among several possible business models:

- *Fully fledged architecture for PTV services*. In this case the telecom operator as network provider is also an IPTV service provider and it provides end-to-end QoS support, using either proprietary IPTV platforms or NGN functionalities.
- *IPTV service provider*. The IPTV service provider offers IPTV streams to different network providers (e.g. NGN providers). For this purpose there are requirements for NGN implementation on both sides, IPTV service provider and broadband access provider (i.e. telecom operator).
- *Internet TV streaming*. This is streaming of live TV over the public Internet, in which case there are no QoS guarantees end-to-end. However, most of the video services

Table 6.11 IoT services types and business models.

IoT service types	Device provider	Network provider	Platform provider	Application provider
Critical IoT (high QoS requirements)	Provider 1			
	Provider 1			Provider 2
Massive IoT (medium to low QoS requirements)	Provider 2	Provider 1		Provider 2
	Provider 2	Provider 1	Provider 2	
	Provider 3	Provider 1	Provider 2	Provider 3

over the Internet are VoD via many video sharing sites (YouTube is the largest such example regarding volume of generated Internet traffic). The GUI for such video streaming services is typically the Web; however, there may be developed client applications for higher QoE when accessing via mobile handsets (to better adapt to smaller screen sizes).

Traffic management techniques on the side of telecom operators may provide better QoE even for Internet-based TV streaming. Video traffic, including OTT video streaming and IPTV over managed IP networks, contributes to two thirds of the overall Internet traffic globally.

Further, the IoT builds a huge ecosystem around itself, which is composed of different business players that include device providers, network providers, software platform providers, and IoT application providers. For QoS support of IoT services the main role in the IoT ecosystem belongs to the network providers (e.g. telecom operators), which have network functions to provide end-to-end QoS required for critical IoT services.

Generally, single players in the IoT ecosystem may play one or more provider roles, such as telecom operator (e.g. can provide IoT devices, network, service platforms, and IoT services). Different business models for the IoT are shown in Table 6.11, including critical and massive IoT service types, depending on the ownership of different providers in the IoT ecosystem.

6.9.2 Regulatory Challenges for Broadband Services

Broadband services impose certain regulatory challenges. One should distinguish between services provided via managed IP networks, which use broadband access, and OTT services provided over the same broadband access (fixed and mobile) for the public Internet network.

So, on one side regulation of traditional services telephony and television has been well established since the end of the twentieth century and beginning of the twenty-first century. The regulation of telephony does not change with its transition from digital telephone networks (PSTN/ISDN) to NGN. IPTV regulation is a similar case, which is not different than existing regulation of TV content and its distribution.

However, OTT services are provided over the best-effort Internet access service. Their regulation is simplified through the means of network neutrality, so there are no direct obligations on OTT service providers like those imposed on telecom operators for telephony and TV over managed IP networks. OTT VoIP (Skype, Viber, WhatsApp, etc.) is offered by third parties (telecom operators and their customers are the first two parties) by using proprietary platforms and applications (i.e. they are not standardized and typically not open for further development by other parties than the owner of that application/service) in a deregulated manner. There are several regulatory challenges that may evolve or change regarding VoIP regulation. For example, QoS-enabled VoIP of telecom operators should support all additional service features available in PSTN/ISDN, such as call line identification, call forwarding, call blocking probability, and so on. Also, emergency calls and legal interception are mandatory for telecom operator grade VoIP (based on national legislation), but they are typically not required for OTT VoIP providers in most countries. Regarding the QoS aspects, the PSTNs were designed to meet strict QoS constraints such as end-to-end delay and bitrate, call success rate, which are followed by QoS-enabled VoIP [1]. That is not the case with OTT VoIP services which are provided over the existing Internet global network; however, customers can make their own choice (although such choice is directly dependent on other users with whom they communicate regularly, such as family, friends or colleagues, hence one or two OTT VoIP services are becoming dominant in a given country or region).

Besides legacy services over IP networks, such as VoIP and IPTV, Internet access services are typically used by third parties for provision of OTT services, which are served in the broadband access network in a best-effort and network neutral manner. Network neutrality refers also to contents over public Internet access, so this part is unregulated in most countries. (That means end-users can access every type of content over the Internet, but it is their responsibility to know which content is allowed by laws in that country and which is not. However, each country can apply its own regulation for access to certain content and sites which provide such content (e.g. prohibit certain type of content for given user groups.)

For the Internet access service (over broadband and ultra-broadband access), the national regulators may monitor OTT services (e.g. with HTTP QoS metrics, based on measurements with HTTP file upload or download, etc.). To prevent degradation of the Internet access service (e.g. lower bitrates than those specified in the SLA) [2], the national regulatory agency (NRA) in a given country may choose between several regulatory tools [39], such as setting minimum QoS requirements, and market mechanisms (e.g. through support of competitive markets for the Internet access service providers in that country). In the case of competitive markets, which have no barriers for entry of new ISPs, QoS regulation could be less strict. For cases with barriers to entry (e.g. entry of new mobile operators is dependent upon spectrum availability), more strict QoS regulation can be applied (e.g. mobile operators are required to obtain licenses on frequency spectrum and provide coverage with specified QoS parameters, such as call or message success ratio, voice quality, minimum sustainable bitrates for nomadic and mobile users).

The IoT regulatory challenges are primarily targeted to security enforcement and privacy and data protection. However, in this chapter we have defined two main types of IoT services: massive IoT and critical IoT. Massive IoT services (which have existed since the appearance of IP-based access networks, especially with mobile-based Internet access) typically do not require strict QoS regulation, and such services are provided via public

Internet access based on best-effort principles. One may say that massive IoT is market driven. However, with the development of mobile technologies targeted specifically to the IoT (e.g. NB-IoT, and M-LTE in LTE-Advanced-Pro, and also 5G IoT network slicing approach), there are possibilities to have dedicated capacity for IoT services which can improve the QoS metrics (e.g. delay). Since the middle of the 2010s, there is also a trend to provide a telecoms single market (TSM) approach (e.g. in Europe enacted in 2016) [24], which puts all ICT/telecom/digital services in the same market, under the same legislation and regulation "umbrella." That offers the possibility to deliver IoT services as specialized services, which may use specific traffic management techniques in access and core networks. That will be particularly important for critical IoT, such as driverless vehicles control, industry automation, digital healthcare, and smart city services [40]).

QoS regulation is also being considered as part of customer protection. However, one should note that customer protection is broader than QoS regulation – for example, for Internet services it also includes sales activities, customer complaints, resolution procedures, and user operator switching/disconnection policies.

Regarding the regulation of cloud computing services, for example, within the telecoms (or lately digital) single market strategy for Europe is noted the key role of cloud computing services for building the digital economy. That has led to legislation of a free flow of non-personal data regulation, which facilitates portability of data and the possibility to switch between CSPs. The cloud computing services therefore have become crucial for development of the digital economy. For example, cloud services in the European cloud market have increased five times in the period from 2013 to 2020 (i.e. from €9.5 billion to €44 billion) [41]. Most uses of cloud computing by companies in Europe are for hosting email systems and storing files in the clouds, and also for financial and accounting software applications (in these cases the dominant type of cloud service category is the SaaS). The efficient delivery of cloud services contributes to the development of the cloud computing market, which depends on the ability to build economies of scale. The establishment of a digital single market (that is, all ICT services to be on the same digital market with the same rules) is aimed toward unlocking the scale which is necessary for reaching the full potential of cloud computing services everywhere.

Overall, QoS regulation for broadband services is targeted to help customers to make informed choices, to check claims by operators, and to understand the state of the digital market. However, the main challenges are to maintain or improve QoS and QoE in the presence or absence of competition in the ICT/digital services market.

References

1 Janevski, T. (2014). *NGN Architectures, Protocols and Services*. Chichester: Wiley.
2 Janevski, T. (2015). *Internet Technologies for Fixed and Mobile Networks*. Norwood, MA: Artech House.
3 ITU-T Recommendation Y.2261, PSTN/ISDN Evolution to NGN, September 2006.
4 ITU-T Recommendation Y.2262, PSTN/ISDN Emulation and Simulation, December 2006.
5 ITU-R Recommendation BT.2020-2, Parameter Values for Ultra-high Definition Television Systems for Production and International Programme Exchange, October 2015.

6 ITU-T Recommendation G.1080, Quality of Experience Requirements for IPTV Services, December 2008.

7 B. Fenner, M. Handley, H. Holbrook, I. Kouvelas, Protocol Independent Multicast - Sparse Mode (PIM-SM): Protocol Specification (Revised), RFC 4601, August 2006.

8 ITU-T Recommendation P.800, Methods for Subjective Determination of Transmission Quality, August 1996.

9 ITU-T Recommendation P.800.1, Mean Opinion Score (MOS) Terminology, July 2016.

10 ITU-T Recommendation P.800.2, Mean Opinion Score Interpretation and Reporting, July 2016.

11 ITU-T Recommendation P.10/G.100, Vocabulary for Performance, Quality of Service and Quality of Experience, November 2017.

12 ITU-T Recommendation P.862, Perceptual Evaluation of Speech Quality (PESQ): An Objective Method for End-to-end Speech Quality Assessment of Narrow-band Telephone Networks and Speech Codecs, February 2001.

13 ITU-T Recommendation P.863, Perceptual Objective Listening Quality Prediction, March 2018.

14 ITU-T Recommendation P.564, Conformance Testing for Voice Over IP Transmission Quality Assessment Models ITU-T, November 2011.

15 ITU-T Recommendation G.107, The E-model: A Computational Model for Use in Transmission Planning, June 2015.

16 Broadband Forum TR-126, Triple-play Services Quality of Experience (QoE) Requirements, December 2006.

17 ITU-T Recommendation E.804, QoS Aspects for Popular Services In Mobile Networks, February 2014.

18 ITU-T Recommendation G.1010, End-user Multimedia QoS Categories, November 2001.

19 ITU-T Recommendation Y.1315, QoS Support for VPN Services – Framework and Characteristics, September 2006.

20 E. Rosen, Y. Rekhter, BGP/MPLS IP Virtual Private Networks (VPNs), RFC 4364, February 2006.

21 Rekhter, Y. and T. Li, A Border Gateway Protocol 4 (BGP-4), RFC 4271, January 2006.

22 ITU-T Recommendation Y.2215, Requirements and Framework for the Support of VPN Services in NGN, Including the Mobile Environment, June 2009.

23 ITU-T Recommendation Y.2811, Framework of the Mobile Virtual Private Network Service in Next Generation Networks, July 2012.

24 ITU-D, Quality of Service Manual, 2017.

25 Ofcom UK, Traffic Management and Quality of Experience, 2011.

26 ITU Internet Reports, The Internet of Things, 2005.

27 ITU-T Recommendation Y.2069, Terms and Definitions for the Internet of Things, July 2012.

28 ITU-T Recommendation Y.2060, Overview of the Internet of Things, June 2012.

29 5G Mobile Network Architecture for Diverse Services, Use Cases, and Applications in 5G and Beyond Deliverable D6.1, Documentation of Requirements and KPIs and Definition of Suitable Evaluation Criteria, 2017.

30 GSMA Industry Paper, Mobile Internet of Things Low Power Wide Area Connectivity, 2018.

31 ITU and Cisco (2016). *Harnessing the Internet of Things for Global Development*. Geneva: ITU and Cisco.

32 ITU-T Technology Watch Report, Big Data: Big Today, Normal Tomorrow, November 2013.

33 ITU Journal – ICT Discoveries, First Special Issue on the Impact of Artificial Intelligence on Communication Networks and Services, 2018.

34 Toni Janevski, Emerging Trends and Technologies in ICT and Capacity Building Challenges for the ITU Academy, ITU Capacity Building Symposium 2018, Santo Domingo, Dominican Republic, 19–21 June, 2018.

35 ITU-T Recommendation Y.3500, Information Technology – Cloud Computing – Overview and Vocabulary, August 2014.

36 ITU-T Recommendations Y.3521/M.3070, Overview of End-to-end Cloud Computing Management, March 2016.

37 Rizvi, S., Roddy, H., Gualdoni, J., and Myzyri, I. (2017). Three-step Approach to QoS Maintenance in Cloud Computing Using a Third-party Auditor. *Procedia Computer Science* 114: 83–92.

38 Meisong Wang, Prem Prakash Jayaraman, Rajiv Ranjan, et al., An Overview of Cloud based Content Delivery Networks: Research Dimensions and State-of-the-art, Transactions on Large-Scale Data- and Knowledge-Centered Systems XX, pp. 131–158, 2015.

39 BEREC Guidelines for Quality of Service in the Scope of Net Neutrality, BEREC, November 2012.

40 ITU-T Focus Group on Smart Sustainable Cities, An Overview of Smart Sustainable Cities and the Role of Information and Communication Technologies, October 2014.

41 European Commission, Digital Single Market Policy – Cloud Computing, https://ec.europa.eu/digital-single-market/en/policies/cloud-computing, accessed in August 2018.

7

Broadband QoS Parameters, KPIs, and Measurements

Broadband and ultra-broadband access networks provide capabilities for delivery of various services toward end-users, which can have different QoS and QoE requirements. To define the requirements for a given service there is required a set of QoS parameters, which are important to all parties in the value chain, including service providers, networks providers, equipment vendors, and end-users. Most important QoS parameters, which describe the service requirements and can be measured and understood by the end-user (i.e. the customer), are referred to as key performance indicators. There are service dependent QoS parameters and hence service dependent KPIs, as well as service independent QoS parameters (e.g. mobile network radio coverage) and hence service independent KPIs. All selected technical QoS parameters as KPIs are subject to measurements, which can be performed by all parties in the value chain.

7.1 QoS, QoE, and Application Needs

In the all-IP world each service includes an application (e.g. VoIP is a service that uses voice application running on the top of RTP/UDP/IP protocol stack). The relationship between network layer QoS and the obtained QoE (this refers mainly to services used by humans) is strongly dependent on the given service and its application. The following are typical examples regarding QoS requirements of some services (Figure 7.1):

- Email is delay tolerant but requires lossless transmission on the application layer (e.g. email client to email server communication). Lossless communication is accomplished with the use of TCP, which is possible due to email delay tolerance.
- The QoS for voice communication (such as in VoIP, i.e. telephony) mainly depends on packet delay, delay variation (jitter), and depends less on packet loss (because human ear is delay tolerant, can accommodate up to several percent of loss). Additionally, voice QoE is impacted by terminal capabilities, various user-related aspects such as current mood, applied tariffs, and other factors that can have an influence. The well-known criterion for ensuring proper experience is that one-way delay for voice services should be below 150 ms for best performances, while delays above 400 ms are unacceptable for general network planning (however, there are exceptions when this upper limit will be exceeded, such as double satellite hop for locations that cannot be reached with terrestrial networks) [1]. With transition from TDM voice to VoIP there is an increase in the delay budget (due to voice packetization, packet buffering, etc.), so

QoS for Fixed and Mobile Ultra-Broadband, First Edition. Toni Janevski.
© 2019 John Wiley & Sons Ltd. Published 2019 by John Wiley & Sons Ltd.

delays of VoIP for inter-regional calls (for terrestrial routes between 5000 and worst case 27 500 km) fall in the range between 200 and 300 ms, which is acceptable. For VoIP connection on routes with lengths below 5000 km, with proper network planning and mandatory prioritization of VoIP over other services, the one-way delays can be kept also below 150 ms [1].

- For interactive gaming, both delay and delay variation (i.e. jitter) are very important, particularly for first-person shooter games. In real-time gaming the one-way and RTT delay requirements are even stricter than VoIP requirements, that is, one-way delays should be kept well below 100 ms. CDNs are required to limit one-way and RTT delay between gaming servers and end-users (i.e. gamers). Although gaming is provided over the public Internet which is network neutral, there is possibility for use of DPI and QoS differentiation on the end-user side, which will be user-specific and determined by the end-user.
- The Internet of Things has divergent QoS requirements. As already discussed in Chapter 6, the IoT can be grouped into two major groups of services (regarding the QoS requirements), they are critical IoT and massive IoT. The critical IoT services require very high reliability and very low delay, while massive IoT can be offered with lower reliability and longer delay.

Only by controlling QoS, it does not guarantee that the transmission or access links will be faster. Network designers and engineers can, however, control the relative priority with which each router processes the IP packets/datagrams waiting to be sent over each transmission link and which packets are to be dropped during periods where more packets are waiting than a given router is able to store or buffer.

Effects similar to prioritization can be achieved by caching (storing frequently used static data close to the user) and by replication (where the same dynamically generated results can be produced in more than a single location in the network – cloud services can represent an example of this kind of distribution or replication function). The use of caching CDNs represents an increasingly common and important means for

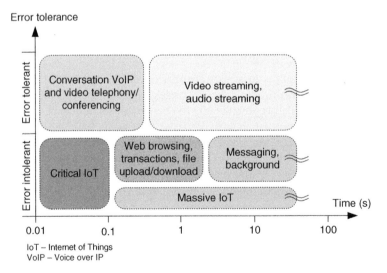

Figure 7.1 User-centric QoS.

improving the QoE. For example, locating YouTube video servers closer to end-users reduces delay and jitter, and hence gives higher QoE. The same approach refers to all services which are based on principle user to content (i.e. Internet native client-server networking approach).

How common is delay-sensitive application traffic? The delay is the most important QoS parameter regarding the real-time traffic. In fact, the real-time services are defined on the basis of low delay requirements, or in other end, non-real-time traffic is more flexible regarding the delays. Hence, typical (now legacy) classification of IP traffic is to real-time conversational and streaming services, and non-real-time interactive (e.g. Web is the most prominent example in this group) and background services (email, file download or upload, etc.). Real-time conversational voice services benefit from limited delay, which can be accomplished in two ways: (i) guaranteed QoS end-to-end (e.g. NGN), and (ii) ultra-broadband access speeds which provide low delay for aggregate bitrates in the access part significantly below the traffic capacity (e.g. below 60% of the ultra-broadband link capacity). The load associated with VoIP is negligible. Video telephony is VoIP accompanied with real-time bidirectional video (e.g. video telephony, or video conferencing), which also requires bounded delay but the requirements are higher due to higher traffic demands of the video (especially for better picture resolutions) than voice and the requirements for the synchronization between the audio and video streams because both media types (audio and video) need to be reproduced with the same time reference. That is accomplished by using timestamp approach in RTP, together with its accompanied control protocol, RTCP, which provides reporting functions between the client and server side of RTP/UDP/UP communication [2].

Additionally, all real-time services including VoIP, video telephony (i.e. video VoIP) or conferencing, video/multimedia streaming (typically by notion of video streaming one assume video with one or multiple audio streams, which can be called also multimedia streaming), all require low jitter (i.e. delay variation). Why? Because, the reproduction of audio or video streams at the receiver's side must be continuous (i.e. without stoppage of the audio or freezing of the video, once the playing has started), which requires playout buffering. The playout buffering should compensate all one-way delay and jitter (for both, real-time audio and video) in order to have smooth reproduction of the content (either that is speech in the earphone for VoIP, or watching video stream on the screen by using a given video player). In the case of only streaming one-way audio or video, the delay plays a minor role (as long as the user is prepared to accept a couple of seconds delay for their playout, while the jitter buffer is filled). The streaming video services have increasing importance as they have become the biggest segment of Internet traffic around 2020.

According to the predictions [3], by 2021 around 68% of the total traffic will be served via fixed broadband Internet access (Figure 7.2), 17% will be served via mobile broadband networks (3G, 4G, and 5G), while 15% will be managed IP traffic (e.g. traffic with QoS guarantees, such as carrier-grade VoIP, IPTV, or VPNs for business customers). However, mobile broadband has a prediction of two times higher compound annual growth rate than fixed broadband. Regarding the type of traffic, the predictions say that consumer video streaming traffic share will further increase and reach 70% of the total IP traffic (managed and Internet) by 2021 [3], which is 82% of the consumer traffic. So, most of the telecommunication infrastructure in the following years will carry video traffic, which is also one of the drivers for the broadband and ultra-broadband demand.

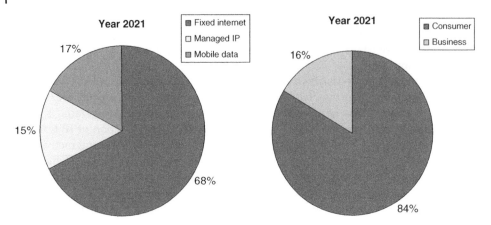

Figure 7.2 IP traffic predictions by type of broadband access and type of customers.

7.2 Generic and Specific QoS Parameters

The QoS parameters are defined to characterize the quality level of a certain aspect of a given service (so, there are many QoS parameters that refer to the same service), and ultimately influence the customer satisfaction (either directly – when the end-user is human, or indirectly when the services are based on machine-to-machine communication).

What is the purpose of defining and using the QoS parameters? The QoS parameters are defined to be used by service and network providers, to manage and improve the offering of their services. Also, the QoS parameters can be used by the customers (as end-users or as partner providers for that service) to check or ensure that they are getting the level of quality for which they have made an agreement (SLA) with the service provider (e.g. telecom operator). So, QoS parameters are also being used to support commercial customer contracts by using them for concrete SLA formulation and verification.

The main purpose of QoS parameters in an operational network is to be measured and then compared with reference values. There are two types of QoS parameters regarding measurement methods:

- *Objective QoS parameters.* They are referring to physical attributes of signals, circuits, links, network segments, and networks.
- *Subjective QoS parameters.* They are obtained via customer opinion surveys based on questionnaires which are designed to reflect the customer's satisfaction from a given service (e.g. by using mean opinion score [MOS] grades scale, from 1 – the worst to 5 – the best).

QoS parameters can bas based on primary metrics that are determined by using direct measurement (e.g. circuit noise, echo path loss, or signaling release cause) or derived from a set of primary parameters (e.g. statistical operations on measured data, applying rating factor for obtaining estimates of customer opinion such as E-model rating, decision thresholds). Particular QoS metrics can be defined to serve specific service types, such as:

- QoS metrics for different types of services (e.g. voice, video, multimedia, various data).
- QoS metrics for one service type (e.g. video), but different types or subtypes of that service (e.g. for video service that can be VoD, IPTV, Web live streaming, video telephony, video conferencing). Different types or subtypes of the same service may have different requirements regarding the same QoS parameter (e.g. video conferencing is very sensitive to one-way delay, which is not the case with video streaming or VoD).
- QoS metrics for the same service type, but different commercial basis for the same service. For example, VoD or TV streaming over Internet access service (IAS) do not have QoS guarantees end-to-end (due to network neutrality principle for the Internet network) and occasionally may enter congestion periods (e.g. when such video traffic is multiplexed with a high volume of other Internet traffic). Meanwhile, paid IPTV service (and included VoD service) has guaranteed QoS end-to-end (e.g. guaranteed bandwidth).

7.2.1 Comparable Performance Indicators

The QoS parameters and metrics must be important and understandable to the end-users, as well as comparable among different service and network providers. The agreed QoS values are also referred to comparable performance indicators [4]. Such QoS values (CPI examples are given in Table 7.1) enable customers to be aware of the actual variety of service quality to help them make informed decisions when choosing a provider (e.g. fixed or mobile telecom operator).

7.2.2 Standardized QoS Parameters

QoS parameters (or metrics, which are used interchangeably in this book) are essential for efficient management of the service quality by the service providers (with the

Table 7.1 Comparable performance indicators.

Telecommunication activities	Objective QoS indicators	Subjective QoS indicators
Service provision	Supply time for contracted service, percentage of orders completed by the date stated in the SLA	User satisfaction grade (e.g. in range 1–5)
Service reliability	Number of faults, number of customers affected with interrupted service due to the faults	User satisfaction grade (e.g. in range 1–5)
Restoration/repair	Average time period for clearing the reported faults and restoration of the interrupted service	User satisfaction grade (e.g. in range 1–5)
Billing	Number of user complaints received per number of customers that are billed (e.g. number of complaints per 1000 customers over a period of 1 mo)	User satisfaction grade (e.g. in range 1–5)
Complaint handling	Percentage of user complaints that are resolved in predefined number of working days (which is dependent upon the service)	User satisfaction grade (e.g. in range 1–5)

Table 7.2 Examples of non-utilization QoS parameters.

Generic QoS parameter type	Example generic QoS parameter from the given type	Description
Preliminary information on ICT services	Parameter 2: Pricing transparency of an ICT service	It is characterized by clarity, conciseness and unambiguity in every tariff structure for all usage conditions for every service provided by the service provider. It is measured as user opinion rating.
Contractual matters between ICT service providers and customers	Parameter 5: Integrity of contract information	True and fair view of pertinent information on supply, maintenance and cessation for a telecommunications service provided by a service provider.
Provision of services	Parameter 10: Time for provisioning	This refers to time period of time between the scheduled provisioning time and the actual provisioning time.
Service alteration	Parameter 17: Time for alteration of service	This is time elapsed from the instant alteration notification is received by the user to the instant the alteration is completed.
Technical upgrade of ICT services	Parameter 30: Outage time due to technical upgrade	Time duration when the service in part or in full is unavailable to the customer for use due to the technical upgrade process.
Documentation of services	Parameter 38: Integrity (correctness and completeness) of documentation	Correctness, completeness, and user friendliness of pertinent information associated with the use of all features of a service and its maintenance. It is measured as user opinion rating.
Technical support provided by service provider	Parameter 44: Number of attempts before successful solutions	Number of attempts before the technical request was successfully resolved.
Commercial support provided by service provider	Parameter 48: Accessibility of the commercial support	Ratio of the number of successful access attempts to the commercial support to the total number of attempts to reach this support.
Complaint management	Parameter 58: Customer perception of the complaint management	The service provider's exhibition of the combination of assurance, empathy and responsiveness in dealing with complaints from reporting to satisfactory resolution. It is measured as user opinion rating.
Repair services	Parameter 62: Successful repairs carried out within a specified period of time	Ratio of the number of repairs successfully carried out to the total number of repair requests accepted by the service provider within a specified period.
Charging and billing	Parameter 71: Accessibility of the account management	Ratio of the number of successful attempts to the total number of attempts to reach the account management.
Network/Service management by customer	Parameter 78: Outage duration	The total time a network/service management facility was not accessible to the customer during a specified reporting period.
Cessation of service	Parameter 88: Contractual cessations achieved	The ratio (%) of the number of contractual cessations requested to the total number of such requests made within a specified period.

means that are available to them). They should be simple to use and measure, to provide accurate representation of the customer perception. In some cases QoS parameters are defined with standards.

One may distinguish between QoS parameters standardized for different service types. For example, regarding the voice services there are several ITU-T standards that define voice-related QoS parameters (ITU-T G.107, G.108.2, G.109, P.862/P.863, and others). Also, the E-model (also discussed in Chapter 6) provides a scalar rating of transmission quality, but it requires knowledge of the end-to-end configuration (e.g. networks, terminals).

Another classification of QoS parameters is regarding whether they refer to service utilization or non-utilization. So, besides service utilization (or one may name them as operational QoS parameters) such as packet delay, loss, jitter, etc., there are non-utilization QoS parameters such as those standardized in ITU-T E.803 [5], which lists 88 different generic QoS parameters. Examples of non-utilization generic QoS parameters are given in Table 7.2.

Generally, QoS parameters are also called QoS metrics, QoS indicators, QoS measures, or QoS determinants. Regardless of that, all QoS parameters (including those that refer to service utilization and those that are non-utilization based) have a goal to characterize the quality level of the service being offered and the level of customer satisfaction (the latter refers to human end-users). In cases when services are used by humans as end-users, the QoS parameters represent subjective and abstract user-perceived "quality" in terms of numeric (i.e. quantified) values. The QoS parameters can be obtained with two main methods: (i) objective measurement, and/or (ii) subjective measurement (based on subjective evaluation of service quality by human end-users). One should note that different QoS parameters may also be interrelated (e.g. higher variation of the link capacity influences the delay variation for the affected services that use that link).

7.3 Interconnection and QoS

Interconnection relates to connection links between autonomous network domains (i.e. different autonomous systems), which may belong to different network providers that provide access and/or transit. The interconnection by itself is an important part of QoS, which is needed to be properly dimensioned regarding the traffic requirements for all service types that go through the interconnection points. For example, the QoS support at the interconnection for voice traffic over legacy fixed and mobile networks (i.e. TDM-based networks such as PSTN and PLMN) is based on traditional QoS voice parameters, such as limited end-to-end delays provided by guaranteed bitrates with TDM in each of the two voice directions, admission control (accepting new call when enough resources end-to-end, i.e. voice channels in both directions, can be established end-to-end, thorough all networks and interconnection points), etc.

Besides voice, the interconnection is also important for the best-effort and network neutral Internet traffic. However, the interconnection of the best effort Internet traffic is harder to design than traditional interconnection for voice traffic. That is due to the variable bitrates of the aggregate Internet traffic, which results in Internet traffic being self-similar and bursty by nature [6], and hence unpredictable at a given time (unlike legacy voice traffic). Of course, interconnection at the Internet traffic exchanges depends upon the volume of traffic that goes out of the originating network domain (the egress traffic), as well as number of interconnection links in parallel with other network domains and their capacities (for higher reliability as well as for ingress and egress Internet traffic balancing).

So, regarding the interconnection there are two main types of interconnection, and they are: (i) TDM interconnection (made initially for PSTN/ISDN interconnections, but also used for Internet traffic exchange later), and (ii) IP/Internet interconnection (over all-IP networks on both sides of the interconnection points, including the traffic originating/terminating from/to public Internet access and managed IP networks).

7.3.1 QoS Aspects for TDM Interconnection

The telecom/ICT world is fully converging onto all-IP networks, but there are still many TDM transport networks in use that were initially designed for legacy telephony (e.g. synchronous digital hierarchy). Such TDM transport networks and their interconnection points nowadays are also used for carrying IP traffic from the public Internet and managed IP networks.

The main purpose of interconnection is to ensure end-to-end network service connectivity with desired QoS, and at the same time to enable customers of interconnected network operators to establish connections with one another.

Table 7.3 defines several important QoS parameters (i.e. key performance indicators) for TDM-based points of interconnection (POIs) [7], and their possible thresholds values. Such KPIs should be measurable and realistic to ensure effective interconnection regulation and mitigation of eventual disputes.

Each interconnection between networks of different telecom operators is mainly a matter of mutual agreement between the interconnecting parties. However, the interconnection capacity (e.g. number of channels/circuits, or bitrates), its expansion, and/or modification (including closing a certain POI) need to be mutually agreed. This can be driven by market forces in a given country (e.g. the POI capacity is expanded when the traffic increases over a given threshold at POI), but in less developed ICT markets (e.g. markets with dominant market position of the incumbent telecom operator) it is desirable for national administrations to establish regulatory guidelines (which should be based on fair, reasonable, and non-discriminatory approaches) for establishing proper interconnection (e.g. referent interconnection offer) in order to facilitate effective and expeditious interconnection. In all cases the concrete details of the interconnection agreement needs to be mutually decided by both interconnecting telecom operators, but in a time limited manner. Alternatively, if both parties fail to establish an agreement within a predefined time frame then an interconnecting agreement may be prescribed by the authority (e.g. national regulatory agency), for the benefit of the customers which require affordable and available ICT services which end-to-end traffic may span across two or more networks and hence across several POIs.

Every interconnection is based on a bilateral agreement. In order to have effective interconnection, one possibility is to establish an interconnect exchange, which will provide interconnection ports to different telecom service providers and at the same time will reduce the number of needed POIs. In such case all interconnections may continue as before (between two connecting peer operators), but any required increase in the number of ports is done through the interconnect exchange.

Table 7.3 KPIs for TDM interconnection.

Key Performance Indicators (KPI)	Description
POI congestion	This refers to number of failed call requests over the POI vs. total number of calls (e.g. benchmark threshold may be set at <0.5%).
Subscriber attempts-seizure success ratio	This parameter refers to the ratio between actual seizures at the interconnection and the total number of call attempts.
Inter-operator POI efficiency	Gives comparative performance indicators of particular inter-operator POI with respect to the other inter-operator POIs.
Time-frame for activation of a new POI	The time limit for providing POI connectivity (one benchmark for this KPI is 90 d).
Time-frame for POI capacity enhancement	This is time frame for enhancement of capacity of the POI. Benchmark value is 60 d from the date of acceptance (maximum); however, the time can be shortened in the case of need.
Interconnection route utilization parameter	This parameter refers to the traffic dimensioning of the POT route. Considering the queuing theory analysis [6], the prescribed benchmark for the route utilization parameter is 70% [7].
Mean time to repair for POI ports	When there is an alternate router for traffic that traverses the POI, then the port failure time duration should not be greater than 72 h [7]; however, when the traffic is completely interrupted due to the failure then the restoration much target much shorter intervals (e.g. up to 1 h; however, the best value is zero time to repair).
Time to repair interconnection route	The interconnection route should be repaired in time shorter than 1 h.
Dual seizure ratio	This refers to simultaneous seizure of the same channel (circuit) from both ends on the different sides of the POI. The best value for this parameter is zero.
Signaling link utilization	This KPI value should not exceed 40% of handling capacity in case of a failure. To avoid signaling congestion there are required sufficient number of signaling links provided between POIs.
Unit of time measurement and recordings in CDRs (Call Data Records)	The measurement time unit for billing purposes in CDRs should be in either seconds or milliseconds, to avoid any billing disputes between telecom operators.
Clock synchronization and accuracy of switch time	The TDM interconnection is based on SDH, where SDH requires strict synchronization which is independently done in each of the interconnecting networks. Therefore is recommended synchronization with provision of reference to the clock of the interconnection provider (or other network provider).
Acceptance testing and monitoring	For establishing a new POI there is required uniform testing procedure (it should be reviewed after given time or upon a need), with a regular monitoring mechanism. It is recommended do be done by a governmental agency or accredited third-party.

7.3.2   Internet Traffic Interconnection

With the transition to all-IP networks, the interconnection points are also becoming IP-based (i.e. they are used for exchange of IP traffic). Most Internet traffic exchange is done with peering and transit agreements, as discussed in Chapter 2 of this book. The peering is an agreement in which two interconnecting ISPs carry each other's traffic. In such case there is no obligation to carry traffic from a third-party ISP. On the other hand, transit refers to business and technical arrangement in which a given ISP agrees to carry Internet traffic on behalf of another ISP. In most of the cases the transit provider carries traffic to and from every destination on the public Internet, which is typically part of the transit arrangement.

While transit is always based on certain payment for the IP traffic transport, peering can be paid or free. Both paid peering and partial or mutual transit were widely established from the privatization of the Internet around 1995. However, peering and transit business agreements are not publicly available, hence the traffic data referred to them is also unavailable.

The number of ASs, which have autonomous routing and traffic management mechanisms (the reader may refer to Chapter 2 for more details regarding ASs), continues to grow constantly (Figure 7.3). This implies that the ASs are more richly connected than in the past, while the number of ASs that IP traffic traverses end-to-end span across 3–5 ASs (as noted in Chapter 2), which directly influences the end-to-end QoS (lower number of ASs and generally network nodes end-to-end results in lower delay and delay variation as well as lower loss probability).

In practice, large network operators typically exchange Internet traffic with comparably large network operators, based typically on nondisclosure peering arrangements and direct connections between the network operators.

Regarding the Internet traffic interconnections, there is growing influence of the trends which have appeared with the appearance of the broadband and ultra-broadband access (they have increased the expectations regarding the quality experienced by the customers, including the bitrates and delay). So, the trend is that the global service providers (Google, Apple, Amazon, Facebook, etc.) have started to deploy and operate

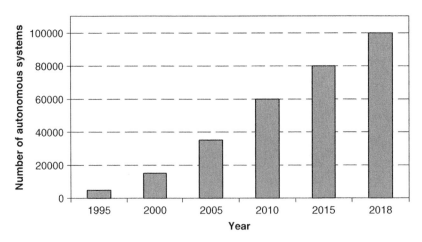

Figure 7.3 Growth of number of autonomous systems.

their own globally distributed networks and/or service centers to increase QoE (quality for their services experienced by their customers), by using CDNs also. Traffic forecasts show that more than 70% of all Internet traffic will cross CDNs by 2021 on a global scale, which is up from 52% in 2016 [3]. Additionally, large content and/or application providers (e.g. Google and Apple) operate some of the largest data transport networks in the world. However, they typically lack local Internet access networks (fixed or mobile) which are provided by national telecom operators based on national regulations in each country. However, large global service providers which build their own networks act also as network operators (i.e. network providers) and hence can negotiate their own peering and transit arrangements by using their role as network operators which is in favor of delivery of their global services with higher QoE. The high QoE is one prerequisite for the global service providers to keep and extend their customer base and to develop new innovative services provided over the global Internet network.

7.3.3 End-to-End QoS and IP Networks Interconnection

The QoS by definition is always considered end-to-end, for each service. For services used by human users the whole concept of QoS is targeted toward satisfaction of user needs and expectations from the given ICT service. So, the big picture for QoS is always end-to-end, which includes all networks and networks segments on the path as well as all interconnection points that the IP traffic traverses (i.e. IP packets from the service flows).

End-to-end communication across the Internet is also noted as UNI-to-UNI (where UNI stands for user to network interface). In such QoS big picture (with the end-to-end concept), different IP network sections may be represented as network clouds with edge routers on their borders toward other networks with which they interconnect. Also, each network cloud has many different routers and switches, where some of them are used only to route (or switch) IP packets (i.e. the data) while others have various functionalities (e.g. control and management functionalities).

When considering end-to-end IP communication, there can be one or multiple IP network sections on that path. Since the Internet is a global network, where clients can be connected through a fixed or mobile access network in one country and servers can be hosted in another country or region, IP connections typically span national or regional boundaries. So, UNI-UNI performance (i.e. end-to-end QoS) for a given service can be estimated or measured by knowing network performances of all network sections across different IP networks on the path of the data.

Generally, each network section may have different policies and different QoS approaches, so end-to-end QoS in IP-based platforms and networks is usually difficult due to their heterogeneity. So, when IP is used as networking transport technology that in practice does not guarantee that various networks are the same or compatible one to another. An extreme case of various configurations among different IP networks will be an end-to-end scenario where each of the networks has different QoS approach (e.g. ISP-A uses DiffServ, Transit provider 1 uses over-provisioning, Transit provider 2 uses multi-protocol label switching, and ISP-B uses Carrier Ethernet QoS, as it is given in Figure 7.4). When there are different QoS mechanisms in each network segment it is hard to provide guaranteed end-to-end QoS for the Internet traffic (which is by default aggregated traffic consisted of different originating Internet applications).

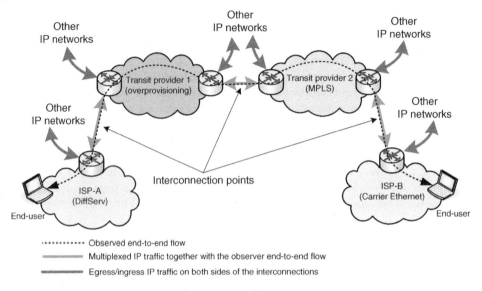

Figure 7.4 End-to-end QoS provision for Internet traffic.

End-to-end performance is influenced by the QoS solutions in different network sections on the path including fixed and/or mobile access networks. One of the main QoS parameters regarding the Internet traffic is capacity (i.e. bitrate). The access networks capacity for the IAS covers all traffic from/to end-users (most of Internet traffic is bidirectional, due to the request-response principle of both client-server and peer-to-peer Internet communication). However, some of the traffic may end (i.e. sink) on the servers which are hosted by the same telecom operator (that provides the IAS). Unlike the telephone networks where most of the traffic resides in the same network or on a national level, most of the Internet traffic goes through an IP interconnection point toward servers on a global Internet network. So, most of the Internet traffic traverses over multiple autonomous networks which may have different QoS solutions on the path between the client and the server. In such case, the interconnection points must be designed to carry all Internet traffic (egress and ingress) because in the opposite case they may become bottlenecks on the way end-to-end (Figure 7.5).

To avoid the appearance of bottlenecks at the interconnection points of a given telecom operator (i.e. ISP), they should be dimensioned to be able to carry all Internet traffic from all egress and ingress Internet traffic during peak hours (e.g. the time period during the day when volume of Internet traffic in that network is highest). So, In this regard, peering and transit agreements of a given operator (i.e. ISP) influence end-to-end QoS and may require regulatory attention to ensure end-to-end QoS performance of services.

With the transition of telecom operators' networks to all-IP networks, similar to TDM POI exchanges, IP eXchanges (IPXs) have been established in most countries. However, the IPX connects national telecom operators among themselves (typically with peering agreements) as well as connecting telecom operators with IP transit providers (e.g. with transit agreements). In many cases QoS regulation at the IPX is not present, especially because for transit traffic the IPX is also a "border" between two jurisdictions (e.g. each

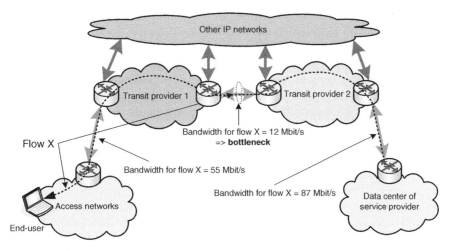

Figure 7.5 Internet interconnection points as possible bottleneck.

country has its own jurisdiction over regulation of ICT services on its own territory). However, there are regional and global agreements between national authorities (e.g. NRAs and/or ministries for ICTs) regarding the QoS because overall the IP communication (including public Internet and managed IP networks) is end-to-end and in most cases spans across multiple networks under different ICT jurisdictions. QoS parameters at the IPXs can be regulated and monitored within a country by the NRA. In general, end-to-end QoS can be achieved only through regional and global harmonization, including the IPXs and the networks attached to them.

IP traffic includes also QoS-enabled VoIP, traffic from business users (e.g. VPN services), as well as IP traffic from IAS and traffic to/from CDNs and other data centers of the global service providers. Regarding the interconnection of different types of IP traffic, one should note that IP packets are not individually transferred through the interconnection points (i.e. IPXs). Typically, a given telecom operator (as network provider) first aggregates IP packets (based on previous packet classification) into VPNs. So, there can be separate VPN tunnels created for carrier-grade VoIP, for signaling (i.e. control) traffic, for business services traffic, for network neutral Internet traffic, for traffic between CDN points of presence (CDN POPs), etc. Such VPN tunnels are established between the two end-peers which can be gateway nodes (or edge routers) placed on the border between each of the two networks that interconnect.

So, all traffic to/from telecom operator networks traverses through national and international IPXs. Therefore, the IPXs are the most appropriate place to enforce end-to-end QoS provisioning, besides the need for QoS encouragement and/or enforcement at the broadband/ultra-broadband fixed and/or mobile access networks.

7.4 KPIs for Real-Time Services

The most important QoS parameters which are important to all parties (customers, service providers, telecom operators, NRAs, and the public) are referred to as KPIs. However, they are primarily performance indicators which are derived by using network performance metrics (when KPIs are related to service utilization). Such indicators are measured on the basis on various network counters (e.g. in routers, switches, user equipment) and are essential for operation and maintenance of services and for support of the ICT service business models (e.g. selected KPIs optionally or mandatory may be specified in the contractual agreements between the services providers and the customers). Defining KPIs is targeted to help in reporting, auditing and other efforts related to QoS (and for long-term QoS sustainability of ICT services sold to the customers). However, KPIs must be used in context with a certain service or part of its delivery, otherwise they may become meaningless. Besides KPIs based on service utilization there are KPIs derived from the set of generic QoS parameters (e.g. as given in Table 7.2).

End-users require and expect the QoS, and they experience it during the utilization of the given service. Therefore, the telecom operators design the networks to meet the QoS requirements for given services used by given number of end-users. However, the planned QoS by the service providers will not always or everywhere match the delivered level of QoS. For example, the QoS delivered can be lower than predicted (or contracted in the user agreements) in certain parts of the fixed networks or in certain cells or coverage areas of the mobile networks due to various causes (bad network design, increase of traffic load without needed increase of the network capacity, network reconfiguration, etc.). So, in certain cases the QoS level experienced by the end-users may be lower than expected. Then, from the end-users' point of view, achieving satisfactory QoS depends on the appropriate combinations of the following elements:

- *Network performance.* Capacity of network links in the access, core and transit networks, as well as capabilities of the networks nodes (processing capabilities, available QoS mechanisms and policies, etc.).
- *Terminal performance.* Different terminals may have different capabilities and that has influence on the QoS metrics and on the QoE. Typical example are smartphones, which develop with fast pace in each mobile generation. So, different smartphones (i.e. mobile devices) in the same network may support different network interfaces (e.g. some smartphones may support 3G/3.5G only, while others may have support for 3G, 4G, and 5G). Also, different mobile devices may have different processing power, memory capacity as well as support for different modulation and coding schemes and different number of antennas (e.g. for MIMO communication over the radio access network). In such case, mobile terminals with better capabilities will experience better QoS (higher bitrate, lower delay including network delay and application delay aspects, etc.) and hence provide higher QoE to mobile users with high end mobile equipment than the QoE experienced by mobile users with older or less capable mobile devices in the same mobile network. Hence, mobile operators need to organize provision of lower cost mobile devices with necessary capabilities in order to offer a high quality experience. Of course, the price of services matters, so higher prices may provide better QoS (e.g. higher bitrates in the same access network) than the same service obtained for a lower price.
- *Retail channels.* There are different channels to sell services to end-users, which may span from service providers' selling points to digital retail channels (e.g. buying or upgrading/downgrading a service online or via the end-user equipment).
- *Customer care.* This is the part of the service providers which is typically combination of human support for service management and configuration/reconfiguration

and automatic voice machines (or Web-based service assistants which are becoming more common with the development of the Artificial Intelligence use in the ICTs).

7.4.1 KPIs for Voice Over LTE Services

The legacy telecommunication/ICT service which has transferred to VoIP (with the NGN deployments by telecom operators, as replacement of PSTN/ISDN) is the main and most known real-time service that includes human users on both sides of the connection. Operator grade (i.e. carrier grade) VoIP is provided with QoS guarantees end-to-end in both fixed and mobile IP access networks. Typically, the end-users are not aware of transition of the digital telephony to VoIP with QoS support, as they were not aware of transition of analogue to digital telephony in the last decades of the twentieth century. The VoIP provided in mobile environments is more prone to QoS degradations than VoIP provided by telecom operators via fixed access networks (e.g. optical access such as FTTH), due to lower predictability of the radio interface. Hence, the KPIs for voice services in mobile networks cover all KPIs for fixed VoIP as well as additional ones that refer to mobile networks only. The first standardized VoIP in mobile networks is Voice over LTE (VoLTE), defined for 4G mobile networks (i.e. LTE/LTE-Advanced/LTE-Advanced-Pro), which started with deployments in the late 2010s (toward 2020) and will continue in the 2020s (i.e. in the 5G era). However, the transition of PSTN/ISDN to NGN-based VoIP first happened in fixed access networks due to mobility management issues in mobile networks (that is, the handovers, which require strict synchronization to accommodate mobile voice services, and that is more complex in all-IP environment than in TDM, e.g. TDM by default has synchronization end-to-end as well as there is no buffering of data in the network nodes – something that adds additional packet delay and jitter to the VoIP).

Considering VoLTE, the relevant KPIs are specified in ITU-T G.1028 [8]. The VoLTE is provided over a managed IP network, so it is also referred to as a "managed" VoIP service (which is provided with its prioritization over the other IP traffic, such as Internet traffic). That is in contrast to OTT VoIP applications/services (Skype, Viber, WhatsApp, etc.) which use the public Internet access that is network neutral and it has no prioritization over the rest of the Internet traffic (video, Web, etc.), although telecom operators may apply traffic management techniques to the aggregate Internet traffic (with the use of DPI) with the aim of offering appropriate QoS to different types of public Internet traffic. Additionally, VoLTE uses the NGN control functionalities based on SIP and Diameter [9], used in IMS (IP multimedia subsystem) in NGN deployments in fixed and mobile networks.

In 4G mobile networks (LTE/LTE-Advanced/LTE-Advanced-Pro), the quality classification identifiers (QCIs) that can be used for VoLTE services are given in Table 7.4. However, the KPI target values are dependent whether the VoIP is established between customers in two LTE access networks, between LTE and PSTN, or between LTE and 3G mobile network. KPI targets for VoLTE services for various scenarios (LTE – LTE, LTE – 3G, or LTE – PSTN voice calls) are given in Table 7.5. As one may note, the delay threshold is set to 400 ms (which is the maximum value for terrestrial voice connections, according to ITU-T G.114 [1]) because the PSTN/PLMN delay threshold at 150 ms is impossible in IP-based mobile networks for voice services such as VoLTE due to large delay budget due to packetization of voice (waiting more voice samples to be gathered in

Table 7.4 LTE QCI for VoLTE.

QCI	Resource type	Priority level	Packet delay (ms)	Packet error loss rate	Service
1	GBR	2	100	10^{-2}	Conversational voice
5	Non-GBR	1	100	10^{-6}	IMS signaling

GBR, Guaranteed Bit Rate; QCI, Quality Classification Identifier; IMS, IP Multimedia Subsystem.

Table 7.5 KPI targets for Voice over LTE (VoLTE) services.

KPI	VoLTE – VoLTE in the same LTE network	VoLTE – VoLTE between inter-connected LTE networks	VoLTE – 3G	VoLTE – PSTN
Registration success rate	99.9%	99.9%	99.9%	99.9%
Service availability	99%	99%	98%	99%
Post dialing Delay (PDD)	LTE-LTE: 3.5 s CSFB: 6 s	LTE-LTE: 4 s CSFB: 6 s	4.5 s CSFB: not defined	4 s CSFB: not defined
Voice QoE (MOS grade)	4	4 (HD voice) 2.8 (otherwise)	3.8 (HD voice) 2.8 (otherwise)	3.1
Mouth-to-ear delay	400 ms (Note 1)	400 ms (Note 1)	400 ms (Note 1)	400 ms (Note 1)
Call drop rate	2%	2%	3%	2%

CSFB, Circuit Switched Fall Back; MOS, Mean Opinion Score.
Note 1: For the best experience for voice the delay is targeted to a preferred maximum value at 150 ms [1]; however, it is impossible to reach for VoLTE although some network operators are able to provide national VoLTE calls with delays below 250 ms. In all terrestrial voice delivery cases the one-way delay should be kept below maximum value at 400 ms.

single IP payload for higher utilization of resources), buffering of voice packets, rerouting of voice packets at handovers from old to new (target) base station, as well as playout buffering at the receiving side (due to the jitter in IP networks). The service availability is in the range of 99% due to variable conditions in the mobile access networks as well as mobility of the users (the exact network capacity cannot be predicted with the same accuracy as in fixed networks).

7.4.2 KPIs for IPTV and Video Services

IPTV as another real-time service provided over managed IP networks (with guaranteed QoS end-to-end), has also requirements to provide satisfactory QoE. While VoIP is less sensitive to packet loss rate and more sensitive to delay and jitter, IPTV is less sensitive to delay and jitter (because it is unidirectional stream, so playout buffering solves all issues regarding these two QoS parameters) and more sensitive to packet loss rate (due to source coding of the video, where data loss in a given video frame influences other frames in the same MPEG Group of Pictures). The main KPI targets for IPTV service with SD and HD [10] are given in Table 6.3 (in Chapter 6). The same KPIs and targets refer also to VoD services that are typically provided through the same system that is used for IPTV delivery.

7.5 KPIs for Data Services and VPNs

Most of the broadband and ultra-broadband IP traffic is coming from data (i.e. OTT) services provided over the public Internet network. With the spread of mobile broadband networks, the same data services used in fixed access networks are transiting to the mobile data service segment, which is constantly increasing with deployments 4G and 5G technologies. However, in all cases the data traffic is carried through core and transport IP networks by using packet tunneling approaches such as VPNs tunnels. Due to their interrelation, this section covers KPIs for data services including mobile ones, as well as KPIs for VPN services.

7.5.1 KPIs for Data Services

The data services are provided by using the network neutrality principle through the public IAS (the reader may refer to Section 6.6 in Chapter 6). So, data services refer to all data that is requested by IAS subscribers/end-users from various content providers and application/service providers globally on the public Internet which are referred also as OTT services (they include various websites, YouTube, Facebook, Twitter, BitTorrent, OTT VoIP services such as Skype or Viber or WhatsApp, OTT cloud services such as Amazon clouds or Google docs, OTT video streaming such as YouTube). The most important QoS parameters for data services are referred here as KPIs for data.

Main KPIs for data services together with the most popular data service/application types are given in Table 7.6 [11]. The table gives the importance (or in other words, relevance) of a given KPI for given data application/service type. The importance of KPI goes from very low (not so important) to high (very important). The wording "application/service" is used, where typically application is a standalone implementation of a client application and a server application which communicate over the public Internet (either they are standardized or proprietary applications), while service refers to the applications which also include control and signaling data exchange (e.g. with different servers or network nodes) for the provision of a given service (e.g. each OTT VoIP is a service because there is mandatory required control data exchange, at least transfer of user's IP address to bind it with his/her username for delivery of terminating calls to that user by other users of the same OTT VoIP service).

Considering the importance of the KPIs for data applications/service, high importance of packet loss means having zero losses, which are provided by using TCP/IP protocol stack where TCP uses retransmissions to recover from any packet losses or errors. So, with TCP on the transport layer given service/application receives error-free data in both directions. In the cases where the delay KPI has high importance (e.g. for gaming services) then application may provide retransmissions when needed and utilize

Table 7.6 Popular types of OTT applications and quality parameters.

Data application/ service type	Importance of KPI for data applications/services (Very low, Low, Medium, or High)				
	Data transmission speed (bit/s)		Delay	Delay variation	Packet loss
	Downlink	Uplink			
Browsing (text)	Medium	Very low	Medium	Very low	High
Browsing (media)	High	Very low	Medium	Low	High
File download	High	Very low	Low	Very low	High
File upload	Very low	High	Low	Very low	High
Transactions	Very low	Very low	Medium	Very low	High
Streaming media	High	Very low	Low	Very low	Low
Voice over IP (VoIP)	Low	Low	High	High	Low
Video telephony	High	High	High	High	High
Gaming	Low	Low	High	High	High

UDP on the transport layer. For real-time services such as VoIP or streaming media (e.g. video streaming with accompanied audio streams), the bitrate in downstream is the most important KPI because video streaming requires high bitrates (which is dependent upon the picture resolution used for the given video stream). Delay and delay variation (i.e. jitter) are less important KPIs for streaming media because they can be compensated (and with that eliminated) by the playout buffering at the receiving media player (one should note that the receiving device that plays the video should have enough capabilities to play the video stream on the screen, such as processing power for real-time decoding of media streams and enough memory for buffering the media in order to have smooth reproduction without the picture freezing or lost media data).

What can be defined as the main KPIs for data services? The availability of broadband IAS is the most influential KPI for data services (that is percentage of time the broadband/ultra-broadband access to Internet is available to the user). Although the availability of PSTN over fixed networks has reached availability of 99.999% of the time, availability of services in mobile networks for voice is in the range 98–99%, and the availability of data services tends to be in the same ranges. However, availability of Internet connectivity depends also on the functionalities and performances of DHCP and DNS functions, which are prerequisites to access any application/content transferred over the Internet (e.g. DHCP is essential for allocation of an IP address to the network interface of the user device, while DNS is crucial for mapping the domain names used by end-users into IP addresses which are used for delivery of the IP packets from a given source to a given destination). Assuming that the IAS is available and considering the fact that 2/3 of the total Internet traffic is video (i.e. media streaming) and its most important KPI is data transmission speed, the straightforward conclusion is that the most important KPI for data services is data transmission speed (i.e. bitrates in downlink and uplink directions). One should note that the delay is also influenced by the bitrates. For example, the transmission delay for transfer of 2 MB with 160 Kbit/s will be $2 \times 8 \times 10^6$ bits/160×10^3 bit/s $= 100$ seconds, while for bitrate of 160 Mbit/s the transmission time delay for the same 2 MB will be 0.1 seconds. So, the KPIs are not independent from each other; however, the bitrate (i.e. data rate) is the most important one (after the availability of broadband IAS). Therefore deployment of ultra-broadband access and transport networks for data services (i.e. for IAS) has crucial importance for further development of data services and for development of digital society based on various data services (provide over IAS).

Considering the above discussion, when the IAS is available, then the most important KPIs will be:

- *Data transmission rate* (i.e. bandwidth). That is the maximum number of bits that a given transmission path can carry. Each end-to-end path consists of many links which may have different bandwidth and different capacity for the given data service traversing that link (flow bitrate is normally below the link capacity). Then, the link with the lowest transmission data rate on the path is called bottleneck link and it defines the end-to-end transmission rate for the given service stream (e.g. media stream).
- *Propagation delay.* This refers to the time that an IP packet requires to be sent by the sender over a given link and received by the node on the other end of the link. Due to maximum speed limit (that is speed of light, approximately 3×10^8 m s) longer link results in bigger propagation delay. That is one of the main reasons for use of CDNs for OTT data services (to limit the propagation delay between the user clients and the servers).
- *Queuing delay.* This refers to the time that packet waits in a queue in network nodes (e.g. routers, switches) before it is transmitted over the next link (i.e. next hop). At high traffic intensity the queue length is longer and hence the queuing delay and delay variation (i.e. jitter) have bigger values. Both the average delay and the jitter matter for real-time conversational services (the highest importance) and for interactive services (e.g. Web browsing) with medium to low importance.
- *Packet loss.* This parameter refers to the probability that an IP packet never reaches its destination. There are several potential reasons for packet loss to occur which also include the transmission errors, but one should note that they appear with very low probability in fiber fixed access and transport networks (e.g. 10^{-8}–10^{-9} error probability, which results in the time period between consecutive loss periods being measured in hours). In practice, in fixed networks IP packets are mainly lost when the queues are full due to the number of packets waiting for transmission is becoming greater than the available storage capacity in the buffers. Additionally, TCP always results in losses when TCP congestion window reaches the bottleneck capacity for the given TCP connection (however, all lost packets due to any causes are retransmitted by using TCP retransmissions). One should note that the error probability in mobile access networks is much higher (e.g. 10^{-2}–10^{-3}) in the macrocellular environment, which may influence the data transmission rate for TCP-based services (e.g. Web services) or real-time services (e.g. VoIP, video streaming) which use UDP/IP. That may result in lower bit rates at cell edges in mobile networks, so mobile networks are transiting to micro and pico cell environment to improve the QoS by lowering both the packet loss and the packet delay in the radio access networks (due to shorter distances between the mobile terminal and the serving base station, which gives less attenuation of radio signals and lower delay due to shorter path).

7.5.2 KPIs for VPN Services

The VPNs are typically used in core and transport networks. The most common use of VPN is in MPLS-based core and transport networks. Also, VPNs can be used in Carrier Ethernet, which is becoming widely spread also in metro, wide area networks, as well as on the regional and global levels. There are layer 2 and layer 3 VPNs, depending on whether the tunneling is done on the layer 2 Ethernet frame (the Ethernet, including WiFi as its wireless "extension," has become the unified LAN technology, which is spreading its use to larger and global networks) or the layer 3 IP packet (all packet networks nowadays are IP-based, hence on the network layer, i.e. layer 3, is always the Internet Protocol).

When Ethernet is LAN technology (as it is in most of scenarios) then QoS management is based on class-of-service (CoS) use. The CoS provides layer 2 differentiation between different types of services such as best effort IAS, business services, voice services (i.e. VoIP with QoS support), video services with QoS support (e.g. IPTV, VoD with QoS). Each of the service types typically uses separate VPNs for its transfer in core and transport networks. One should note that 3GPP mobile networks use GTP (GPRS Tunneling Protocol), which is standardized by 3GPP (unlike other tunneling protocols, such as IP-in-IP, which are standardized by the IETF) [12].

There are mainly two models (Figure 7.6) for provision of VPN services regarding QoS [13]:

- *VPN pipe model.* In this case the VPN service provider needs to provide adequate service bandwidth between any pair of VPN endpoints (e.g. routers, switches), along the path of that VPN, which is referred to as a "pipe." In this model, if the customer for VPN services wants to reach N destinations, then QoS requirements should be specified for N paths (i.e. N pipes). The pipes can be established in the network by using MPLS (e.g. though label switched paths). The QoS requirements for pipes provide isolation between the QoS metrics of each pipe.
- *VPN hose model.* In this case are specified QoS requirements per VPN endpoint and not for each pair of VPN endpoints. This model is targeted for multi-point to multi-point VPN services. In such case, unlike targeting the SLA on point-to-point basis, i.e. for VPN pipe basis, the SLA is defined in terms of "hose" between each site (on customer side) and the provider-edge (on VPN provider network side). Each such "hose" is defined with two KPIs:
  - o *Ingress committed bitrate.* The agreed VPN hose bitrate from the customer toward the provider network.
  - o *Egress committed bitrate.* The agreed VPN hose bitrate from the provider toward the customer.
- For example, VPN pipes are replacement (in IP transport networks environments) of leased lines (in circuit switching transport networks). Typical implementation of VPN is any-to-any connectivity using IP or MPLS technology based on the Border Gateway Protocol (BGP)/MPLS VPNs [11]. Overall, the main QoS parameters (i.e. KPIs) for VPN services include the following:
- *Packet delay, jitter, and packet loss.* These are not specific VPN QoS metrics, but refer to all IP-based services. However, their importance is related to the type of traffic carried in the VPN. So, these KPI reflects in fact the KPIs for the service type carried within the VPN (voice, IPTV streams, Internet data, etc.).

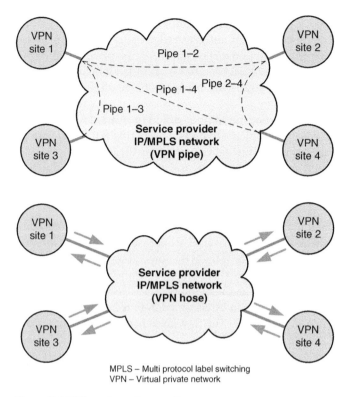

MPLS – Multi protocol label switching
VPN – Virtual private network

Figure 7.6 VPN services: pipes and hoses.

- *VPN sustainable/committed ingress/egress bitrate.* That is the average guaranteed bitrate for a given VPN pipe or VPN hose. These parameters are dependent upon the SLA for a given VPN or traffic management decision as well as the number of network providers that particular VPN traverses.
- *VPN service availability.* The percentage of time when the VPN is available is comparable with the transport network's availability (e.g. 99 999% of time).

Regarding KPIs for bitrates used for the VPN, there are different constraints in pipe and hose models. For example, a full mesh of N sites that should be connected requires N × (N−1)/2 VPN pipes. But, VPN pipe model can be inefficient due to low resource utilization in the network. For example, if site-1 provisions 100 Mbit/s to each of the sites 2, 3, and 4 (that is 300 Mbit/s in total), and there is low traffic to site 2 that unused capacity cannot be allocated for carrying traffic between site 1 on one side and sites 3 and 4 on the other (so, 100 Mbit/s network capacity will be idle in this case). However, VPN pipe model can be used for business services where both endpoints of the pipe are in the network of the VPN service provider, while in cases where VPN endpoints belong do different networks (of different providers) then hose model can be more convenient.

7.5.3 KPIs for Mobile Services

Mobile networks have different characteristics than fixed access networks. They differ in two main aspects:

- Mobile networks have radio-based access which cannot be completely isolated and controlled, so there is interference between different radio signal in the same spectrum, higher attenuation of signals due to many obstacles on the path, signal reflections and scattering, multipath propagation (due to signal reflections) which result in time dispersion of signals, so the bitrates are harder to be guaranteed in mobile networks than in fixed networks.
- *Mobility of the users.* There is different mobility of different users, from nomadic mobile users (at home, office, or public place) to mobile users that are in transportations vehicles (e.g. cars, buses, trains). The higher mobility is the lower is the bitrate over the same frequency spectrum resources due to usage of more robust modulation and coding schemes in such cases (which transfer less bits per symbol, hence result in lower bitrates).

To combat radio interface and mobility of users (for higher QoS and QoE), one solution is to use smaller cells in the mobile access network (that results in smaller attenuation of signals) which may support very high bitrates in the radio access part and provide higher QoE for mobile users (due to better QoS, that is, higher bitrates and smaller delays in smaller cells, especially for users with nomadic mobility).

The QoS parameters in mobile networks can be classified in four layers, according to the ITU-T E.804 [14], as shown in Figure 7.7. The first layer is called "network availability." It defines QoS from the viewpoint of the service provider. The second layer is "network access," which provides the basic requirements for all other service-related QoS parameters. The third layer in this model contains the other three QoS aspects from the service point of view: service access, service integrity, and service retainability. Finally, the fourth layer defines service-specific QoS aspects and parameters, which differ from one mobile service to another.

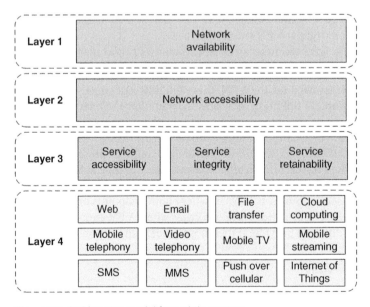

Figure 7.7 QoS layering model for mobile services.

Regarding their implementation mobile services can be grouped into the following two main groups:

- *Direct mobile services.* These services are provided over mobile IP-based networks (e.g. 2.5G/3G/4G/5G) by using the IP client-server or peer-to-peer principles. Such examples are Web browsing, email, FTP, mobile broadcast, push-to-talk over cellular (PoC), streaming video, mobile VoIP, video telephony, group/conference calls, Web radio, and similar others.
- *Store and forward mobile services.* This group refers to every non-real-time service (sometimes called "background"), in which two clients are assumed and one or more servers in the middle in the following manner: A-party uploads a message to a server; the server optionally forwards the message to another server; the server notifies (optionally) the B-party that a new message is available; and the B-party downloads the message. The QoS parameters for this type of mobile service can be divided into several groups, which include generic parameters (e.g. message upload session failure ratio (%)), email parameters (for SMTP, POP3, IMAP4), MMS parameters, SMS, and SDS parameters. Overall, the group of store-and-forward services includes various types of messaging services which have less stringent QoS requirements regarding delay and jitter (on the other hand, they require lossless transmission).

Besides service-specific QoS parameters, there are service-independent QoS parameters which include mobile technology dependent parameters (e.g. for radio access network), or general IAS availability parameters (e.g. DHCP or DNS parameters). Service independent QoS parameters [14], include the following: radio network unavailability, network non-accessibility, attach failure ratio, attach set-up time, Packet Data Protocol (PDP) context (in 3G) or Evolved Packet System (EPS) bearer (in 4G) activation failure ratio, PDP context/EPS bearer activation time, PDP context/EPS bearer cut-off ratio, data call access failure ratio, data call access time, DNS host name resolution failure ratio, as well as DNS hostname resolution time [14].

One of the most important service-independent KPIs for data services are referred to DNS availability and DNS response time, which refer to almost all Internet-based services (e.g. Web-based services, email), including mobile and fixed networks. Why? Because the DNS connects the two name spaces in the Internet, IP addresses (IPv4 and IPv6) and domain names. Without proper functioning of the DNS (as well as DHCP because the DHCP sets the IP addresses of the DNS servers at the end-user hosts), even in the case there is available IAS (i.e. through an access network) there would be in practice no access to data services over the IAS, either through fixed or mobile access.

Considering the end-user's point of view, there are no differences between mobile broadband access and fixed broadband Internet access (i.e. access to data services). So, the KPIs for data services are those given in Table 7.6.

5G mobile networks have novelties with the use of network virtualization and definition of network slices. KPIs in 5G mobile networks are directly related to the three main 5G network slices: eMBB, URLLC, and mMTC. These also define the three main service types regarding QoS requirements [15]. In that manner the KPIs are targeted separately to each of the service types. However, although URLLC and mMTC are expected to lead further the innovations in mobile services, one may expect IP traffic growth in mobile networks in the 5G era (i.e. the period 2020–2030 and beyond) to continue to be Internet traffic (that is, traffic from all OTT applications and services) via ultra-broadband mobile access (i.e. eMBB). URLLC and mMTC are targeted for IoT services, i.e. making different things smart via development of various smart services (e.g. smart homes, smart cities).

7.6 KPIs for Smart Sustainable Cities

With increasing use of the ICTs in all segments of living, one of the most remarkable use cases of smart services is through the development of smart sustainable cities (SSC) services. According to the ITU-T definition [16], an SSC is an innovative city that uses ICTs as well as other means to improve the quality of life of citizens, efficiency of urban operation and services, and competitiveness, while ensuring that it meets the needs of present and future generations with respect to economic, social and environmental aspects.

The ICT architecture for the SSC aims to provide certain services for the citizens, which include intelligent transportation services, e-government, e-business, smart health services, smart energy services, smart water supply services, etc. For provision of such services, SSC is based on multi-tier architecture from the communication point of view, as shown in Figure 7.8.

The ICT architecture for SSC consists of the following layers (bottom – up): sensing layer (it connects various terminals, such as sensors, RFIDs, cameras, etc.), network layer (it consists of different fixed and mobile networks), data and support layer (it stores the collected data from the sensing layer into clouds), and application layer (it includes applications for the SSC services, such as smart healthcare, smart governance, public safety, environmental protection, etc.). Overall, SSC architecture is integrating many emerging ICT technologies based on IoT services including massive and critical ones. The main target is to provide a sustainable quality living environment for the citizens based on use of the ICTs for provision of "digitalized" city services.

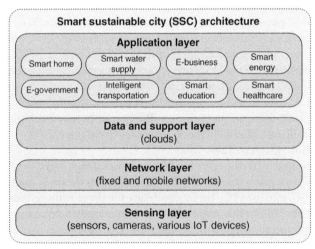

IoT – Internet of Things

Figure 7.8 Smart sustainable cities (SSC) architecture.

Regarding the scope of KPIs for the SSC [17], they can be categorized in six dimensions: ICT, environmental sustainability, productivity, quality of life, equity and social inclusion, and physical infrastructure. Then, each dimension has its subdimensions. For example, subdimensions of ICT are network and access, services and information platform, information privacy and security, and electromagnetic field. For the dimension quality of life, the subdimensions are education, health, safety/security, and convenience and comfort. If we look (as an example) into the subdimension access and networks (from the dimension ICT) [18], the KPIs are as follows:

- *Availability of computers or similar devices.* Refers to proportion of households with at least one computer or similar device (e.g. smartphones).
- *Availability of Internet access in households.* Refer to proportion of households with Internet access via a fixed or mobile network.
- *Availability of fixed broadband subscriptions.* Refers to fixed broadband subscriptions penetration (number of broadband subscriptions per 100 inhabitants).
- *Availability of wireless broadband subscriptions.* Refers to wireless-broadband penetration including mobile and wireless terrestrial networks and satellite networks (number of broadband subscriptions per 100 inhabitants).

Regarding the practical implementations, SSC services are all based on the ICT dimension, including fixed and mobile/wireless access networks as well as deployed SSC services.

Considering the focus of 5G mobile network to network slicing, one may expect wider deployment of SSC services with the 5G deployments in the 2020s and beyond. However, different SSC services may use different network slices which are dependent upon their QoS requirements as well as end-points (e.g. human users or machines/devices) [15]. eMBB is convenient for using tourist guide SSC service based on video services with different video resolutions (e.g. HD, Full HD, 4K) and augmented or virtual reality. Similarly, V2I (vehicle to infrastructure) infotainment services based on video streaming to passengers or vice versa (by using cameras on vehicles) may utilize eMBB in 5G (or mobile broadband in 4G or 3G). V2I assisted driving has strict QoS requirements such as guaranteed bitrate, very low delay, and very high reliability. This service belongs to critical IoS services, for which are the most convenient URLLC network slices in 5G. However, as one may assume, most of the SSC services will be based on non-real-time machine type communications, which will include many devices per square kilometer (massive deployments) connected via the mobile network such as 5G (Table 7.7). The mMTC services require minimum connectivity bitrates, have higher tolerance to delays (e.g. delays can be in range of seconds), and are served in best-effort manner through the mMTC network slices.

Overall, the SSC services can be provided by using different standardized fixed and mobile access technologies. Having in mind that SSC will typically use mMTC communication (for most of SSC services) the mobile networks are primarily in the focus for such use. However, WiFi access networks (best-effort or IEEE 802.11e for QoS support) are also convenient for mMTC services or generally for best-effort SSC services. In most of the cases backhaul and sometimes fronthaul networks for SSC services will be based on Ethernet or Carrier Ethernet technology, depending on the QoS requirements of the given service. So, all existing ICT technologies can be used in different parts for the SSC

Table 7.7 Performance requirements for future SSC services over 5G mobile networks.

5G network slice	SSC service	Minimum required bitrate	End-to-end delay (ms)	QoS aspects
eMBB	Video applications (Tourist guide, city Virtual Reality, etc.)	10 Mbit/s downlink or uplink	<100	Best effort reliability
	Vehicle to Infrastructure (V2I) infotainment	10 Mbit/s downlink or uplink	<100	Best effort reliability
URLLC	V2I assisted driving	0.5 Mbit/s downlink and uplink	<10	High reliability
mMTC	V2I – driver information service	0.5 Mbit/s downlink/uplink	<100	Best effort reliability
	Environmental monitors, waste management, and ITS	Minimum connectivity uplink (e.g. 100 s of Kbit/s)	In range of seconds	Best effort reliability
	Smart meters – sensor data, meter readings, individual device consumption	Minimum connectivity uplink (e.g. 100 s of Kbit/s)	In range of seconds	Best effort reliability
	Smart grid sensor data and actuator commands	Minimum connectivity uplink (e.g. 100 s of Kbit/s)	In range of seconds	Best effort reliability
	Logistics sensor data for tracking goods	Minimum connectivity uplink (e.g. 100 s of Kbit/s)	In range of seconds	Best effort reliability

eMBB, enhanced Mobile Broadband; ITS, Intelligent Transportation System; mMTC, massive Machine Type Communication; URLLC, Ultra-Reliable Low-Latency Communication.

services, and QoS should be applied accordingly in each of them (either that is Ethernet, WiFi, or 3GPP mobile networks such as 4G or 5G).

7.7 QoS and QoE Assessment Methodologies

The main objective of the quality measurements (i.e. QoS/QoE) is to provide information to customers, potential customers, and service providers. Quality measurements may be used for generation of reports to customers about the delivered services, or for attracting new customers by focusing on service quality. For service providers and third-parties the quality measurements can be used for services design and their offering, for troubleshooting, for marketing purposes, as well as capacity planning and further service development.

7.7.1 QoS/QoE Measurement Systems

The QoS measurement system aims to provide QoS measurements which can be compared to given targets and used for network/service monitoring or design/optimization purposes. Such a system should be easily understood by service providers and customers, and should be useful for QoS monitoring of services used by customers. Also,

QoS measurement systems provide input data for network dimensioning for capacity enhancements and its optimization. With the aim of making measurement results comparable, the measurements should be independently repeatable by the same party or another party. Also, given QoS measurement system should be widely applicable for measurements of different traffic types, different link-layer technologies (e.g. Ethernet, WiFi, mobile access networks) and any IP network. For the purposes of QoS enforcement the QoS measurement system must be reliable and comparable with the SLA targets.

There are certain requirements that quality measurement systems should comply with [19], which include the following:

- *Accuracy.* Measurement results should be reliable, reproducible and consistent over time, including both QoS and QoE indicators.
- *Comparability.* This refers to having comparable KPIs (which may be enforced by the NRA), so the services from different service providers or from the same service provider in different regions or different customer packages can be compared.
- *Trustworthiness.* The measurement system components should be reliable over time, based on agreed procedures (e.g. between service providers and NRA, or according to regional QoS guidelines [19]) and also have to be protected against security attacks which may jeopardize the integrity of the results (e.g. during their storage or transmission from a probe station to a centralized database).
- *Openness.* This refers to availability of details about the measurement (e.g. open source code). However, the quality measurement system should be preferably based on specifications, standards, recommendations as well as best practices.
- *Future-proof.* The measurement system should have flexibility, be extensible and scalable (e.g. regarding introduction of new test scenarios). In this manner the system will also be more cost-effective.

The QoS/QoE measurement systems can be based on either software applications (that can be downloaded and installed on end-users' devices such as computers and smartphones) or dedicated probe stations/nodes which are deployed on the side of the network provider or the customer's network. In all cases the QoS measurement system can be based on two following measurement methods [20, 21]:

- *Passive measurement method.* This type of measurements is based on observation of user data packets on a network link in a non-intrusive manner (e.g. monitoring/sniffing regular IP packets that carry user data, control or management information, and their statistical processing afterwards for calculation of various QoS parameters such as average or maximum bitrate, packet delay, or jitter, etc., which can be per type of service, per flow, and so forth). However, the drawback is that this may result in significant processing load of network nodes or requirement for large storage (on disks) of the collected data, which increases significantly with availability of ultra-broadband access. Therefore, purely passive measurement method is unlikely to be used for estimation of QoS metrics such as KPIs in nowadays broadband and ultra-broadband networks. However, passive measurements may be also based on test packets (e.g. marked IP packets, in the IP payload) sent from the QoS measurement systems, which are passively monitored by the network nodes and further used for statistical calculation of the QoS parameters such as delay,

delay variation (i.e. jitter), and packet loss. Passive methods are also convenient for troubleshooting purposes.
- *Active measurement method.* In this case probe stations are deployed which can be installed software agents or deployed hardware devices (small computers or network devices with installed measurement software) on network elements on the operator's site or at users' premises (e.g. connected to home gateways). The traffic sources and sinks of the probe packets can be separate measurement devices (probe stations), or network nodes (routers or switches) that are enabled for measurement tasks or network elements that simultaneously carry user data traffic and perform QoS measurements.

Most of the QoS measurements by telecom operators are based on the active measurement method. To enable measurement of QoS parameters across multiple provider networks, each provider can agree to use a common QoS measurement probes and make probe nodes to be available to other providers, or each network provider can use own measurement methods with its own probe devices and based on that estimate the end-to-end performance (e.g. due to absence of QoS measurements data from other network providers along the path of the traffic).

There is a need to carry out the QoS measurements in such a way that they can be proven to reflect a service provider's "realistic" performance. Truly representative QoS measurements are not only critical for regulatory purposes (e.g. for QoS enforcement), but they are also important for all end-users (e.g. consumers of ICT/telecom services, service or network providers) to have confidence in the accuracy of performed QoS or QoE measurements. For example, where packet samples are used instead of real user traffic, the QoS measurements should provide a precision of ±10 at a maximum, with a confidence level of 95% [21]. This confidence interval requires 96 sample measurements to be carried out. However, if there is required precision level of $\pm3\%$ for the same confidence level of 95% there will be required 1067 samples from the measurement of the same parameter under the same conditions (Table 7.8).

7.7.2 Basic Network Model for Measurements

In an ideal scenario QoS measurements would be made between the same endpoints for each type of customer's traffic. These endpoints are the customer's terminal (TE – terminal endpoint), customer equipment such as edge router (CE – customer edge), or provider edge router (PE – provider equipment) [22].

Generally, the IP network is partitioned in segments, which include access network and core networks (ingress and egress segments) and transit networks. Hence, a given telecom operator (i.e. provider) can be access provider and/or transit network provider. The point of interconnection (POI) of access network and regional network (e.g. core

Table 7.8 Measurement sample sizes and confidence intervals.

Confidence level		95%		
Degree of variability		0.5		
Precision level	$\pm3\%$	$\pm5\%$	$\pm7\%$	$\pm10\%$
Required number of samples	1067	384	196	96

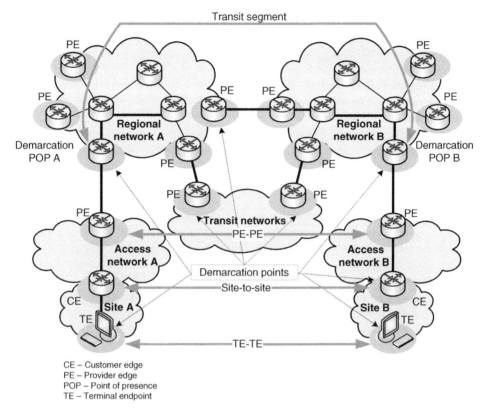

Figure 7.9 Network models for QoS measurements.

network of the telecom operator) is referred to as demarcation POP. The providers offer services between different endpoints which include the following cases:

- *Edge-edge services.* In this case the demarcation points (DPs) are PEs (provider edge routers).
- *Site-site services.* In this case the service extends to the customer edge point (managed CEs).
- *TE-TE services.* In this case the services extend to the customers' terminals.

The types of offered services by providers in fact define the positioning of the measurement points, based on the network model for QoS measurements shown in Figure 7.9. In this model network is partitioned on network segments, where each segment is being monitored independently according to the three service cases (edge-edge, site-site, and TE-TE). Typically, telecom operators are interconnected with peering over some transit networks. A given service provider can act as an access provider for some traffic and as a transit provider for some traffic, or can have both roles for a given type of traffic.

7.7.3 Quality Assessment Methodologies

The quality can be assessed with parameters that can be measured on network segments or end-to-end and compared with target reference values (for those QoS parameters), or they can be assessed on the basis of the users' experience from service usage (e.g. it refers to services used by human users). Hence, there are two main methodologies for QoS/QoE assessment:

- *Subjective assessment method.* In this case the quality of audio and video media that are consumed by human users is evaluated in subjective terms. However, "real" subjective quality assessment, where human subjects evaluate the quality of various services under different testing conditions, can be time-consuming and expensive. Also, for such testing scenarios there are required professional audio-visual devices or soundproof chambers. Therefore objective quality assessment is preferred as a means for estimating subjective quality solely from performed objective quality measurements. As already discussed, MOS is used as a measurement grade (from 1 to 5) of the average subjective opinions on a given quality metric. One should note that non-utilization QoS parameters (Table 7.2) are subjectively assessed by using questionnaires (filled by end-users on voluntary basis) given to customers (after completion of certain non-utilization service) via Web form, email, SMS, or other digital (or paper) forms.
- *Objective assessment method.* In this case objective quality assessment is used to estimate subjective quality (i.e. QoE) by using solely objective quality measurements (i.e. QoS). In practice, many times it is more convenient to estimate QoE based on objective testing and using statistical estimation models for the QoE [11]. There are three modes of objective testing to evaluate QoE: intrusive mode (a signal is injected into the system under test in order to generate a degraded output signal), non-intrusive mode (the quality assessment is performed on the "live" user traffic is carried by the network, without generation of active test signals), and planning mode (it is not used in real-time environment, but it is used for the planning and design of ICT systems). Of course, all QoS parameters are measured with objective measurements.

Certain quality indicators such as call set-up time, call success ratio, etc. can easily be measured by installing probe stations in appropriate locations. Then, quality measurements can be made on real user traffic or by using artificially generated traffic from/to probe network stations/nodes. Such approaches can be used on both public Internet network or managed IP networks.

The purpose of quality assessment via measurements is to perform analysis and evaluation of the service or network performances and to take corrective or proactive actions where it will be found necessary (based on QoS/QoE analysis with performance measurements). The same service provided by a telecom operator can be measured by the service provider and perceived by the customer (two views), as shown in Figure 7.10.

In addition to definitions and standards regarding QoS parameters and their assessments, measurements need to be defined so they will be practical for operators (e.g. measurements are typically the basis for networks and services design and optimization) and important to customers (e.g. can be used for comparison of QoS provided by concurrent operators in a given country, or can be used to check/validate whether contracted QoS level in the SLA is supported by the service provider).

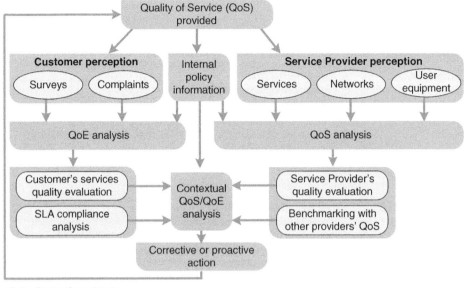

QoE – Quality of experience
QoS – Quality of service
SLA – Service level agreement

Figure 7.10 Customer and service provider QoS perceptions.

7.8 Broadband QoS Measurements

Current telecommunication networks have become all-IP networks with broadband/ultra-broadband access. All ICT services are delivered over the same access, including legacy telephony (as VoIP with QoS support), legacy TV (as IPTV), and IAS. On the other side, the QoS is always considered end-to-end. That means that for any service the QoS measurements are performed over broadband access, core (i.e. regional) and transit networks, either end-to-end QoS/QoE measurement or network performances measurement on different network segments.

7.8.1 Framework for QoS Measurements of IP Network Services

Considering that all networks and services in the modern telecommunication/ICT world in the twenty-first century are IP-based, there is necessity of QoS measurements of network services offered by service providers such as fixed and mobile telecom operators. For the purposes of operation and maintenance, networks and services design and optimization as well as regulatory monitoring of the provided QoS by the telecom operators (as ISPs), it is necessary to define a minimal set of QoS parameters (which will be used by all parties, including providers and regulators) for evaluating the quality of the IP network service. Such minimum set of parameters for evaluating the quality of the IP network service is defined with ITU-T 1545.1 [23], as given in Table 7.9.

Table 7.9 Minimum set of QoS parameters for IP network service.

Parameter	Definition
IP network service activation time	This parameter refers to the allocation of IP address and configuration settings by the DHCP server in the attached network (which includes setting of DNS servers IP addresses via DHCP).
DNS response time	It defines average Round-Trip-Time (RTT) delay metric for DNS response in IP networks, which is directly dependent upon the distance between the DNS client and the DNS server, and DNS server average domain name resolution time.
Number of IP network interconnection points	This refers to number of interconnection points to other Autonomous Systems (ASs).
RTT to IP network interconnection points	This metric measures the RTT between subscribers' service demarcation points and the interconnection points to other ASs. Typically, such interconnection points are located at the Internet eXchange Points (IXPs). The IP packet delay is defined as average of $(t_2 - t_1)$ over all successful and errored packet outcomes between occurrence of two corresponding packet reference events, ingress event at time t_1 and egress event at time t_2.
One-way IP delay variation [24]	This is one-way delay variation to IP network interconnection points.
One-way IP packet loss [24]	This is one-way packet loss to IP network interconnection points.
Mean data rate achieved (separately for downlink and uplink directions)	This is the average of the data transfer rate achieved for transfer of a given number of samples (e.g. download or upload of file samples with a given size – the higher the bitrate, the larger the sample file should be in order to provide realistic measurement of the data rate because TCP with its Slow Start needs many RTTs until it reaches the available data rate).
Percentage deviation of the mean data rate (i.e. mean bitrate of Internet Access Service – IAS)	This denotes the deviation between the data rate contracted (advertised) to the achieved data rate, obtained as: $$Deviation[\%] = \frac{|contracted\_rate - achieved\_rate|}{contracted\_rate} 100\%$$
Internet IP network service availability	Represents the fraction of time probability that the end-user is able to access services over the Internet Access Service (IAS).
Radio coverage availability	This parameter refers to received signal strength in mobile networks (e.g. better than −95 dBm) which is however mobile technology dependent parameter.

Measurement tools can be hardware-based or software-based. Hardware-based measurement tools can have at least the following options of implementation:

- *Option 1.* The probe stations replace the end-user's equipment and no other equipment can be connected to the IAS during QoS measurements (applicable for both fixed and mobile IAS).
- *Option 2.* The probe stations share the Internet access with other user traffic (e.g. probes connected to home gateway node on the end-user's side). In this case probes can do both, passive QoS monitoring of the end-user's traffic behavior (with approval

of the end-user) and perform active measurement tests only when there is no user traffic.

- *Option 3.* In this case the application programming interface is embedded into the customer's home gateway, to act as a probe and test the Internet connection (typically for testing fixed broadband connections).

Besides the hardware measurement tools listed above, at least three types of software-based measurement tools can be distinguished:

- *Web-based tool.* In this case the QoS measurement is executed via the user's Web browser by accessing a specific Web page for that purpose (this is the most common approach for testing the available bitrate in downlink and uplink direction on a global scale, such as Speedtest tool [25]).
- *Dedicated software client.* In this case the measurement software is permanently installed on the end-user's equipment.
- *Testing API.* In this case an API is included in a website in order to perform the test transparently every time when users access the website from their Web browsers.

Regardless of the chosen tools for the QoS measurements, any two independent implementations of the tools should measure with high confidence equivalent performances of the same network path under the same conditions.

7.8.2 QoS Evaluation Scenarios

For QoS measurements on the IP network services there are different scenarios that can be applied. However, all of them can be grouped into two main types:

- QoS evaluation scenario at the national level (e.g. test server located at the local Internet eXchange Point);
- QoS evaluation scenario at the international level (test server located at an international IXP).

In the QoS evaluation scenario at the national level, the position of the test server (used for the measurements) is at the local IXP while probes are installed at the end-user premises. These measurements can be done by the NRAs with or without explicit participation of the ISPs. With the scenario on the national level, the measurement path includes all links and network nodes from the customer terminal to the test server located at the local IXP (Figure 7.11). So, the test is started by the probes (as one end of the measured Internet connection) and targeted at the local IXP (as the other end of the Internet connection) when measuring/testing local KPIs (e.g. mean data rate, packet delay, losses).

The evaluation scenario at the international level is similar to the evaluation scenario on the national level, with the main difference being the location of the test servers (Figure 7.11). In the international case the test server is typically located at an international IXP (e.g. it may be located in another country). As usual, higher bandwidth available to the given Internet connection end-to-end results in better QoS and vice versa. The international QoS evaluation scenario provides possibility to NRAs to test international data KPIs (e.g. download/upload data rate, delay).

The QoS evaluation of IP network services at the national and international levels allows comparison of QoS levels locally and globally.

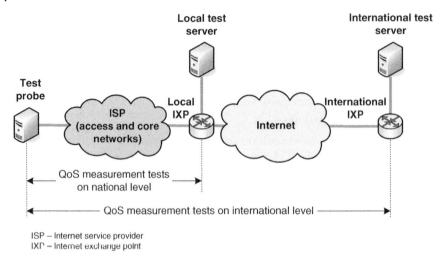

ISP – Internet service provider
IXP – Internet exchange point

Figure 7.11 Measurements of Internet QoS at national and international levels.

7.8.3 Discussion About the Sampling Methodology

Regardless of the evaluation scenarios and the types of tools used in the measurements (e.g. hardware-based or software-based), the crucial aspect is having confidence in the results. For the purposes of a high level of confidence in the QoS measurements, the number of probes to be used in testing should be large enough to cover all parts of the network and collected data from measurement is sufficient from a statistical point of view. Also, the data collection should be designed in a manner that reflects the QoS as it is being perceived by the end-user.

Further, the measurements with the sampling methodology should reflect the market share regarding the number of selected access lines for each speed package, with aim the sampling methodology to be confident [23]. As usual, there are differences between QoS measurements in fixed and mobile access networks. For mobile networks, for example, probe stations should be positioned at different locations around the country and be proportional with the installed capacity in a given region (e.g. more probes are required in urban areas than in suburban or rural areas). For broadband QoS measurements in fixed access networks (e.g. DSL, cable, or fiber access), the selection of access points for QoS measurements can be challenging because one endpoint of the measurement must be located at users' premises (QoS is always considered end-to-end). In such cases (for fixed broadband), cooperation between NRAs, consumers, and ISPs is required, with the aim of attracting sufficient number of end-users volunteers which will want to participate in the QoS measurement campaign by allowing installation of either hardware or software probes at their end.

Even in the case when there are enough probes, the broadband QoS measurements are not continuous events, but discrete events, either periodical (with installed probes) or triggered by certain decision (e.g. drive tests for QoS measurement in mobile networks). Since the twentieth century with telephony as the main telecommunication/ICT service the measurement moments were targeted to peak hours, the hour in the day (on average) with highest traffic intensity. A similar approach can be applied for Internet traffic.

However, due to the heterogeneous nature of the Internet traffic and globally spread Internet content, the moments for the measurements should in principle cover both high and low traffic, which will include also peak hours (for the Internet traffic). But when simplicity is required, the measurements may be targeted only at time intervals with high Internet traffic intensity. The logical reasoning for measurements in the peak hours only is that if the IAS is working properly in peak hours (i.e. hours with highest traffic intensity), then the quality of Internet-based services/applications in hours with lower traffic intensity should be even more acceptable. However, Internet QoS measurements should be based on sample sizes and confidence intervals as given in Table 7.8.

7.9 Quality Measurement Tools and Platforms

The broadband and ultra-broadband access to Internet requires also efficient and easy to access tools for QoS measurement. Since all parties (customers, service providers, network operators, and NRAs) require a certain level of quality experience which is based on certain values of the QoS parameters (bitrates in downlink and uplink, RTT delay, etc.), there was increasing need for measurement tools for QoS of broadband/ultra-broadband Internet access (including fixed and mobile one). Some of them are developed by specialized companies for such tools, some are developed by SDOs (standards developing organizations), some are developed by NRAs (e.g. by contracted third-party for development of such tools for the given NRA), so many such QoS measurement tools and platforms are available. This section gives several QoS measurement tools and platforms; however, such a list is not exhaustive.

Some QoS measurement tools include the following:

- *Network diagnostic tool (NDT).* This is a Web-based client/server program (based on M-lab measurement platform) that provides measurement of download and upload speed, RTT delay, as well as jitter (including minimum, maximum, and average values from the consecutive measurements) [26]. Besides NDT there are other measurement tools available on the M-lab platform.
- *SpeedTest tool by OOKLA.* Similar to NDT this measures RTT time, downlink and uplink speeds, and it is claimed to be used by the most ISPs and users globally [27]. It also offers commercial a measurement solution to ISPs (based on the SpeedTest technology), with customizable broadband performance measurement tools.

All QoS measurement tools are based on accompanying measurement platforms, which provide the required client-server infrastructure needed for active measurements. Such platforms may provide a pre-defined set of measurement tools or offer an open infrastructure where any measurement tool may be implemented. The tools used can be either hardware or software based. Some of the most popular measurement platforms for performing broadband Internet measurements include the following:

- *Measurement Lab (M-Lab).* This is an open, distributed server platform which aims to provide Internet measurement tools developed by different parties [28]. M-lab is not a complete measurement platform by itself, but it provides the necessary infrastructure to set up a measurement platform by deploying QoS measurement tools over the platform. However, the measurement data collected by such tools is made available in the public domain on the basis of an open data approach.

- *RIPE Atlas.* This is the Internet data collection system of the main RIPE Network Coordination Centre (NCC) [29], where RIPE is one of the five regional Internet registries (RIRs), for the Europe region (each RIR manages IP addresses and autonomous system numbers in its own region). The RIPE Atlas is in fact a global network of hardware probes (small hardware devices that are connected to an Ethernet port on the router at users' premises) and anchors (enhanced RIPE Atlas probes that have bigger measurement capacity), which perform active quality measurements. Anyone can access this data via Internet traffic maps, streaming data visualizations, and via an API. RIPE Atlas measurement nodes (probes and anchors) perform built-in measurements which include ping for measuring of RTT delay, traceroute for measuring the hop distance (and possibly number of ASs on the packets' path, based on IP addresses of router interfaces), DNS, and HTTP (with or without SSL/TLS – secure sockets layer/transport layer security). As typical for a measurement platform, users can also define their customized measurements.
- *SamKnows – Internet Performance Measurement.* This is a distributed network of whiteboxes (i.e. probes, as outlined in Figure 7.12) for Internet QoS measurements [30], which are placed in consumers' homes to measure the performance of fixed broadband access. They are controlled by servers that do scheduling of measurement tests and send the data to centralized databases. Besides the built-in performance measurements, the SamKnows platform is designed to provide flexibility for future modifications or enhancements (when they will be required). Primary quality metrics for SamKnows are given in Table 7.10.

Overall, all measurement tools and platforms typically use active measurements, where the probes (hardware or software-based) generate additional traffic (for QoS measurement purposes) in addition to the user traffic. For example, the measurement of the bitrate (by using HTTP file download and upload, for measurement of downlink and uplink speeds, respectively) requires larger file sizes for higher bitrates. For example, SamKnows for fixed broadband access requires file sizes of 6 MB for bitrates below 30 Mbit/s and file sizes of 12 MB for bitrates in the range 30–50 Mbit/s, while for higher speeds above 50 Mbit/s the HTTP connection for download/upload is required to last at least 10 seconds, which requires a file size of approximately 60 MB (per test)

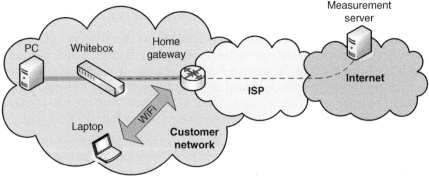

ISP – Internet service provider
PC – Personal computer

Figure 7.12 SamKnows measurement probe.

Table 7.10 Primary quality metrics in SamKnows platform.

Target for measurement	Primary quality metrics
Web browsing	The total time to retrieve a reference Web page from one of the popular websites.
Voice over IP (VoIP)	Uplink and downlink packet loss and jitter, and RTT delay.
Speed	Downlink and uplink bitrate (Mbit/s), measured separately in each direction with several (e.g. three) concurrent TCP connections for more realistic measurement (based on user experience, where single broadband access is concurrently used for more several Web client-server connections).
UDP delay	Average RTT obtained through a series of randomly transmitted UDP packets.
UDP packet loss	Percentage of UDP packets lost (obtained from UDP delay tests).
DNS resolution	The time measured for the recursive DNS resolvers (used by the ISP) to return IP address from resolving a popular website domain name (e.g. www.google.com)

or more [31]. However, the data cap (maximum volume of traffic included in the IAS package on daily, weekly, or monthly basis) can be an obstacle for such measurements if the Internet traffic due to measurements is not differentiated (and hence not included in the data cap, i.e. not charged) by the ISP.

7.10 Discussion

The QoS measurements are influenced by different methods used. While passive QoS measurements (based on user traffic sniffing) may provide valuable statistical data, QoS measurements are typically provided with active measurement because the measurement procedures can be specified or standardized and the results are repeatable and comparable. However, even with active QoS measurements, for example when there are multiple flows over the same Internet access, they all share the available bandwidth (i.e. bitrates) in the downlink and uplink respectively. So, to have accurate results (e.g. for fixed broadband access) the link should be idle at the time when the installed probe generate test packets toward the servers (of the measurement platform). For example, if the user has 100 Mbit/s of bandwidth (i.e. link capacity) in downlink, and concurrently downloads 10 files with HTTP or FTP over the given Internet access link (with file sizes large enough to exists concurrently for period of several tens of seconds), then each TCP connection (i.e. each HTTP or FTP connection) will have less than 1/10th of the downlink bandwidth, i.e. less than 10 Mbit/s in this concrete example. That is due to the fairness in the design of the TCP because both HTTP and FTP (which typically are used for bitrate measurements) use the TCP/IP protocol stack. The TCP by itself is designed to provide congestion control on the two ends (TCP client and TCP server) by adapting its speed to the available bandwidth, which results in available access link capacity being split equally among different TCP flows over the same link. It is possible that some of the TCP flows will experience lower bitrates due to bottleneck link on the end-to-end path between the client and the server.

What about the QoS measurement of real-time services (i.e. services sensitive to delay) over broadband Internet access? Real-time Internet services such as OTT VoIP (Skype, Viber, WhatsApp, etc.) use RTP/UDP/IP for transfer of the user real-time data, while TCP can be used also in this case for control and management traffic (because signaling is important to be lossless, and TCP provides that for the applications running above it). However, UDP/IP flows do experience packet losses because UDP does not have retransmission mechanisms of the lost packets/datagrams (as TCP has).

The QoS measurements may be influenced by the traffic management approaches by the service provider (e.g. telecom operator). For carrier-grade services such as operator-grade VoIP, IPTV, business services, or IAS, there is explicit bandwidth allocation in the access links and certain traffic dimensioning in core and transit networks. Network slicing and network virtualization further contribute to introduction of different slices with different levels of QoS support for service provided over them (e.g. critical IoT services).

Regarding all QoS measurements, they need to be established end-to-end. Such approach requires Internet measurement tools and platforms (e.g. M-Lab, RIPE Atlas, SamKnows). Some NRAs have also developed their own tools for Internet QoS measurements in line with their national legislation regarding the telecom/ICT sector.

Overall, there are many different factors that can influence the measured data, including (but not limited to) other networks on the path, background traffic from other applications, the type of network (fixed or mobile), end-user equipment and its capabilities (e.g. processing power, memory, operating system, available network interfaces), traffic management techniques that are applied by the ISPs for the given tariff package, etc. So, QoS measurements are not a "linear game" and knowledge about how broadband and ultra-broadband networks and technologies is required, as well as knowledge about different types of IP-based services, their traffic characteristics, and QoS requirements.

References

1 ITU-T Recommendation G.114, One-way Transmission Time, May 2003.
2 IETF RFC 3550, RTP: A Transport Protocol for Real-Time Applications, July 2003.
3 Cisco, Cisco Visual Networking Index: Forecast and Methodology, 2016–2021, June 2017.
4 Geza Gordos (ed.), Telecommunication Networks and Informatics Services, 2003, http://w3.tmit.bme.hu/thsz/onlbook.pdf
5 ITU-T Recommendation E.803, Quality of Service Parameters for Supporting Service Aspects, December 2011.
6 Janevski, T. (2003). *Traffic Analysis and Design of Wireless IP Networks*. Norwood, MA: Artech House Inc.
7 ITU-T Recommendation E.847, Quality of Service Norms for Time-Division Multiplexing Interconnection Between Telecom Networks, March 2017.
8 ITU-T Recommendation G.1028, End-to-end Quality of Service for Voice Over 4G Mobile Networks, April 2016.
9 Janevski, T. (2014). *NGN Architectures, Protocols and Services*. Chichester: Wiley.

10 ITU-T Recommendation G.1080, Quality of Experience Requirements for IPTV Services, December 2008.

11 ITU-T Recommendation G.1011, Reference Guide to Quality of Experience Assessment Methodologies, July 2016.

12 Janevski, T. (2015). *Internet Technologies for Fixed and Mobile Networks*. Norwood, MA: Artech House.

13 ITU-T Recommendation G.1315, QoS Support for VPN Services – Framework and Characteristics, September 2006.

14 ITU-T Recommendation E.804, Quality of Service Aspects for Popular Services in Mobile Networks, February 2014.

15 5G Monarch project 5G Mobile Network Architecture for Diverse Services, Use Cases, and Applications in 5G and Beyond, Deliverable D6.1, Documentation of Requirements and KPIs and Definition of Suitable Evaluation Criteria, September 2017.

16 ITU-T Focus Group on Smart Sustainable Cities, Setting the Framework for an ICT Architecture of a Smart Sustainable City, May 2015.

17 ITU-T Recommendation Y.4900, Overview of Key Performance Indicators in Smart Sustainable Cities, June 2016.

18 ITU-T Recommendation Y.4901, Key Performance Indicators Related to the Use of Information and Communication Technology in Smart Sustainable Cities, June 2016.

19 BEREC, Monitoring Quality of Internet Access Services in the Context of Net Neutrality, September 2014.

20 IETF RFC 7799, Active and Passive Metrics and Methods (With Hybrid Types In-Between), May 2016.

21 ITU-T Recommendation E.802 Amendment 1, Framework and Methodologies for the Determination and Application of QoS Parameters, March 2017.

22 ITU-T Recommendation Y.1543, Measurements in Internet Protocol Networks for Inter-domain Performance Assessment, June 2018.

23 ITU-T Recommendation Y.1545.1, Framework for Monitoring the Quality of Service of IP Network Services, March 2017.

24 ITU-T Recommendation, Y.1540, Framework for Monitoring the Quality of Service of IP Network Services, July 2016.

25 Speedtest by Ookla, http://www.speedtest.net, visited in August 2018.

26 NDT (Network Diagnostic Test) by M-Lab, https://www.measurementlab.net/tools/ndt, accessed in August 2018.

27 SpeedTest by Ookla, www.speedtest.net, accessed in August 2018.

28 Measurement Lab (M-Lab), www.measurementlab.net, accessed in August 2018.

29 RIPE Atlas, https://atlas.ripe.net, accessed in August 2018.

30 SamKnows – The Global Platform for Internet Measurement, https://www.samknows.com, accessed in August 2018.

31 European Commission, Quality of Broadband Services in the EU, October 2013.

8

Network Neutrality

Network neutrality is directly related to the appearance of the Internet. It has played a consistent part since the beginning of the modern Internet, which started with the privatization of the network around 1995. The main idea of independent applications running over unified networking based on the IP in the middle of the protocol hourglass has made the Internet a success and Internet technologies have become a unified packet-switching technology on the global telecom/ICT scene [1]. The independence of applications and services from network providers (generally speaking) and from national regulations has created possibilities for innovative services and applications globally without requirements for a consensus about them among standards developing organizations, researchers, or policy makers. That has allowed the number of services and applications in the telecom/ICT world to increase exponentially, including some standardized and some proprietary ones. But all of them are created for the benefit of the end-users as the customers of service providers that offer such services, noted as over-the-top services in previous chapters of this book. The same relates to the contents that are transferred by various OTT services and applications. This is regarded as network neutrality of the Internet (as a global network) to all services and applications running over it, including those that exist now and future "killer" services.

8.1 Introduction to Network Neutrality

Network neutrality refers to the principle that all Internet traffic should be treated equally [2]. However, as in everything, network neutrality has proponents and opponents. The proponents claim that telecom companies without network neutrality may try to impose a tiered service mode in order to control the Internet traffic via certain pipelines (per traffic type of per flow) and with that to remove the competition and oblige end-users to buy particular services from them. That may result in the creation of the "dirt road" effect. This refers to the concern that Internet traffic management and prioritization may motivate network operators to degrade non-prioritized traffic by applying prioritization to traffic from selected OTT services and applications, thus turning the best-effort-based Internet into a "dirt road" of poor capacity and quality for many other services.

The opponents of network neutrality regulation argue that the best solution for the broadband providers is to encourage greater competition among service providers,

QoS for Fixed and Mobile Ultra-Broadband, First Edition. Toni Janevski.
© 2019 John Wiley & Sons Ltd. Published 2019 by John Wiley & Sons Ltd.

which is currently limited in many areas by having no possibility for agreements between telecom operators and third-party service providers for differentiated provision of their services over the public Internet network (the telecom operator has the means to provide differentiated QoS in its broadband and ultra-broadband fixed and mobile access networks, as well as in regional core networks, but with network neutrality that is limited to managed IP networks).

Network neutrality has taken on various meanings. For example, in the European Union [3] it is defined as the ability of all Internet end-users to access and distribute information or to run applications and services of their choice. Another definition is that all Internet traffic should be treated equally, without any discrimination, restriction, or interference, independent of the sender, receiver, type, content, device, service, or application. Another possible definition of network neutrality is the absence of unreasonable discrimination on the part of network operators for the transmission of Internet traffic. However, not all definitions are exactly the same. For example, "all traffic to be treated equally" may refer to prohibition of any form of differentiated QoS provision. But if the definition refers to all Internet traffic to be treated equally, that does not prohibit differentiated QoS (e.g. QOS differentiation can be applied to different Internet traffic types such as voice, video, torrent, email).

There are two general types of services over broadband access networks [4]:

- *Internet access service (IAS)*. This is in fact access to the public Internet over fixed or mobile access networks, which provides connectivity to the Internet and with that to all OTT (i.e. data) services.
- *Specialized services*. They refer to all services provided over the managed IP networks by using the same broadband access as IAS. These managed IP networks for specialized services use strict admission control and are typically optimized for specific content, applications, or services (e.g. operator's grade VoIP, such as VoLTE, IPTV, critical IoT services) based on extensive use of traffic management techniques to meet requirements of the content, applications, or services for a specific QoS level.

The monitoring and assessment of both types of services, IAS and specialized, (i.e. managed) is performed by both telecom operators and NRAs. Typical QoS parameters which are the subject of interest for objective quality assessment are bitrates, packet delay (i.e. latency), jitter (i.e. delay variation), and packet loss (e.g. it is important for real-time services such as voice). The important quality criterion is to avoid degradation of IAS for provision of specialized services, which can be accomplished by having sufficient network capacity (for carrying the highest volume of traffic during the day). What is sufficient network capacity? It is network capacity on each link and on each path that can transfer IP traffic with no more than 70% load in the peak hour (due to statistical multiplexing of bursty Internet traffic), without noticeable congestion. Sufficient capacity for IAS is assessed by performing QoS measurements with and without specialized services over a given broadband/ultra-broadband network.

8.2 Degradations of Internet Access Service

Network neutrality refers directly to IAS and therefore the main regulatory issues concerning QoS are related to detection of its potential degradation. For that purpose

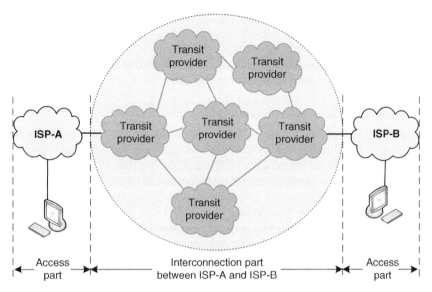

ISP – Internet service provider

Figure 8.1 End-to-end Internet communication model.

appropriate tools for monitoring the IAS are required, to know its quality [4]. Internet communication in general refers to end-to-end communication over the public Internet, which includes ISPs on both ends and an interconnection leg between them which may consist of one or more transit networks (Figure 8.1).

Considering the end-to-end Internet communication model, there are two main types of degradation of IAS:

- *Degradation of IAS as a whole.* This refers to IAS provided by telecom operators to the end customer, which however is the basis for all OTT (i.e. data) services used by the end-user. To identify possible cases of IAS as a whole degradation requires monitoring the service quality of the IAS in fixed and mobile networks by measuring the main IP network quality parameters (as given in this chapter).
- *Degradation of IAS considering individual OTT applications/services.* This case refers to individual applications (Skype, Viber, Facebook, YouTube, etc.), which may experience degradation due to different causes that may be put into the following three main groups:
  - *Congestion.* This is the situation in IP networks when IP traffic increases to a level at which routers run out of buffer space and are forced to start dropping some IP packets. Congestion appears randomly for various reasons, including not enough capacity in certain parts of the network (e.g. access, core, or interconnection), or it may be the result of temporary network link or segment failures. However, congestion affects bitrates, delays, jitter, and losses and with that all OTT Internet applications that are traversing that path.
  - *Traffic management.* Different approaches applied by ISPs in their own networks may include both limiting measures (e.g. application traffic blocking and/or throttling) and enabling measures (e.g. routing and traffic forwarding).

○ *Network security and integrity.* This refers to degradations of applications caused by various security attacks (e.g. a denial of service attack on a given website), as well as degradations due to network failures (network integrity in fact includes measures to maintain the quality level during network failures, as well as measures to prevent network failures such as using redundant nodes and links in the network).

Certain degradations of OTT services, such as service blocking (e.g. blocking of OTT VoIP services such as Skype, Viber, or WhatsApp), are easier to detect. Meanwhile, throttling of certain services (e.g. torrent traffic throttling by the ISP) is harder to detect and requires measurement (e.g. by the NRA). Congestion management is always required in IP networks; however, the ISPs should not misuse that for degradation of chosen OTT applications, i.e. congestion management should be application agnostic. Network security and network integrity management are also necessary, but again, this should not go beyond the actual requirements.

Network neutrality means that all traffic is treated equally over the IAS. However, the ISP may apply traffic classes by using DPI then differentiating and classifying IP packets belonging to different types of applications or traffic types (e.g. audio, video, data) for their transfer over the core or transit networks (e.g. by using separate VPNs). In such cases certain Internet traffic classes will be prioritized over others (e.g. in core and transit networks), which may lead to degradations of OTT services that are classified into lower priority classes in this case, especially in time periods when there is high volume of high(er) priority traffic in the network. However, proper packet scheduling at network nodes (e.g. routers, gateways) as well as network dimensioning (regarding the capacity of links on all paths over the operator's IP networks) may prevent such degradations for OTT services provided over the IAS (i.e. through the public Internet network).

Other degradation of both IAS and specific OTT services (provided over IAS) can be due to high volume of traffic of specialized (i.e. managed) services over the same IP networks. Both types of services, Internet-based and specialized (managed), can be provided over the same broadband and ultra-broadband IP networks, with applied traffic engineering for sharing of the total bandwidth (e.g. dedicated bandwidth for IAS and for managed IP services, or shared bandwidth between IAS and managed IP services, with priority applied to managed IP services in the ISP's network) [5]. However, in peak hours congestion may occur for specialized services which may use certain bandwidth of the IAS and with that degrade the IAS and individual OTT services provided through the IAS. However, if the bandwidth for the IAS is strictly reserved, specialized services may have lower QoS than specified or contracted (e.g. more VoIP calls rejected or dropped in mobile networks). What is more important, IAS or specialized services? This is a sensible question. When the specialized services are critical (e.g. critical IoT services such as remote control of vehicles or drones), then their importance is normally much greater than that of the IAS (which uses a best-effort approach). There are also services of the same type, such as VoIP, provided as OTT services through the IAS (Skype, Viber, WhatsApp, etc.) and as specialized services over managed IP networks (e.g. NGN, VoIP as PSTN/ISDN replacement, VoLTE), by using the same broadband access (either in fixed or mobile networks). In such cases both voice services may be treated with the same importance. However, not all voice calls have the same level of importance. For example, emergency calls (e.g. 112 in Europe, 911 in the United States) must have the highest priority (e.g. they should be possible even without a SIM card in a smartphone), regardless of possible degradation of other non-critical services.

So, there are different reasons for possible degradations of the IAS or specific OTT applications/services. In such cases, when the degradation of services provided by an ISP is noticed, the NRA may proceed to an intervention by imposing minimum QoS requirements [6]. The main approach would be to impose requirements on the ISP to improve QoS until degradation is eliminated. The ISP may need to perform maintenance (in the case of malfunctioning of certain equipment or links), optimization (for better usage of available network and service resources), and network design (to deploy capacity that can serve all contracted users, considering the projected growth in offered bitrates for IAS, the data cap for IAS, number of customers per IAS and specialized service, and number of specialized services offered in parallel with IAS over the same broadband fixed or mobile networks).

The need for intervention from the NRA will depend on the quality of the IAS and therefore monitoring of IAS and assessment of the measurement results are necessary to detect eventual degradation, which may include the following cases [4]:

- ISPs prioritize specialized services at the expense of the IAS as a whole.
- Internet traffic load grows faster than the increase in available capacity in the given access and core (i.e. regional) networks, or at the interconnection points.
- IAS of sufficient quality is accessible to only a limited number of users, or in limited regions (e.g. in mobile access networks).

When there is degradation of the IAS as a whole (Figure 8.2), the IAS-providing ISP (that is ISP-A in Figure 8.2) will not be able to directly control the performance beyond its own interconnection points through which they interconnect to neighboring networks (ISP-B and ISP-C). However, ISPs first perform traffic dimensioning and network design and then also complete interconnection agreements, after investigating network performance needs and negotiating with peering and transit partners.

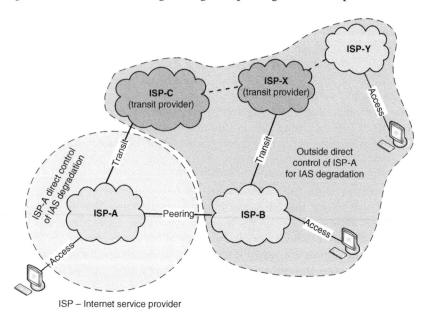

ISP – Internet service provider

Figure 8.2 Degradation of internet access service (IAS) as a whole.

Examples of degradations of individual OTT applications and services and applications include (but are not limited to) the following:

- blocking of OTT VoIP (e.g. Viber, WhatsApp, etc.) on mobile IASs;
- blocking or throttling of the OTT peer-to-peer file sharing (e.g. BitTorrent) applications on mobile or fixed IASs;
- differentiation of OTT traffic that is generated from different content and service providers by giving priority to one OTT service over the others.

Network neutrality does not allow degradations of individual OTT applications and services by their blocking and throttling. However, reasonable traffic management should be applied on the Internet traffic. In the ideal case, traffic management may contribute to better QoS of OTT services and applications (e.g. by giving priority to real-time services and applications over non-real-time but avoiding bandwidth starvation for any OTT service). One possible approach for better QoS support on a network neutrality basis on the Internet is to provide means to end-users (e.g. via fixed access networks) to specify preferences for certain OTT services (e.g. priority of gaming traffic over Web traffic); however, that should refer only to the customer's own network (either home or business) and can be supported by fixed service providers (e.g. via settings on the home gateway/router).

8.3 Main Regulatory Goals on Network Neutrality

Bearing in mind the possible causes of IAS degradation as well as degradation of individual OTT services and applications, the question is, when is regulatory intervention needed in respect of network neutrality? The market situation should be evaluated to determine whether the problem requires a certain type of regulatory intervention, which is dependent upon the availability of such offers and the ease of customers switching ISP (that provides the IAS), which includes all burdens faced by customers (e.g. price difference between the offers for IAS of the old ISP and the new ISP).

In general, there is no need for intervention when there is good availability of IAS offers on the market with satisfactory quality at a reasonable price, and at the same time there exists sufficient possibility and ease to switch providers (i.e. ISPs), as shown in Figure 8.3. When switching ISPs is hard and availability of IAS offers on the market (e.g. in the given country) is low, there is a need for regulatory intervention by the NRA.

Such regulatory intervention should be targeted toward imposing minimum QoS requirements to ISPs for their IAS offers. That can be completed in a regulatory process [4], which starts with the NRA identifying degradations. That is followed by evaluation of QoS indicators for the purposes of verification of IAS degradation (or degradation of individual OTT service) and analysis of the measurement results to decide whether the intervention is needed. If the QoS assessment results in necessary regulatory intervention, then the NRA chooses a regulatory tool for that purpose. The NRA makes a decision about the requirements imposed on the ISP and notifies the ISP about them. Finally, the NRA makes a final decision for the set of minimum QoS requirements on the ISP, which is followed by monitoring of QoS indicators for compliance purposes.

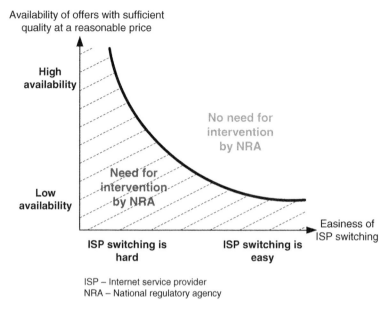

Figure 8.3 Regulatory interventions in regard to network neutrality.

Overall, there is no mandatory explicit regulation of network neutrality in all countries. Some countries may decide that no specific action is required regarding network neutrality. Some might argue that the regulatory approaches already taken go beyond what is reasonable. Overall, there is no clear global consensus on how to deal with network neutrality, so NRAs make make their own reasonable judgments on the basis of their national legislation and telecom/ICT market development.

One specific approach regarding network neutrality is zero rating, which refers to the ISP denoting a price of zero to the Internet traffic associated with a particular application or group of applications (in such a case the data does not count toward any data cap applied to the IAS for a given customer). This may refer to OTT VoIP (e.g. zero rating Viber or WhatsApp in mobile networks), social networking (e.g. Facebook), or video streaming (e.g. zero rating Netflix or YouTube). The lower data cap (Internet traffic allowance for a given user in a given period) is likely to result in stronger influence of price differentiation. Additionally, such price differentiation between individual applications which belong to a same type or category (e.g. zero rating for one video streaming site only, and regular charging or counting in the data cap for all others) has an impact on competition between service providers that offer that type of service (in this example, video streaming). Price differentiation is generally not forbidden (or not allowed) everywhere and in every case. However, the goal is to continue functioning of the Internet application ecosystems as an engine of innovation on the basis of network neutrality, which may be jeopardized by price differentiation of OTT applications. On the other side, practices which apply a higher price to the data from a specific application or group of applications are likely to impose restriction of choice on the end-user (the price on average influences user choice) [7]. Additionally, price differentiation of different applications offered through the IAS (i.e. over the public Internet) may discourage innovation of applications and services.

8.4   Network Neutrality Business Aspects

Considering IAS market developments, there is continuous growth of Internet traffic in both fixed and mobile networks. However, the growth in fixed Internet traffic is driven more by the increasing volumes of Internet traffic by each user rather than due to an increase in the total number of fixed broadband users. In many developing countries penetration of fixed broadband is very low. In all countries user penetration of mobile broadband is higher than fixed broadband. But most of the Internet traffic (about two-thirds of total traffic, according to Figure 7.2 in Chapter 7 of this book) is coming from fixed broadband access. This is due to higher bitrates of fixed broadband, but also due to the fact that many users access the Internet by using mobile devices (smartphones) via WiFi which is typically connected via fixed broadband access lines to the Internet. So, in most, WiFi access is in fact fixed broadband Internet access, although the WiFi technology has its influence on the experienced QoS because it uses unlicensed spectrum, which is crowded, and it experiences much higher deviation of signal strengths and with that, deviation in the available bitrates between the user WiFi interface and the access point on the side of the network connected to the Internet.

The growth in mobile data traffic is not driven only by an increase in the number of subscribers. It is influenced by different factors such as availability and affordability of powerful mobile devices (e.g. leading-edge smartphones), deployed mobile networks with high capacity per mobile user, and the OTT applications and services ecosystems developed for mobile devices (resulting in increased use of download or upload of various mobile data). With increased broadband penetration the broadband market is saturating because all human users have either fixed or mobile Internet access, or both. That is why although the overall volume of Internet traffic is increasing, its growth rate is declining over time.

The growth rate is most important for business because it is important how much a given business can grow in the near future. Higher growth attracts investments in that sector and vice versa. The revenues of fixed and mobile operators are saturating regarding their growth, due to saturation of the market and the already developed competition (e.g. in each country there are multiple fixed and mobile operators that share the ICT/telecom market). For example, fixed voice revenues have started to decline around 2005 when the number of subscribers showed a downhill trend (which was caused mostly by higher penetration of mobile voice and also by the availability of OTT voice services such as Skype at that time). In mobile networks, SMS revenues have been threatened by OTT messaging applications which are typically included with OTT voice services, thus making them attractive to end-users (think of Skype, Viber, WhatsApp, Messenger). Although the revenues from SMS and mobile voice provided by mobile operators decline over time, they are being compensated by increased revenues from mobile data services (that is the IAS provided over mobile access networks) because mobile users still have to pay the cost of Internet access (for instance, a flat fee price or volume-based fee – e.g. per MB or GB).

To provide the required QoS the telecom operators (which operate the access and core networks) must invest in capacity by deploying additional bandwidth for the growing Internet traffic. So, traffic growth needs infrastructure investment and also traffic management tools to address congestion issues. To remain sustainable and profitable, telecom operators (which are also ISPs) may change existing business models or introduce new ones, including the following:

- Differentiation of IAS offers with different data caps are a standard approach in fixed and mobile networks (the data caps are many times higher in fixed networks than in mobile networks for the same price). For example, mobile operators can manage traffic growth (and with that capital investments in new equipment in access and core networks) by applying appropriate data caps in their mobile IAS packages to manage traffic growth. The smaller data cap slows traffic growth, and vice versa.
- Price differentiation and time differentiation can appear in different offerings – for example, free voice calls during the night for packages targeted at younger users, or faster broadband speeds for gamer tariffs (higher speeds also result in lower latency, which is essential for online gaming applications). Possible differentiation is also provision of mobile services through a managed IP network part in the mobile network, such as VoLTE or mobile IPTV (both are provided with assured QoS, via a managed IP network, i.e. not via mobile IAS).
- ISPs that make wholesale purchases of access networks from the network providers (e.g. other fixed and mobile operators in the given country) may have less control over traffic management and with that the QoS provided to their customers.
- Various content and application providers naturally want to prioritize their data delivery where possible. Such global providers build their own content delivery networks; however, their services are provided over fixed or mobile IAS on the basis of network neutrality. So, placing the CDN nodes as close as possible to ISPs' networks increases the QoS for their services. ISPs can also have revenues from wholesale offering of bandwidth on the side of CDN data centers (which are on the opposite side of the access part from the viewpoint of the ISPs).

One may conclude that traffic management in respect of QoS and network neutrality is directly related to telecom operators' businesses. To optimize resource utilization in different part of the networks (e.g. core network, interconnection points), the possible tool is differentiation of applications regarding traffic management, which can be done in two main ways [4]:

- *Application-agnostic traffic management*. All applications are treated in a similar manner (e.g. IP packets from all OTT applications put in the same forwarding queue).
- *Application-specific traffic management*. Individual applications are treated differently. For example, torrent traffic is throttled or video traffic is shaped, while voice traffic is not.

In developed ICT/telecom markets, the prices for both fixed and mobile broadband and ultra-broadband services are not fixed at any particular level, but rather they appear to respond to normal forces of supply and demand by operators. So, in broadband markets where competition is developed and effective, "dirt road" effects (with lower QoS for IAS-based services due to higher volume of specialized services over the same broadband and ultra-broadband) are unlikely. Why? Because the operator would face a higher probability of losing its market share and with that would become less profitable.

In broadband markets where competition is ineffective or not developed, "dirt road" effects also are unlikely because the network operator can exploit its market power in more effective ways. So, in all cases the "dirt road" effect is not likely to happen to the network neutral traffic, hence network neutrality is not threatened by such business factors. Overall, one may expect that network operators which provide IAS (and with that they are ISPs) will be motivated to continue provision of best-effort IAS of acceptable quality.

Let's discuss QoS and accompanied traffic management from the IAS business perspective. If all packets were of high priority (or all of low priority), then prioritization would have no effect. If only a few packets were of high priority while most of them were low priority, then the low priority packets would experience only a small additional delay (due to serving high priority packets first at network nodes). The third case in this discussion refers to the case if most packets were of high priority and few were of low priority. In such a case packet prioritization would only slightly accelerate the small number of high priority packets and the low priority packets would experience significant additional delay. So, considering the business logic, prioritization of traffic seems weak if almost all traffic is high priority in any case. What will be the upper limit of the capacity dedicated for the prioritized traffic depends on the type of traffic – is it regular voice or emergency voice, or is it massive IoT or critical IoT? Of course, emergency voice service as well as critical IoT must have high priority regardless of the non-critical traffic over the same access network. To avoid the "dirt road" effect, proper capacity planning is required by using traffic engineering techniques [5], or by using overprovisioning (as the simplest approach, which is more costly due to the requirement for more resources deployed in the network).

8.5 Role of NRAs in Regulation of Network Neutrality

Regarding the regulation of network neutrality, there are similar approaches with certain differences in other countries. For example, India's network neutrality approach forbids price differentiation of OTT services (e.g. free Viber or Facebook over the mobile network) because it favors one service provider over the other and hence may jeopardize network neutrality [6]. Meanwhile, price differentiation for OTT services is not forbidden in Europe in certain cases [8].

Network neutrality regulation in different countries may include different approaches, such as the following:

- *Cautious observation.* This refers to cases where countries and their NRAs have chosen not to take any specific measures regarding network neutrality because they think that no specific action is required and the regulatory framework is sufficient.
- *Tentative refinement.* This is the case when a country has adopted a light-handed approach, which is typically based on disclosure and transparency, lowering switching barriers between ISPs, imposing minimum QoS on IAS (e.g. minimum supported bitrates in the access networks), including light-handed network neutrality measures (without prohibition of specific ISP behaviors).
- *Active reform.* This is the case when a given country has taken specific regulatory measures which are targeted to prohibit specific behaviors by ISPs, which is typically related to network and traffic management practices.

The main role of the NRA regarding network neutrality is to establish common rules to safeguard equal and non-discriminatory treatment of Internet traffic in the provision of IAS and at the same time to protect end-users' rights. In that manner, the role of NRAs typically includes the following activities:

- *Supervision.* This involves monitoring of contract information between the ISPs and customers in the given country, analyzing commercial practices, traffic management practices, and specialized (i.e. managed) services. That is done by means of assessment of practices in the market, technical measurements, and gathering of information from customers.
- *Enforcement.* There are different approaches in enforcement which may span from naming and shaming (in countries with a developed telecom market, with enough competition, and applied legislation) to financial penalties (in countries with less developed telecom markets, and significant QoS degradations of IAS or violation of the network neutrality rules). For example, the NRA may require ISPs to deal with degradation of IAS, or to cease or revise specific problematic traffic-management approaches in their networks, to cease providing specialized services (i.e. services with QoS guarantees) in the absence of sufficient capacity for the IAS, and finally imposing fines on ISPs.

The main tools of each NRA for QoS enforcement are measurement systems [9]. A reference measurement system to be used by an NRA is shown in Figure 8.4. The concrete implementation is based on connecting the measurement system (through a router which acts as a gateway – with added functionalities regarding QoS measurements) to a national IXP where the ISPs are connected within the NRA's jurisdiction. The router/gateway node of such a reference measurement system is responsible for the

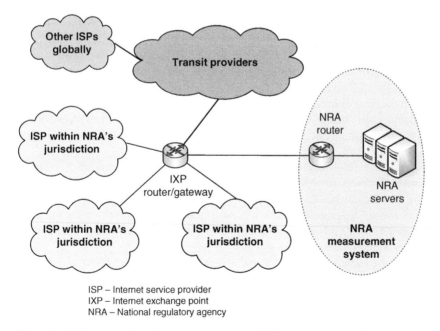

ISP – Internet service provider
IXP – Internet exchange point
NRA – National regulatory agency

Figure 8.4 Reference national measurement system for Internet access services.

logical configurations considering that different networks connected to IXP are different autonomous systems, hence it should facilitate IP connectivity via BGP, which is used for inter-AS routing, or DNS, which is used to "connect" the two name spaces in the Internet, IP addresses and domain names. Such measurements system of the NRA also needs to use different servers to set up and control Internet measurements, as well as to collect, store, process, and present the results (Figure 8.4).

8.6 Network Neutrality Approaches

There is no clear global consensus or one best approach to deal with network neutrality. Countries have taken different approaches to the regulation; however, the birthplace of the Internet, the United States, and the home of the 3GPP mobile networks, Europe, in fact define the shape of network neutrality in the massively dominant IP-based telecom world. Therefore this section further covers network neutrality approaches in Europe and the United States.

8.6.1 Network Neutrality Approach in Europe

In 2013 the European Commission proposed a Telecoms Single Market (TSM) Regulation to the European Parliament, where network neutrality was an important part of the proposed legislation. Later, in November 2015, the European Parliament and the Council enacted Regulation EU 2015/2120 [10]. This regulation (2015/2120) provides the first European network neutrality rules, which have been applied since April 2016. The key elements include the following:

- *Rights of end-users.* End-users have the right to access and distribute information and content, use and provide applications and services, and use terminal equipment of their choice, irrespective of the end-user's or provider's location or the location, origin, or destination of the information, content, application, or service, via their IAS.
- *Traffic management.* Providers of IAS are not prevented from implementing reasonable traffic management measures. However, to be reasonable, such measures should be transparent, non-discriminatory, and proportionate. Also, the measures should not be based on commercial considerations, but on objectively different technical QoS requirements of specific traffic categories.
- *Prohibited practices.* Network operators may not block, slow down, alter, restrict, interfere with, degrade, or discriminate between specific content, applications, or services, or specific categories thereof, except as necessary, and only for as long as necessary. However, this should be done in such manner that preserves the integrity of the network and/or end-user equipment, and deals with network congestion by treating equivalent traffic categories equally.
- *Specialized or managed services.* Network operators, as well as content and service providers, are free to offer services (other than Internet access) which are optimized for specific content, applications, or services, where such optimization is necessary in respect of QoS. However, these services can be offered only if the network capacity is sufficient for their provision in addition to any IAS. Overall, this is closely related to network planning and design.

Overall broadband quality is assessed on a periodical basis or assessment is triggered by customer requests. If the results show significant discrepancy, continuous or regularly recurring, regarding IAS performance such as speed (i.e. bitrates) in uplink/downlink or other QoS parameters, that will be decided (e.g. by the NRA) as non-conformity of performance and may trigger the remedies available to the consumer which must be in accordance with national laws.

In European countries, ISPs are required to ensure that any contract which includes IAS specifies at least the following aspects [7]:

- Information about how traffic management measures applied by that provider could impact on the QoS of the IAS, on the privacy of customers, and protection of their personal data.
- Explicit information about any volume limitation (i.e. data cap), speed, and other QoS parameters that may impact the IAS and any OTT service or content provided over the public Internet access.
- Clear explanation about the minimum, normally available speeds (i.e. bitrates) as well as maximum possible and advertised download and upload bitrates of the IAS for fixed broadband networks, or an estimation of maximum and advertised download and upload bitrates of the IAS in mobile broadband networks, and also information about how the deviations from the advertised download and upload bitrates can impact the services used by end-users.
- Clear explanation of the remedies available to the customer according to national legislation for cases of continuous or regularly recurring discrepancy between the actual and advertised/contracted performance of the IAS in respect to bitrates or other QoS parameters.

In Europe, QoS is considered as being part of consumer (i.e. customer) protection. In that regard, consumer protection provisions include QoS supervision and enforcement concerning IAS, which entitles NRAs to monitor the IAS and promote the availability of non-discriminatory IAS with the QoS which reflects current advances in the technologies (e.g. individual bitrates provided in the 2020s are ten times higher than average individual bitrates provided in the 2010s, and so on). For such purposes, NRAs may impose minimum QoS requirements and other necessary measures on ISPs for IAS provision (as well as for other services). Also, there are obligations imposed on ISPs (and in general, on telecom operators which provide any ICT/telecom services) to make available to the NRA all required information relevant to QoS and network neutrality [7].

The rules for network neutrality in the TSM in Europe have been established to avoid any fragmentation in the single market, and at the same time to create legal certainty for businesses which are using IAS and to make easier their work on an international level. According to the European viewpoint, network neutrality ensures that the Internet will remain a driving engine for innovations, including the recent new technologies (IoT services, 5G services, etc.) and those that will appear in the future.

According to the network neutrality rules enacted in Europe in 2016 [10], the European NRAs are obliged to closely monitor and ensure compliance with the provisions on open Internet, as well as to publish annual reports and share with other NRAs in the European region. In parallel with network neutrality regulations, the European countries (e.g. the EU) in 2015 introduced the Digital Single Market strategy, which is targeted toward removing national barriers in the European region to transactions that take place online, over the public Internet access. The main pillars for such a Digital Single Market are improving access to digital goods and services (through public Internet), creating an environment where digital network and services (which are all-IP based) can prosper, and promoting digital as a driver for growth (it is targeted to maximize potential growth of the digital economy, enhancing digital skills [11], and creating an inclusive digital society). For development of digital economy and digital society in Europe and worldwide, the crucial aspects are having open Internet access, based on network neutrality, as well as protection of digital consumers' rights.

8.6.2 Network Neutrality Approach in the United States

According to regulation in the United States, telecommunication services are subject to numerous regulatory obligations while information services are subject to fewer explicit obligations as they were felt not to be subject to market power so long as basic services were available on a non discriminatory basis.

This distinction between telecommunication and information services historically enabled the U.S. Federal Communications Commission (FCC) to avoid regulation of the Internet core. Until 2002 Internet access was regulated, then in the period 2002–2005 the FCC classified broadband access when it is bundled with IAS to be in fact an information service. In 2005 [12], the FCC issued its "Internet Policy Statement" to ensure the existence of broadband networks that are widely deployed, open, affordable, and accessible to all consumers, as well as to preserve and promote the open and interconnected nature of the public Internet network. The policy statement reflected the views of the FCC commissioners at that time, without adoption of specific rules or enforcement mechanisms.

Further, the FCC issued an open Internet ruling in December 2010 [13], which includes the following open Internet rules targeted to preserve free and open Internet:

- *Transparency*. Fixed and mobile broadband providers must disclose the network management practices, performance characteristics, and terms and conditions of their broadband services.
- *No blocking*. Fixed and mobile broadband providers may not block lawful content, applications, services, or block certain applications (e.g. OTT voice and messaging) that compete with their own services (e.g. fixed or mobile telephony, video services).
- *No unreasonable discrimination*. Fixed broadband providers may not (unreasonably) discriminate transmission of lawful Internet traffic over the broadband Internet access service.

Such ruling (in 2010) was challenged in U.S. courts, although it imposed fewer restrictions on mobile broadband Internet access than on fixed broadband access. The increase of mobile broadband in the 2010s with the deployment of LTE (long term evolution) has influenced the FCC's open Internet ruling with a new open Internet order in 2015 [14]. The main idea was to prevent the threats to open Internet by broadband providers which have tools to deceive consumers, degrade content, or disfavor the content they do not like. While in Europe at the same time (2015) network neutrality regulation which introduced specialized services was proposed, in the U.S. such (specialized) services are defined by what they are not, that is, broadband internet access services (BIASs). In the FCC open Internet order of 2015 [14] BIAS is defined as "a mass-market retail service by wire or radio that provides the capability to transmit data to and receive data from all or substantially all Internet endpoints, including any capabilities that are incidental to and enable the operation of the communications service, but excluding dial-up IAS." With such a definition BIAS also does not include enterprise services, VPN services, hosting services, or data storage services. The 2015 order goes further in protecting open Internet than the 2010 ruling, by specifying the following rules:

- *No blocking*. ISPs shall not block lawful content, applications, services, or non-harmful devices, subject to reasonable network management.
- *No throttling*. ISPs shall not impair or degrade lawful Internet traffic on the basis of Internet content, application, or service, or use of a non-harmful device, subject to reasonable network management.
- *No paid prioritization*. ISPs shall not engage in paid prioritization (here "paid prioritization" refers to the management of a broadband provider's network to directly or indirectly favor some Internet traffic over other Internet traffic).

Additionally, the 2015 open Internet order allowed the FCC to enforce the rules through investigation and processing of customer complaints. The 2015 order does not directly regulate Internet interconnection but allows interventions on a case-by-case basis. The 2015 order fully includes mobile broadband networks together with fixed ones, unlike the 2010 order which was stricter toward fixed networks (due to lower availability and lower penetration of mobile broadband technologies before 2010). Also, prohibition of paid prioritization over the Internet was explicitly noted in the 2015 order unlike that of 2010. The open Internet order of 2015 has faced various court challenges [8].

After debate started at the end of 2016, the FCC published new rules at the beginning of 2018, which have returned to a light-touch framework for broadband Internet access by reclassifying the IAS as an information service (which is less heavily regulated) rather than a telecommunication service (which is more heavily regulated) according to the U.S. Communications Act. The FCC's restoring Internet freedom order took effect in June 2018 [15], and it is targeted at providing a framework for protecting an open Internet while paving the way for better, faster, and cheaper Internet access for consumers. This framework for protecting Internet freedom has focused also on consumer protection (against ISPs for anticompetitive acts or unfair and deceptive practices), and transparency about ISPs' business practices (to publicly disclose information regarding their network management practices, performance, and commercial terms of service). The main obstacle with this order was to remove unnecessary regulations to promote broadband investment (including fixed and mobile).

Generally, in the U.S. the main issue which changes from one open Internet order to another is the following: does IAS qualify as an information service or a telecommunications service? While the open Internet order of 2015 classified IAS as telecommunication service, the order of 2018 reclassified it as an information service. Fundamentally, all Internet orders in the U.S. are targeted toward an open Internet; the difference is in the legislation rules which refer to it.

Someone outside the U.S. may note that telecommunications in the twenty-first century have become ICT, which inseparably include both telecommunication and information services.

8.7 Challenges Regarding QoS and Network Neutrality

From a regulatory perspective, QoS management creates numerous challenges. Differentiated management of QoS potentially offers benefits not only to network operators but also to content and application providers, and also to consumers and other end-users. However, OTT service providers (Google and its YouTube, Netflix, Amazon, etc.) are not entitled to QoS provision for their services due to network neutrality rules (when and where they are in force). However, OTT services also tend to provide higher user experience because that gives certain guarantees that users will remain customers of that service and hopefully the customer base will increase (if it has not reached saturation point where almost all possible users have already become service customers). On the other side, network neutrality prohibits OTT service differentiation in the access networks for the IAS. However, Internet traffic is transferred over the same links with traffic from managed IP networks (e.g. carrier VoIP, IPTV, critical IoT services), hence QoS differentiation is certainly applied in the core and transit networks.

Overall, striking a sensible balance in QoS approaches is not always easy. For example, considering the business side of national telecom operators, QoS-enabled services require higher investment and higher operational costs due to QoS functions in the networks. One should note that IAS is also an QoS enabled service, while individual OTT services and applications (provided over the IAS) are not. Additionally, there is a trend to invest in ultra-broadband networks, and targeted individual speeds in the 2020s are at least 100 Mbit/s over both fixed and mobile access networks. Most of that capacity is in fact for IAS (two-thirds of that is for video traffic over the Internet), then for IPTV (over managed IP networks), and the least for carrier-grade VoIP and other specialized (i.e. managed) services (over managed IP networks).

Network neutrality is often considered to be the key driver of innovation in the application and service space. It appeared with the public Internet, so from the outset the Internet was based on a network neutrality principle and best-effort approach for service provisioning. However, since the 1990s, with the convergence of telecommunication networks and services toward all-IP networks and services based on Internet technologies, there is a need for end-to-end QoS provisioning in the new all-IP environment. That is necessary for legacy telecommunication services which are now provided over IP networks (managed IP networks) such as carrier grade VoIP and IPTV, VPN business services, as well as for emerging critical IoT services (all of them belong to the specialized services).

So, all services are being transferred over IP-based networks, but network neutrality refers only to IAS and not to every IP-based network and service. For example, NGN VoIP as a PSTN/ISDN replacement is not part of the public Internet and network neutrality does not refer to it. OTT voice services such as Skype and Viber are provided via IAS, over the public Internet, based on network neutrality principles.

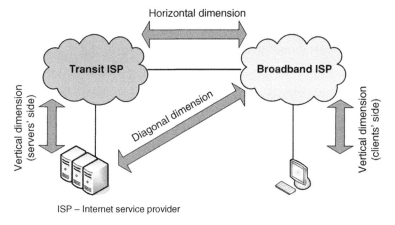

ISP – Internet service provider

Figure 8.5 Three dimensions of network neutrality.

In general, one may distinguish three main dimensions regarding the relation between QoS differentiation and network neutrality [16], which are typically accompanied with separate policies (Figure 8.5):

- *Vertical dimension*. This dimension refers to connections to an ISP for transfer of Internet traffic to/from end-users and from/to service providers' data centers. Hence there are two sub-dimensions vertically:
  - o *Users' vertical subdimension*. This sub-dimension refers to the client's side (considering the client–server Internet communication), that is the uplink/downlink relation between the ISP and its customers who use the IAS provided by that ISP.
  - o *Service providers' vertical subdimension*. This refers to the server's side (of the Internet client–server communication), i.e. this dimension refers to the relation between the ISP and service/application or content providers.
- *Horizontal dimension*. This refers to interconnections between the ISPs as the same level entities considering the network value chain. The horizontal dimension also covers relations between different types of ISPs, such as broadband ISPs (which provide broadband access) and commercial ISPs (which provide transit of Internet traffic). Typically, their relations are based on peering or transit agreements.
- *Diagonal dimension*. This refers to the relation between entities that are in different but interlinked value chains. Such common diagonal dimension refers to the relation between the content and application/service providers on one side and broadband ISPs on the other side.

Considering the dimensions given above regarding IAS, QoS, and network neutrality, one may note that the vertical dimension denotes the network provider (e.g. the ISP) versus the end-user or service provider. Then, the vertical subdimension between the ISP and the end-user as well as the diagonal dimension (between ISP and OTT service providers) are the two (sub)dimensions to which network neutrality refers on the global scale. That means that network neutrality prohibits QoS differentiation for Internet traffic between the ISP network and end-users (as clients) as well as prohibiting business relations (in respect of QoS differentiation of the service providers' Internet traffic) between content and application/service providers and ISPs (which provide IAS toward end-users, on the client's side). On the horizontal dimension, QoS differentiation due to reasonable traffic management should be allowed and sometimes it is necessary for the purposes of network planning and dimensioning.

8.8   Network Neutrality Enforcement

Network neutrality is essential for innovation in the OTT services space, which gives freedom to end-users to choose content and applications they want to consume, from local (i.e. national) or global (i.e. international) content and application providers (i.e. OTT service providers).

Considering the network neutrality ruling in different parts of the world (e.g. Europe, the U.S.), typically non-network-neutral practices refer to blocking, throttling, and preferential treatment of certain OTT services (by the ISP, i.e. telecom operators that provide IAS) [17]. Regarding the duties of NRAs, there are three types of actions to monitor and ensure compliance to network neutrality rules:

- *Supervision.* It includes network neutrality monitoring by the NRA by gathering information from ISPs on monitoring of restrictions of end-user rights, contractual conditions and commercial practices, traffic management, assessment of IAS performance, and impact of specialized (i.e. managed) services on the general quality of the IAS, as well as transparency requirements on the ISPs.
- *Enforcement.* It includes a variety of interventions and measurements on the IAS provided by different ISPs in the given country.
- *Reporting.* This is typically done by the NRAs based on the results from their monitoring of the IAS and specific Internet-based OTT services.

In order to ensure availability of non-discriminatory IAS at quality levels that reflect advances in technology (e.g. newer fixed and mobile broadband technologies provide higher individual and aggregate bitrates), the NRAs may decide to:

- oblige the ISP to remove all factors that are causing IAS degradation;
- define technical characteristics of the IAS (e.g. via QoS parameters and their target values) that can be the subject of regulation (e.g. for removal or revision of prohibited traffic management practices);
- impose and enforce minimum QoS requirements on IAS, as well as any other necessary measures (e.g. the obligation of the ISPs to ensure sufficient network capacity for the offered IAS packages to their customers);
- enforce IAS with the desired level of QoS (e.g. continuous availability of contracted minimum and advertised bitrates in uplink and downlink) by monitoring QoS of the IAS and making the results publicly available (for the light touch approaches) so customers can make informed choices;
- impose ceasing orders for particular specialized services (which are not critical) when there is no sufficient capacity available for the IAS. However, such an approach is not a primary one, but a possible one. For example, emergency calls cannot be subject to ceasing, neither can critical IoT services such as control of driverless vehicles and drones;

- impose fines for various QoS degradations of IAS or degradations (blocking, throttling, etc.) of particular OTT services and applications, such as periodical penalties (e.g. on daily, weekly or monthly basis), in accordance with national legislation. However, one should note that fines will not change the QoS of the IAS in the short term when there is not enough capacity to support the desired bitrates as the main QoS parameter (after IAS availability in the first place) because such a case will require a longer time period for network capacity upgrade. The other possible measure to improve IAS QoS unpopular and unwanted: increasing prices of IAS services (or decreasing data caps for IAS) which will decrease Internet traffic due to lower affordability of IAS.

In the cases of OTT services blocking and/or throttling, discrimination or similar measures applied by the ISP, the NRA could prohibit such restrictions which clearly violate network neutrality regulation. Such measures include the following:

- prohibit the ISP from blocking and/or throttling specific OTT services;
- prohibit discriminatory (in regard to network neutrality) traffic and congestion management practices in regard to individual OTT services (e.g. prohibit noticed QoS differentiation within the IAS);
- set a minimum required QoS level in accordance with the development of the ICT/telecom technologies at the given time (e.g. European countries have set a target of at least 50% of their population to have IAS with minimum 100 Mbit/s after 2020) [6]. Also, available IAS speeds in downlink and uplink should be comparable to advertised/maximum speeds.

Requirements and different measures in regard to network neutrality can refer to one or several ISPs or can be applied generally to all ISPs in the given country. However, the imposition of these requirements regarding IAS and network neutrality principles should be judged on the basis of their effectiveness, necessity, and proportionality.

Effectiveness refers to efficient use of the measures regarding the results obtained from putting them into practice (e.g. prevention or removal of QoS degradation of IAS). Necessity refers to selection of the less troublesome solutions in network neutrality enforcement – for example, penalties should not be first on the list of regulatory tools, but rather, naming and shaming of ISPs or warnings issued to ISPs with IAS degradation or network neutrality infringements. If a light touch approach is deemed insufficient or ineffective, more drastic enforcement tools can be applied (e.g. financial penalties). Proportionality refers to adequate application of enforcement measures, and that the imposed obligations are appropriate to the goals and cannot be replaced with other, less interfering and equally effective alternative actions.

8.9 Discussion

Network neutrality is part of the core philosophy of the Internet network, so it would not disappear but typically would change its definition or positioning on the telecom ("digital") market. In fact, network neutrality refers to an Internet network which does not offer any guarantees to any particular flow or content. It has sped up innovation in the telecom sector with the appearance of the Internet available to everyone since the middle of the 1990s. Why? Because OTT service providers do not need to adhere to strict national regulations for services provided with QoS guarantees. To prohibit censorship of content and applications or their differentiated treatment, network neutrality (as a native feature of the Internet) was accepted on the global scene by most countries worldwide. That allows Internet traffic to traverse over terrestrial networks through different countries without blocking or filtering. Why? It is related to the freedom of the individual to choose what they want to use as a service and which content they want to consume. Such an approach can be called "freedom" in the telecom/ICT space, in the cyber world. The cyber world by itself is mimicking the real world and all of its functions, so freedom in the real world (to move freely, to act freely, etc.) is translated on the Internet via the principles of network neutrality (it can also be called Internet freedom or Internet equality). As in the real world, there are legislations and laws in different fields (e.g. financial or banking sector laws, criminal law, human rights laws) which are obligatory for Internet users (when they are using OTT services/applications) as they are obligatory for citizens in the real world. However, laws can differ from country to country (e.g. online betting is allowed in some countries while it is forbidden in others) and hence such rules are mapped into the Internet cyberspace although it is network neutral (it does not mean that it is neutral to law enforcement).

Is traffic management by ISPs contrary to network neutrality? The clear answer is that it is not. Traffic management is a tool to provide better IAS, which directly influences all OTT services that are in fact provided on a network neutrality basis. In practice, traffic management not only potentially offers benefits to network operators (i.e. telecom operators), it provides better QoS and QoE to OTT content and application/service providers (either on the side of their customers or on the side of their servers in data centers connected to the Internet), and also to end-users.

Due to network neutrality, OTT service providers cannot establish agreements with network operators (e.g. for provision of a higher level of QoS). If they could, then well established and big global OTT service providers might have higher QoS (e.g. guaranteed bitrate for certain services or priority over other Internet traffic) and with that they would further be able to monopolize their position in the global market, especially because there are no global laws (including the ICTs) but consensus to support (or not) certain network policies such as the network neutrality approach. That does not forbid service providers from establishing contracts with network providers (e.g. telecom operators) for delivery of services over managed IP networks (e.g. for interconnection of different data centers in the service provider's CDNs). But when the data is transferred over the public Internet (by using the IAS), the service providers (local or global) cannot have any prioritization regarding QoS. So, it is possible (and typical) that Internet traffic over broadband and ultra-broadband is delivered on the basis of network neutrality and the same traffic between different data centers of the service providers be exchanged over managed IP networks for which strict QoS can be guaranteed. That provides the possibility of service providers placing CDN data centers as close as possible to end-users in different regions, and then applying differentiation of the traffic, then performing aggregation per traffic type (e.g. voice, video, and other data) and transferring such different traffic types between their own data centers with different QoS (e.g. by using different VPNs). In this manner, network neutrality does not prevent service providers from improving the QoS of their services, especially with broadband and ultra-broadband Internet access offered by national telecom operators over both fixed and mobile access. Telecom operators directly influence the QoS of the OTT services by the bitrates (and delays) provided over their own networks as well as the data cap offered to end-users. Overall, sustainable ultra-broadband bitrates for IAS benefit all parties, including service providers (better QoS on average for their services over the network neutral Internet), network operators (satisfied customers, who will remain users of their services), and finally the most important (regarding network neutrality), end-users (they benefit from the higher QoS level for IAS as a whole, and for the OTT services that they use over the IAS on a network neutral basis).

Overall, network neutrality provides a greater possibility for service innovations by local (i.e. national) and global OTT service providers, and that is innovation without permission. That means that anyone can develop an application (e.g. server and client applications) and offer them on the public Internet. However, the success of such OTT service will depend on its usability (is it useful for end-users?), popularity, and sustainability (e.g. investments in data centers and CDNs, if the service provider is global). Similar to QoS-enabled services over managed IP networks (provided by telecom operators), higher QoS requires more investment and imposes greater operational costs on OTT service providers.

Finally, network neutrality does not mean no QoS for OTT services, just changes to the approach for how to deliver it to end-users. Different players in the value chain have different roles, which are important for all sides. Network operators (which in fact are the telecom operators running under national legislations) benefit from selling IAS to end-users, while service providers benefit from the usage of their services and making revenues from different channels (e.g. by selling ads and commercial space to different businesses). The society also benefits from the network neutral IAS for its continuous digitalization and development of public digital services and digital markets.

References

1 Janevski, T. (2015). *Internet Technologies for Fixed and Mobile Networks*. Norwood, MA: Artech House.

2 ITU, Trends in Telecommunication Reform 2013, April 2013.

3 Regulation (European Union) 2015/2120 of the European Parliament and of the Council, Measures concerning open Internet access and amending Directive 2002/22/EC on universal service and users' rights relating to electronic communications networks and services and Regulation (EU) No 531/2012 on roaming on public mobile communications networks within the Union, L310/1-18, November 2015.

4 BEREC, Guidelines for Quality of Service in the Scope of Net Neutrality, November 2012.

5 Janevski, T. (2003). *Traffic Analysis and Design of Wireless IP Networks*. Norwood, MA: Artech House Inc.

6 ITU, Quality of Service Regulation Manual, 2017.

7 BEREC, Guidelines on the Implementation by National Regulators of European Net Neutrality Rules, August 2016.

8 J. Scott Marcus, New Network Neutrality Rules in Europe: Comparisons to Those in the U.S., May 2016.

9 BEREC, Net Neutrality Measurement Tool Specification, October 2017.

10 Regulation (EU) 2015/2120 of the European Parliament and of the Council, Measures concerning open Internet access and amending Directive 2002/22/EC on universal service and users' rights relating to electronic communications networks and services, November 2015.

11 Toni Janevski, Emerging Trends and Technologies in ICT and Capacity Building Challenges, ITU Journal on Capacity Building in a Changing ICT Environment, June 2018.

12 FCC 05-151, Policy Statement, 23 September 2005.

13 FCC 10-201, Report and Order, 23 December 2010.

14 FCC 15-24, Report and Order on Remand, Declaratory Ruling, and Order, 12 March 2015.

15 FCC 17-166, Declaratory Ruling, Report and Order, 4 January 2018.

16 European Parliament, Network Neutrality Revisited: Challenges and Responses in the EU and in the US Study, December 2014.

17 James Allen, Andrew Daly, J. Scott Marcus, David de Antonio Monte, Robert Woolfson, Study on Net-Neutrality Regulation, Final Public Report for BEREC, 18 September 2017.

9

QoS Regulatory Framework

The quality of service is important for each network and for each service in the tele-com/ICT world. The importance and the need for telecommunication services impose requirements on their delivery. In general, quality is important in many areas, such as food quality, roads quality, bridges quality, airplanes quality, etc. Also, quality is required for many public services (e.g. electricity, cleaning, water supply system) which improve the lives of citizens. Telecoms provide communication services (such as telephony or messaging) and information services (such as TV, the Web, etc.). Certain services are less important than others (e.g. ordinary telephone calls vs. emergency calls). With the development of smart services that are based on massive machine-type communication (e.g. smart cars, smart homes, smart factories, smart cities), the importance of QoS increases. However, to maintain the desired level of QoS, it should be regulated. This chapter covers the QoS regulatory framework and its applicability in different networks (fixed and mobile) and in different countries (regarding the development of their ICT/telecom markets).

9.1 Scope of QoS Regulation

Quality of service is being monitored in most countries worldwide [1]. From a historical point of view, QoS requirements have been applied to voice services (provided by telecom operators), but more recently regulators have been incorporating minimum QoS requirements also for data services (that is, for IAS). These imposed QoS requirements (i.e. by NRAs) can span from high-level transparency guidelines (without stricter obligations to ISPs) on how the information on traffic management techniques is disclosed to end-users, toward requirements to be satisfied given quality indicators for data network performance (i.e. IAS performance) for fixed and mobile broadband/ ultra-broadband providers.

The main scope of QoS regulation is to ensure that customers experience services of good and sustainable quality. What level of QoS do customers want or need? It depends on the price-quality tradeoff from the customer's viewpoint because higher QoS tends to cost more than lower QoS for the same service (e.g. higher bitrates in fixed ultra-broadband network cost more than lower bitrates, as bigger data cap in

QoS for Fixed and Mobile Ultra-Broadband, First Edition. Toni Janevski.
© 2019 John Wiley & Sons Ltd. Published 2019 by John Wiley & Sons Ltd.

mobile networks costs more than the smaller one). The question is also, what QoS are customers willing to pay for for the service? Then, what measures (if any) will be conducive enough to customers to have their services available at the QoS level that they want and need? If there were no QoS regulation, would telecom operators be motivated to provide services with satisfactory QoS at prices that reflect the service provisioning costs? Would it be possible that certain NRA interventions would make QoS worse for customers? If not, the next question is, what kind of applied policies interventions can be beneficial for higher QoS? In general, different countries may approach such issues in different ways, for different reasons (does the broadband infrastructure need to be built or does it already exist, current national and regional broadband targets, etc.). Simply put, there is no single best way to approach QoS regulation.

One possibility is to impose own QoS standards. However, such cases should be supported by market forces that should be sufficient to ensure appropriate QoS. So, imposing QoS standards is directly connected to the degree of competition in the ICT/telecom market. Thus there are different approaches for QoS regulation between countries with developed ICT/telecom markets and those countries whose markets are developing. The main characteristic in regard to QoS in countries with developing markets is the level of competition and diversity of organization of the market. These questions (regarding QoS regulation) are relevant to both traditional networks and IP-based networks.

In general, the scope of QoS regulation in all countries worldwide includes the following [2]:

- to help ICT/telecom providers' customers to make informed choices;
- to check claims by telecom operators;
- to understand the state of the ICT/telecom market;
- to maintain or improve QoS in presence of competition;
- to maintain or improve QoS in absence of competition;
- to help telecom operators to achieve fair competition;
- to make interconnected networks work well together for end-to-end QoS provisioning.

What are the final results of QoS regulation? Well, the result is having better QoS for targeted services, either mass or critical services (e.g. different QoS targets may be imposed for different types of services, such as open Internet access and critical IoT services). NRAs may impose different types of sanctions on national telecom operators (because NRAs have jurisdiction only on a national level) as well as on global OTT service providers (for services offered to users in the given country, where the NRA also has jurisdiction) which are not meeting the specified QoS and other obligations (e.g. privacy). Such sanctions may range from publicizing failures, fines due to low QoS, license suspension (e.g. for mobile providers), toward criminal prosecution in some cases (Figure 9.1) [3]. Considering the given ICT statistics, many countries have already imposed certain fines in the process of QoS regulation. The second on the list of NRA actions regarding QoS enforcement is publicizing failures, followed by license suspension (Figure 9.1). However, with QoS regulation by the NRA (or related ICT ministry in the government), there is no single measure that fits all cases. Concrete sanctions and measures to deal with QoS in a given country (e.g. applied by the NRA to providers) strongly depend on national circumstances as well as maturity of the telecom/ICT market.

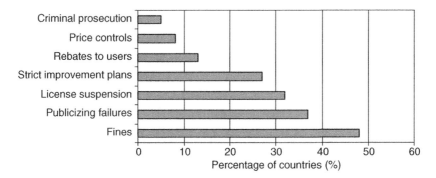

Figure 9.1 QoS measures applied by NRAs to telecom operators.

9.2 Fundamentals of QoS Regulation

The fundamental approach in QoS regulation includes the following steps in the definition of the QoS regulation framework:

- *Step 1.* Development of QoS standards by global SDOs, such as the ITU, ETSI, etc. QoS standards can be implemented with their specification in license conditions, ICT regulation, or industry guidelines.
- *Step 2.* License regulation by specification of QoS standards and/or QoS parameters and their target values in the provider's license conditions, ICT regulation (e.g. telecom/ICT law), or imposed industry guidelines.
- *Step 3.* KPIs measurement techniques should be specified for a given set of measurable parameters. Such QoS parameters are defined for QoS measurements on a national basis (or globally). Examples of technical KPIs include voice call success rate, call drop rate, bitrates in downlink and uplink, messages delivery time, etc. Non-technical KPIs are focused on the customer's subjective opinion and typically include parameters such as billing accuracy, fault, customer care response time, or satisfaction from the response, etc.
- *Step 4.* Monitoring of defined KPIs. This includes measurement of technical KPIs via network auditing, drive tests in mobile networks, etc., as well as subjective customer surveys of non-technical KPIs via questionnaires or customer responses. Different measurement methods (to monitor KPIs) can be used in different countries. Such KPI monitoring can be performed by NRAs, telecom operators, independent organizations, and also by customers (e.g. via installed software or hardware probes in their home networks or on their devices).

The purpose of monitoring defined KPIs (used to define the required QoS in a given country) is to detect degradation of QoS when it appears and then to apply appropriate measures. For example, QoS enforcement can be performed through a regulatory notice (e.g. publishing KPI monitoring results on a public website, through press releases, via directives) to allow customers to make informed choices. But when the applied QoS enforcement measures are not enough to improve degraded or low level QoS, more drastic QoS enforcement measures should be taken, such as financial penalties (e.g. applied to non-conforming operators or service providers) or opening a dispute.

In general, there are four main regulator approaches to deal with QoS matters [2]:

- *Obtaining appropriate information on the level of QoS and identifying problem areas.* This approach is needed to obtain information (from QoS measurements, customer surveys, etc.) which will be needed for a decision on imposing certain QoS measures.
- *Undertaking a constructive dialog with the operator.* The dialog is the first approach by the NRA (directed to operators) to define a set of KPIs, or to point to QoS degradation issues and require responses from the operator (in order to encourage and foster improvements on QoS).
- *Publishing information on telecom operators' QoS performance.* This is the typical approach in developed ICT/telecom markets, which aim to provide QoS measurement results to customers so they can make informed choices.
- *Imposing explicit regulations on QoS.* In this case there are specified and required minimum levels of QoS, and fines for operators with continuous unsatisfactory QoS (over a given time period) or compensation (to the customers) are defined.

QoS regulation has as its main goal provision of QoS at a certain level that reflects the development of the technologies at a given time. There are two main approaches regarding QoS regulation:

- *Regulation-oriented approach.* In this case QoS reporting is performed by the NRA, the QoS targets are defined in national regulations, and fines are specified by the regulator if national operators do not achieve QoS targets.
- *Customer-oriented approach.* In this case QoS reporting is toward customers. The minimum QoS levels are provided in the contracts, while compensation for poor quality performances is payable to the affected customer.

One should note that QoS regulation imposes additional costs (e.g. for the QoS measurement systems and human capacity for performing measurements and analysis) and such additional costs should be weighed up against the benefits of QoS assessment. That can be accomplished by focusing QoS measurements on services and areas where there are known (or detected) quality problems. However, QoS problems can change dynamically over time, including the location of appearance as well as the time of appearance and service being affected, so there needs to be a certain flexibility in the approach to QoS assessment.

Although QoS regulation is based on a selection of QoS parameters, not all existing parameters should be subject to regular measurement action. As already stated, most important QoS parameters are referred to as KPIs. The target values for the same KPIs may vary from one country to another, which is dependent upon the development of the telecom/ICT market and its competition (e.g. call drop rate can be maxim 1% in developed ICT/telecom markets, but up to 10% may be acceptable in some countries with developing telecom/ICT markets [1]). QoS measurements as well as chosen KPIs (for monitoring) need to distinguish between market segments (e.g. private consumers, businesses, wholesale, and retail offerings), reporting areas (e.g. rural areas, urban areas, roads, and railways), operators (considering their market shares), and services (different KPIs are used for voice, video, messaging, IAS, etc.). In all cases the assessed KPIs should be those that have the highest impact on end-users, such as bitrates, delay, jitter, and packet losses for the IAS.

9.3 QoS Regulation Guidelines by the ITU

QoS regulation is needed in both legacy telecommunication networks and IP-based ones. In legacy networks the allocation of resources in access and transport networks was based on a simple but effective concept (allocation of a channel on all network segments end-to-end) accompanied by heavy regulation in place. In IP-based networks there are heterogeneous services offered over the same networks, which makes it harder to define a common set of QoS parameter requirements. Also, due to the autonomous nature of different building blocks of the Internet (the autonomous systems) regarding traffic and QoS management practices, the responsibility for end-to-end QoS has been lost in IP environments. In IP networks all services should be considered as applications that are executed in the end-users' terminal devices. The Internet is built on best-effort principles, hence IP networks by themselves cannot provide self-standing end-to-end QoS. Therefore, a typical approach is performing QoS differentiation for services offered over IP networks (this includes most of the services).

Due to a more complex approach for QoS provisioning and its regulation in IP-based networks, there are different challenges to different parties in the values chain, which include the following:

- *SDOs.* These organizations do not define the QoS standards by themselves only, but they are typically industry stakeholders. If stakeholders wish to not invest resources in globally recognized QoS standards, then there is not very much for SDOs to do in that regard.
- *Network equipment manufacturers.* They typically rely on QoS requests from network and service providers. However, they also include the standardized QoS protocols and mechanisms in their equipment (e.g. switches, routers, gateways), which can be used optionally by different providers.
- *Terminal device manufacturers.* They are directly confronted with the mass end-users' market of today. While in PSTNs the terminal related standards were targeting minimum requirements regarding the network interface, in an all-IP network the terminal standards may also target provision of high-level end-to-end QoS toward the customer. In mobile networks the terminals (e.g. smartphones) have the highest influence on QoS because for many multimedia services QoS performances heavily depend on the terminal capabilities and supported radio interfaces (including supported frequency spectrum, access techniques, and hence bitrates in downlink and uplink).
- *Network operators and service providers.* The network providers (i.e. national telecom operators) make huge capital investments in both infrastructure and access technologies. The network operators react to QoS issues by investing in new capacity or by optimization of deployed network infrastructure. Additionally, network operators use various traffic management tools to provide the desired QoS level to different services.

According to ITU guidelines [2], QoS regulation requires the NRA to perform the following activities (Figure 9.2):

- *Defining QoS parameters.* The QoS parameters which are subject to monitoring and/or enforcement are defined by the NRA with prior consultation with telecom operators in the given country.
- *Setting QoS targets.* The target values for the defined QoS parameters are also set by the NRA, based on consultation with telecom operators and information gathered

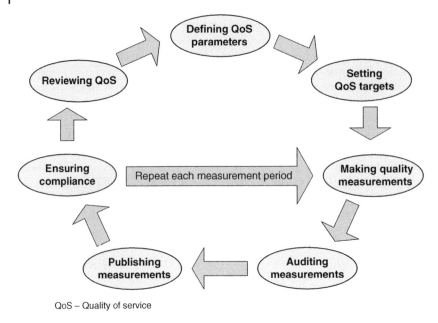

QoS – Quality of service

Figure 9.2 QoS regulation activities.

from monitoring of the collected measurement data (if any) and consulting the national, regional, and global QoS standards and practices.

- *Making quality measurements.* Measurements can be made by the telecom operator, the NRA, external measurement agencies, or end-users.
- *Auditing measurements.* Measurements verification could be completed by the employees of the telecom operators (i.e. "self-certification" process), or they can be audited by external QoS audit agencies, or checked by the NRAs themselves.
- *Publishing measurements.* Typically, QoS measurement results should be published by the NRA (e.g. to offer comparison of QoS levels provided by different telecom operators).
- *Ensuring compliance.* For this purpose there exists a range of approaches that the NRA can use, which typically start with "naming and shaming" practices (the encouragement) toward stricter QoS regulation through financial penalties and similar, more drastic enforcement measures. In general, for ensuring compliance both encouragements and enforcements should be proportional to the QoS goals (that is, achieving better QoS).

9.4 SLA and QoS Regulation

QoS parameters and their targets are specified in SLAs between the service provider and the customer, where the customer can be an individual (residential) user, business user, or service provider (e.g. CDN data centers of OTT service providers are also connected to the Internet via network providers).

According to ITU-T E.860 [4], an SLA is a formal agreement between two or more entities that is reached after negotiating activity, with the scope of assessing service characteristics, responsibilities, and priorities of each part. The SLA may include statements about performance, tariffing and billing, service delivery, and compensations. On a national level, such commercial agreements are typically approved (initially) and monitored by NRAs.

When one considers a multi-provider environment, there are also relationships between interrelated and interconnected providers. For example, one service provider that delivers a service to its own customers must use services provided by other providers (e.g. transit providers). With many service providers in the end-to-end chain, it may become much more complex to guarantee the QoS level as stated in the SLA. Therefore, in IP environments, due to the autonomous nature of different service providers, the SLA typically includes QoS in the access part and in the core network of the telecom operator which provides the access. Most SLAs are established between telecom operators which have local infrastructure and customers who use that infrastructure to access both Internet-based OTT services (i.e. data services) and specialized services (provided via managed IP networks and offered with QoS guarantees). Typically, OTT service providers do not guarantee specific technical QoS to end-users because they offer their services on the principles of network neutrality over the open Internet network and thus they cannot have differentiated QoS provisioning regardless of service types (voice, video, messaging, Web, etc.).

Because end-to-end communication typically may involve different networks (which are operated by different network providers), the standardized approach regarding QoS is one-stop responsibility [4], which is agreed between the service provider and the customer by using an SLA. Such an approach allows the user to have an SLA only with the primary service provider (that is the telecom operator which provides the access network to the user, either to the public Internet or to managed IP networks) as the sole provider responsible for the QoS received. Such an approach can continue between the primary service provider (in regard to the given user) and its sub-providers, though this is not directly related to the SLA signed with the end-user. The QoS agreements with sub-providers (e.g. transit network providers) refer more to the aggregate traffic at the interconnection points between the given provider and its transit providers.

9.4.1 QoS Agreement

The part of the SLA which defines the QoS is referred to as the QoS agreement. Its main goal is to define the agreed QoS with the end-user (when signing the agreement) and to obtain the end-user's satisfaction. For the purposes of the SLA and its corresponding QoS agreement, the definition of QoS parameters (e.g. minimum, average, and maximum supported bitrates are different definitions of the bitrate as a QoS parameter) is essential [5]. In all cases the QoS parameters should be defined in a clear way, which means with "simple" language that can be understood by the ordinary end-user while more technical details in the QoS agreement can be used for QoS agreements with other network and service providers.

The definition of QoS parameters is required for development of the SLA and corresponding QoS agreement (Figure 9.3). Overall service level requirements influence the choice of QoS parameters, including service dependent and service independent QoS

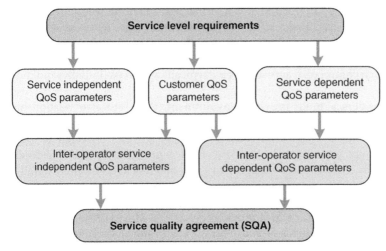

QoS – Quality of service

Figure 9.3 Service quality agreement (SQA) and QoS parameters.

parameters as well as customer QoS parameters. Finally, that results in the definition of the QoS agreement. The QoS parameters for the SLA purposes can be divided into the following groups:

- *Direct QoS parameters.* These refer to a concrete service element and can be determined by direct objective measurements.
- *Indirect QoS parameters.* These are defined as a function of one or more direct parameters. One of the most important indirect QoS parameters is service availability (SA).

For example, QoS parameter service availability can be calculated by using the following simple equation (SA refers to service availability, while UA refers to service unavailability):

$$SA\,[\%] = 100\% - UA\,[\%] \tag{9.1}$$

Further, UA can be calculated by measurement of the outage intervals and their averaging over the total measurement time, given as follows:

$$UA\,[\%] = (\Sigma\,\text{outage interval}/\text{active time}) \times 100\% \tag{9.2}$$

In this concrete example, given with Eqs. (9.1) and (9.2), the indirect QoS parameter is expressed through a direct QoS parameter such as average outage time (i.e. UA).

9.4.2 SLA and QoS Regulation

Definition of QoS parameters in SLA contracts is required for promotion of the desired QoS level, but such agreements are not always effective because the desired QoS targets may not be clearly defined. For example, IAS with bitrates up to 100 Mbit/s does not clearly define whether the user can expect bitrates of 1 or 11 Mbit/s or any number of Mbit/s up to stated maximum bitrate, and whether such provision will be continuous over time. To be effective, a given SLA should include the following:

- The minimum QoS level offered to the customer specified by selected number of QoS parameters which are important to end-users (i.e. KPIs), not the average level to be achieved for all customers or the values (of the KPIs) that are provided only at shorter time intervals.
- The compensation payment for situations when the minimum level is not achieved. The sum should be proportional to the degree of failure. However, one may note that compensation payment is an option, not a mandatory part of QoS agreements (e.g. it is typically not included in SLAs between telecom operators and end-users in developed telecom/ICT markets).

When compensation is used as a measure for non-conformance with the SLA and its QoS agreement, one possible mechanism is compensation to be paid automatically without the requirement that the customer makes a claim. Also, it is possible to enforce penalty clauses for failing to meet QoS targets considering relevant national legislation. The main goal of financial penalties should be more toward encouragement for better QoS. But when QoS degradation persists, the penalties may be charged proportionally to the interruption period for one or more QoS parameters specified in the SLA. Another option (e.g. to avoid penalties) is to provide ease of switching service provider (i.e. telecom operator), which means release from the SLA (with the previous provider) before its expiration.

Overall, the SLA plays an important role in QoS regulation because it includes the main KPI targets between the service provider and the customer (i.e. end-user). Typically, the NRA is the authority which monitors the format of the SLA between the telecom operators (as service providers) and end-users, and provides means for action in cases when the QoS level specified in the agreement is not provided.

An NRA can impose certain requirements on selected KPIs such as bitrates, for example, as part of national broadband strategy. In that manner, by imposing requirements to specify minimum bitrates (where it is technologically feasible, e.g. in fixed broadband networks), the NRA can ensure that all customers in a given country will receive a certain level of broadband and ultra-broadband access. The final packages offered to customers should be driven by the telecom/ICT market and its competition (in developed ICT/telecom markets), while the NRA may define the minimum bitrates that can be offered to end-users to foster penetration of broadband/ultra-broadband access on a national basis because it is of significant importance to society (e.g. for the development of the single digital market), not only a part of the market "battle" (for customers) between national telecom operators.

9.5 Specifying Parameters, Levels, and Measurement Methods

QoS regulation is directly related to specification of QoS parameters, their targets, and measurement methods. QoS measurement methodology should be established in such a manner to be able to provide estimation of the given set of QoS metrics.

Regarding measurements in the 2010s and the 2020s, they are almost all on IP-based networks which have either fixed or mobile broadband access (in the 2010s) or ultra-broadband access (in the 2020s). IP-based networking is based on use of TCP or

UDP as the main transport protocols between the IP (IPv4 or IPv6) on the network layer and the application on the top. In that regard TCP is always based on duplex connection because TCP segments travel from node A to node B and acknowledgements (ACKs) travel in the opposite direction (from node B to node A). Also, the Internet is based on a client server paradigm (in both networking cases, TCP/IP or UDP/IP) where the client sends a request (from the client application) to the server (to the server application) and receives a response from the server (e.g. with the requested data, such as a Web page with its content, video stream, audio stream, etc.). However, outbound and inbound traffic paths (or routes) may differ, so any measurements at the intermediate network nodes (i.e. routers, gateways) may practically cover only one way of the service. Such an approach refers to passive measurements in the network by monitoring (for measurement purposes) user traffic. For such reasons, QoS is assessed mainly by active measurements where the measurement points are in fact on one of the communication endpoints (e.g. client or server side of the application). For example, if the operator, the NRA, independent agency for QoS measurements, or end-users want to perform QoS measurements on the operator's network, they use a client application installed at the end-user's side (or hardware/software probes installed in defined locations for measurement purposes) and the server endpoint (for active measurements) is typically connected to the IXP either nationally (for assessment of QoS in the targeted telecom operator's network) or internationally (for assessment of end-to-end QoS in the more realistic scenario where traffic traverses multiple autonomous networks on the path between the two endpoints, the client and the server).

9.5.1 Defining QoS Parameters

For the purpose of defining QoS parameters, there is typically a need to involve telecom operators in the process. To avoid possible stronger influence (by operators) in such a consultation process, the NRA needs to have strong leadership, taking into the account the various factors that may influence the choice (of QoS parameters). However, the

Table 9.1 List of proposed parameters.

Customer interface	Network infrastructure	Service functionality
Complaint submission rate	Broadband access coverage (fixed or mobile)	Call setup ratio (for voice)
Complaint resolution time	Service supply time	Call retention ratio (for voice)
Customer service call answer ratio	Fault report submission rate	Listening voice quality (mean opinion score – MOS grading)
	Fault repair time	Message transmission ratio (e.g. for SMS, MMS)
		Data transmission capacity (e.g. downlink and uplink for IAS, typically specified in Mbit/s)
		Packet delay (in ms)
		Packet error ratio (probability or %)

measurements should be possible for operators, and there should be the possibility to use independent agencies/entities for audit of the measurement results obtained from different operators. But the most important thing is to retain the aspects of real customer experience for the given services.

A possible list of QoS parameters [2] that are important to end-users and with that to telecom operators is given in Table 9.1. One should note that such a list is not exhaustive, but it can be used as a reference point regarding definition of QoS parameters that should be assessed with objective or subjective methods.

The targets for QoS parameters are defined as a potential value or a range of values that must be reached for QoS to be regarded as satisfactory. In Table 9.1 the QoS parameters are classified into three classes:

- *Customer interface parameters.* They are non-technical parameters for a given service which are subjectively assessed by customers (e.g. by questionnaires).
- *Network infrastructure parameters.* They are parameters which refer to availability and proper functioning of the network infrastructure, including fixed and mobile broadband access coverage. These parameters are typically used by telecom operators for network design as well as operation and maintenance.
- *Service functionality parameters.* They are typically grouped according to the given service type – for example, call setup ratio refers to voice (not to Web sessions), message transmission ratio refers to SMS or MMS (not to voice or video), etc. Use of the same service functionality parameters provides the possibility of comparability of operators within a given country or between countries. Also, these parameters can be measured by external agencies.

Parameters are named according to the same conventions regardless of how they are fully named in different countries. For example, the parameters of type "rate" define the frequency of certain events (e.g. error rate is frequency of error events), "ratio" is used to denote the proportion of successful events (number of successful events vs. total number of events). Further, parameters of type "time" refer to the average time taken by certain successful actions (e.g. packet delay refers to the average delay of all successfully transmitted packets).

9.5.2 Setting Target Levels and Making Measurements

The definition of QoS parameters has a certain goal: assessment of QoS via measurements and audit of the obtained results by using the target levels for those parameters. The targets for QoS parameters are typically set by the NRA; however, they have to go through consultation with the operators, so certain adaptations of the measurement procedures may be required in some cases (e.g. for some new broadband technologies such as 5G in the 2020s). Regardless of the targets set by the NRA, the telecom operators may set their own targets, which may be stricter than those imposed by the regulator.

The formulation of QoS parameters' target levels may involve different approaches. For example, if one states the percentage of new IAS delivered in more than X days (as the service supply time), then X indicates a target level. If one specifies the number of days within which 90% of IAS requests were delivered to customers, then no target level is indicated for the IAS supply time.

The measurement methods should be objective in all cases. In certain cases it may not be possible to specify concrete parameters that can be assessed objectively, such as satisfaction from the customer care center. Then parameters are assessed subjectively by asking the customer to grade the satisfaction from the response obtained from the customer care center of the given telecom operator. However, such subjective assessment is not likely to be used as a basis for enforcement (e.g. by the NRA). Subjective assessment can be realized by third parties as well as by telecom operators themselves. Such measurements are best built into business and operation support systems and should be provided automatically, via SMS (to the mobile user who called the customer care center), Web form (to the Internet user of an OTT service, such as a booking site), via phone call (which is less convenient), via email (when an email address is available), or a combination of these.

The measurements can be either active or passive (as discussed in Chapter 7 of this book). Regarding the practices used in different countries [2], active measurement is most widely used by the NRA for QoS measurements (e.g. for broadband IAS). With active measurement methods NRAs can produce comparable reports from measurements on national telecom operators on a regular basis. Yet active measurements impose additional costs, including the following:

- Costs for deployment of measurement systems for active measurements, such as software and hardware probes and centralized servers and databases for processing and reporting of the collected measurement data.
- For the purposes of comparison of QoS performance of different telecom operators, the sampling methodology can be critical to provide comparable (i.e. benchmarking) results from QoS measurements on different operators' networks.

The measurement results are used by the NRA for auditing QoS. The auditing process is based on aggregated results, not on individual customers. However, there are different possibilities of aggregation of QoS measurements per service (IAS, voice, etc.). The aggregation can be per geographic region, per type of area (urban, rural, highways, and railways), per subscriber types (business and residential). How aggregation of the results will be used can be decided on a case-by-case basis by taking into account various local circumstances and quality problems.

9.6 KPIs and Measurement Methods for Fixed and Mobile Services

Most important QoS parameters are referred to as KPIs, and they are the subject of QoS regulation. KPIs are internal indicators which are based on network performance. Typically, KPIs are based on network counters and are essential for operation and maintenance as well as telecom business models. Also, KPIs can be reported and audited, but they are meaningless if used out of context. In the IP-based networks (that is IAS, VoIP, IPTV, or other services), besides service availability and bitrates (for the IAS), the important KPI is end-to-end delay, which has a major influence on the QoS and QoE (for services used by humans).

Example: End-to-End Delay Calculation in IP-Based Networks

When a flow portion does not contain a satellite hop, its computed Internet Protocol packet transfer delay (IPTD) is (according to ITU-T Y.1541):

$$IPTD \text{ (microseconds)} \leq (R_{km} \times 5) + (N_A \times D_A) + (N_D \times D_D) + (N_C \times D_C)$$
$$+ (N_I \times D_I)$$

where R_{km} represents the route length assumption computed above, $(R_{km} \times 5)$ is an allowance for "distance" within the network portion, while N_A, N_D, N_C, and N_I represent the number of IP access gateways, distribution nodes, core nodes, and internetwork gateway nodes respectively, and finally D_A, D_D, D_C, and D_I represent the delay of IP access gateway, distribution node, core node, and internetwork gateway nodes respectively (typically, nodes and gateways are IP routers).

9.6.1 Audit of QoS and Publishing the Measurements

The QoS audit is targeted at verification of QoS experienced by customers in order to compare results against the obligations which are specified in licenses (e.g. for mobile operators) or in the SLA. There are different methods for audit of telecom operators; however, they are dependent on the types of access network (fixed or mobile) and services provided to end-users. For example, the audit in mobile networks may include a drive test performed on a periodic basis (e.g. weekly, monthly, quarterly), consumer surveys, and measurement data which are submitted monthly or quarterly by telecom operators which are also service providers (e.g. they provide voice, IPTV, and business services over managed IP network as well as IAS). For example, targeted consumer surveys can highlight certain weak elements in respect of QoS and will provide valuable feedback to operators. Such data will also allow customers to compare their QoS experience with other users.

The main goal of publishing the measurements is to help customers to compare QoS from different telecom operators so they can make informed choices. NRAs typically are required to publish information on performances on their websites. Also, the regulator may oblige telecom operators to send such information about the QoS audit periodically to subscribers (e.g. with their bills. QoS information examples that should be published include QoS results from the audit campaign (e.g. drive test, consumer survey).

9.6.2 KPI Measurements in Mobile Networks

Regarding KPI measurements in mobile networks there can be variations in measurement metrics and methodologies in different countries. Such variations may lead to [6]:

- difficulty in comparability of mobile coverage, which may result in difficulties in comparison;
- inconsistency regarding the measurements of different mobile operators;
- inconsistency from the point of view of new and innovative digital services for vertical use cases such as 5G applications and services (e.g. massive and critical IoT services).

Table 9.2 Main KPIs for mobile voice and IAS services.

Mobile voice service KPIs	Mobile Internet access service (mobile IAS) KPIs
Coverage [%]	Coverage [%]
Call success rate [%]	Speed downlink/uplink [Mbit/s]
Call block rate [%]	Data transfer cut-off rate [%]
Dropped call rate [%]	IP service access failure rate [%]
Speech quality [MOS]	IP service access time [s]
Call setup time [s]	Session failure rate [%]
Handover success rate [%]	Session time [s]
	Ping round trip time [ms]

However, in the mobile networks in the 2010s and the early 2020s the main KPIs are in fact service-independent KPIs such as 3G, 4G, or 5G mobile network coverage, as well as KPIs for two main service types in mobile broadband networks, mobile voice (over managed IP networks) and mobile IAS (Table 9.2).

Various KPIs are measured using different methods. The speed (bitrate) measurement is based on multiple parallel HTTP connections (it is also possible to use FTP connections). Regarding coverage, it is also related to type of service in the mobile network. For example, voice coverage can differ from mobile IAS (i.e. mobile data) coverage because voice can be provided also via circuit-switched mobile networks such as 2G (e.g. GSM) and the 3G CS part, while IAS requires IP connectivity over the mobile networks. Additionally, one should note that even in covered areas, services (either voice or IAS) are not available at 100%. This is due to fluctuation in radio conditions (e.g. weather, peak hour which may result in attachments to a worse cell in the mobile network) and sometimes equipment failure. In this regard, service availability is the KPI which can be added to mobile coverage to gain a complete notion about the accessibility of a given service, such as mobile voice or data (as the two most used over mobile networks). For example, for mobile IAS (i.e. mobile data) IP service access failure rate is in fact the opposite of IAS availability. IP access may fail for various reasons, including DHCP failures, DNS failures, congestion in the mobile network, packet errors (which may be caused by interference, low SNR, etc.), long round trip delays in the mobile network (which may be the result of congestion), and so forth.

Generally speaking, service time availability refers to the availability of a given service (e.g. voice, data) during a defined time period (expressed in percentage of time when service is available). Also, it may be expressed as outage time (in minutes, hours) when the given service is not available within a given reference time period (e.g. day, month, or year). Although service availability is an important KPI, it is difficult to measure in many different locations through the mobile networks and at different time periods. Typically, measurements (by the NRA or the mobile operator) are done via drive tests or by using deployed probes; however, these can cover only limited numbers of locations (e.g. drive test can cover only roads, while probe stations can provide information only about the mobile networks regions where they are located). Another approach to obtain information on service availability is for NRA to oblige mobile operators to keep network logs. Otherwise, the NRA may consider service time availability as the percentage of successful measurements provided with drive tests or results obtained from QoS measurement probes. A similar approach can refer to other KPIs in mobile networks.

However, certain KPIs in mobile networks are significantly different from the same KPIs in fixed broadband networks. Such KPIs are downlink and uplink speeds (i.e. bitrates). The bitrates in mobile networks cannot be guaranteed as in fixed (fiber) networks for the following main reasons [7]:

- *Modulation and coding scheme (MCS) being used.* Time varying and location dependent SNR directly influences the MCS that can be used for the given communication. There are different modulation schemes which provide different number of bits transferred per symbol (where the symbol rate on the radio link is dependent upon the frequency carrier width and multiple access technique based on TDMA). For example, 16 QAM (quadrature amplitude modulation) can transfer 4 bits/symbol (because $16 = 2^4$), while 256 QAM can transfer 8 bits/symbol (because $256 = 2^8$). That means with the same resources (e.g. the same frequency carrier and time slot on it), 256 QAM provides two times higher bitrate than 16 QAM (e.g. if 16 QAM provides 100 Mbit/s, then use of 256 QAM will provide 200 Mbit/s). However, MCS depends on the SNR ratio, and SNR ratio depends mainly on signal strength. Signal attenuates with distance (which happens on shorter distances in urban areas due to more obstacles on the path of radio waves), so longer distance between the mobile device and the base station (3G, 4G, 5G or some future mobile generation) results in use of MCS with less bits per symbol, and vice versa. Therefore, ultra-broadband speeds (e.g. in urban areas) require small cells in the mobile network.
- *MIMO (multiple input multiple output).* The bitrates in the mobile networks depend proportionally on the number of antennas used on both ends in the radio link between the mobile terminal and the base station. For example, if SISO (single input single output) gives 100 Mbit/s, then 4×4 MIMO will give 4×100 Mbit/s $= 400$ Mbit/s.
- *Mobility of users.* Higher user mobility results in lower bitrates over the same radio resources and the same mobile device, and vice versa. Therefore, speed measurements in mobile networks should be targeted at different mobility of customers, such as nomadic users (e.g. pedestrians in urban areas) and mobile users in vehicles and trains on highways and railways, respectively.

Due to the above reasons it is not possible to guarantee certain specific speeds everywhere in the mobile networks. Some mobile operators for mobile data services (i.e. mobile IAS) include "up to" mobile speeds (by specifying the theoretical maximum speeds in the mobile network). Unlike mobile IAS, mobile voice has standardized bitrates (e.g. full or half rate GSM codec, adaptive multi rate (AMR) codecs), so the availability of voice depends on the available resources in the given cell. With transition from CS voice (e.g. GSM) to VoIP with QoS guarantees (e.g. VoLTE), there is increased flexibility regarding the capacity for voice services because in such a case they are provided over the same mobile broadband and ultra-broadband access (as mobile data i.e. IAS) and hence in peak hours may use additional capacity at the expense of mobile data capacity if that significantly does not degrade mobile data speeds (which, however, are dependent on different factors as outlined above). VoIP with QoS guarantees (i.e. managed VoIP or specialized VoIP, which are synonyms) is serviced with higher priority than mobile data, hence it is up to network planning and dimensioning what maximum capacity will be dedicated to VoIP (e.g. VoLTE, VoIP over 5G). However, the measurements in the mobile network cannot determine the exact capacity dedicated to specialized VoIP and what is allocated to the IAS (typically the IAS capacity is based on hysteresis, which means in peak hours for voice IAS capacity is at the lower capacity threshold, while in low voice traffic hours it is in its upper capacity threshold in the mobile access network).

9.6.3   KPI Measurements in Fixed Broadband Networks

While mobile networks are more heavily monitored due to their specific characteristics, fixed access networks are less monitored by NRAs. Typically, the QoS measurements for fixed networks can be completed by the operator because the network attachment points of users are at fixed locations (e.g. copper lines, cable networks, fiber access), so monitoring can be set at the intermediate routers in access or core networks, which is usually done by business and operation support systems of mobile operators.

One of the emerging approaches for QoS measurement in fixed networks is customer-based measurement. That can be triggered by certain events of QoS degradation noticed by the customer (e.g. no service at all, or degraded speeds for the IAS), or it can be automated via crowdsourcing tools, such as QoS measurement application (e.g. provided by the NRA). However, one should note that crowdsourcing applications may be used only for IAS due to its network neutrality principles (e.g. they cannot be used for managed IP services). Crowdsourcing applications are developed by NRAs or specialized companies for QoS measurement tools. Typically, users download the client crowdsourcing applications (e.g. Web-based or standalone application). When the measurement test is started [6], the crowdsourcing application client establishes a connection with the measurement server which is placed at the IXP. Such tests are based on multiple TCP connections (e.g. HTTP or FTP) in parallel between the crowdsourcing client (on the end-user's side) and the server. Although such measurement applications are free to use, typically the measurement results may be influenced by possible usage of the user's IAS from other applications at the same time. Also, the speed may be throttled due to exceeded data volume regarding the user's data cap. However, crowdsourcing applications are useful for IAS over fixed networks, enabling end-users to autonomously detect possible QoS degradations (e.g. service unavailability or lower speeds than the contracted ones).

Another aspect of QoS monitoring for fixed broadband is broadband mapping [8]. Possible broadband mappings are given in Table 9.3.

The broadband mapping process is targeted at establishing a single information base for broadband and ultra-broadband services, existing infrastructures, service demand, and ultra-broadband investments. Infrastructure mapping creates transparent access to relevant information. Service mapping provides an insight into the current state of broadband availability in the given country to assist in decision-making processes regarding future public and private broadband investments. For both service and infrastructure mapping, the authority in charge is the NRA.

Table 9.3 Broadband mapping.

Broadband mapping	Description
Infrastructure mapping	Geo-referenced and structured data of physical infrastructure, such as ducts/fiber/nodes, antenna towers, and other needed infrastructures (energy, etc.).
Broadband services/QoS mapping	Maps information on the supply side of broadband service provision, including the available bandwidths and QoS, technologies, operators and service providers.
Broadband demand mapping	Data on actual latency and speeds experienced by users; data usage (per household); expectations regarding QoS/QoE and willingness to pay by different user groups.
Investment and funding mapping	Information related to prospective public and private investment of high speed broadband (i.e. ultra-broadband).

9.7 QoS and Pricing

QoS as an end-to-end feature depends upon all players in the end-to-end communication path of IP traffic for a given service. Such players include national telecom operators, international transit providers, service providers' data centers and CDN, and customers. The telecom operators are driven by market forces in national markets, the international transit providers (also referred to as commercial ISPs) are dependent upon the international market, while the service providers may be dependent upon the national market (e.g. for services provided by telecom operators) or the global market (for OTT services over the public Internet).

From the economical point of view the Internet can be considered as a two-sided market, with the network operators collectively serving as a platform connecting providers of content (e.g. websites) and various OTT services with consumers that use the services by using the access through the telecom operator, which has a role as a multi-sided platform (as shown in Figure 9.4). Such interpretation also may suggest that quality differentiation may not be problematic. With view, some may consider that issues simply refer to how costs and profits should be divided between network operators as enabling platform of the two-sided market and the two (or sometimes more) sides of the market.

QoS and pricing have been related from the beginning in the telecom world. Telephone networks in the twentieth century used differentiated pricing for the same service (telephony, i.e. voice) at different times during the day or different days in the week, in order to discourage callers with lower demand from making calls during peak hours (i.e. the congested period of the day). The idea with such time-of-day pricing was to shift network usage to less congested periods by applying lower per-minute charges. The driver for such an approach is different willingness to pay for the service (from the customer side), which provides higher utilization of the network resources (or lower investments in the network to satisfy traffic demand).

However, with the convergence of telecommunication networks and service on Internet technologies, the approaches have changed to mainly flat fee charges regardless of the volume of voice traffic (on a national basis). Why? Because the access bitrates have increased significantly while the bandwidth required for voice bitrates requirements

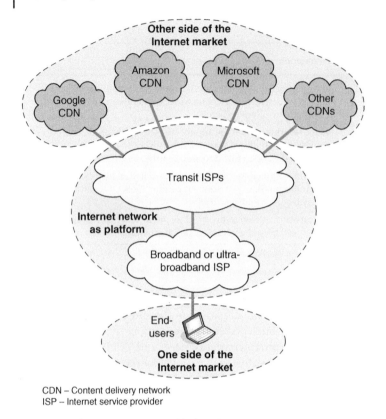

CDN – Content delivery network
ISP – Internet service provider

Figure 9.4 The Internet as a two-sided market.

have not changed over time, hence broadband and ultra-broadband capacities can generally accommodate voice connections (because data services are using most of the capacity in the 2010s and the 2020s). Another reason for voice services provided by telecom operators (with guaranteed QoS) is the increasing competition from OTT voice providers (Skype, Viber, and others). For these reasons, lower cost per bit/s and competition from OTT services, the charging per minute or per second is applied less today in fixed and mobile networks; more typically it is based on a flat fee (e.g. daily, monthly) for a given service for use on the national level.

The pricing of IAS is also related to QoS because the pricing has a direct influence on the demand for a given service. For example, a lower price for a given service (e.g. IAS) increases its consumption by customers, which in turn increases overall traffic in the network and hence increases the probability of congestion appearing. There is also a usage-based pricing approach, which is a tradeoff between price and QoS. With usage-based pricing the service demand (e.g. for IAS) can be shifted from peak to off-peak time periods. High demand (e.g. high traffic intensity for IAS in the given part of fixed or mobile access networks) can also indicate to telecom operators to increase the capacity of their networks to support the targeted level of QoS.

Regarding pricing for IAS, time-differentiated pricing will not be beneficial as in the case of legacy telephone networks The reason is that the Internet consists of many inter-connected autonomous networks, and even when they are in the same time zone, peak usage may well occur at different times in different places, and with that in different tele-com operators' (i.e. ISPs') networks. On the side of the transit provider, an ISP that is providing transit to several broadband ISPs, it may appear that providers will have different Internet traffic peak usage times, which will suggest different prices to be applied at the same time of day to different ISPs and this may raise network neutrality concerns. Also, the Internet traffic is not related only to communication between humans (so, higher traffic appears during the period of the day when most people are awake, such as in legacy telephone networks), but mainly it is used for client-server communication where the server side is always a machine located somewhere on the global Internet network, while the client side can be run by humans on their end-user devices (e.g. smartphones, computers) or machines and various things connected to the Internet (to which time of the day may has no influence on their operations). Therefore, legacy time-of-day pricing approaches are not suited for the IAS.

Overall, the demand for high QoS for services has an influence on customers' WTP [9] for higher QoS (as illustrated in Figure 9.5). If the demand for services that require high

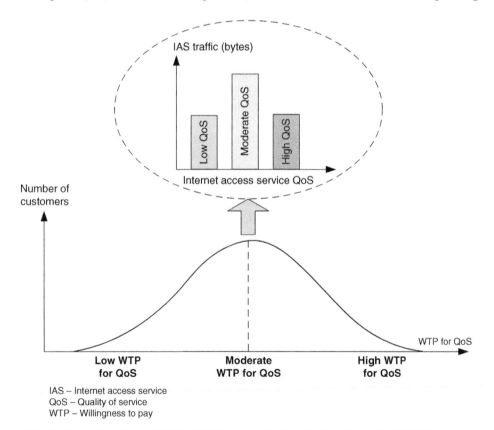

IAS – Internet access service
QoS – Quality of service
WTP – Willingness to pay

Note: For Internet Access Service (IAS), the high QoS refers to ultra-broadband speeds (e.g. 100–1000 Mbit/s per user), low QoS refers to broadband speeds (e.g. up to 50 Mbit/s), while moderate QoS refers to broadband speeds in between (e.g. 50 to 100 Mbit/s). The bitrates for IAS here are given only for illustration purposes and they are relevant in the early 2020s. All IAS bitrates will gradually increase over time with the continuous development of fixed and mobile access technologies.

Figure 9.5 Internet access service users versus willingness to pay (WTP) for QoS.

QoS is very low, then the WTP for high QoS will be also very low. In such a case, telecom operators will have lower interest in providing high QoS. When the demand for services that require high QoS is comparable with demand for services not requiring QoS, then the WTP for QoS is higher. For example, consumers have WTP for QoS for IPTV (as a replacement for traditional TV), VoIP (as a replacement for PSTN/ISDN), business services (e.g. VPNs as replacement of leased lines), critical IoT services (which can be provided without QoS, such as smart transportation or industry), as well as related future smart services (e.g. smart cars, smart homes, smart sustainable cities). So, many services that demand QoS to function cannot be delivered without QoS, such as the best-effort and network neutral approach which is used for IASs. Thus WTP for services that demand high QoS is higher; however, the number of customers that have such higher WTP is lower than the number of customers with lower (i.e. moderate) WTP.

Considering a single service, such as IAS, different WTP by different customers may results in different offers, such as a bigger data cap for customers with high WTP and a lower data cap for IAS customers with lower WTP. Due to lower price per unit in the wholesale, the price per volume of data (e.g. per GB) is lower for bigger data caps than for smaller data caps, so customers with higher WTP at the end will pay less per gigabyte than customers with lower WTP. Similarly, if one considers fixed broadband and ultra-broadband IAS (e.g. based on fiber in the access part), the bitrates can also be used as the main QoS parameter (together with the data cap, e.g. on a monthly basis) for differentiation of users with different WTP. Users with lower WTP will get IAS with lower speeds, while users with higher WTP will get IAS with higher speeds. If there are many customers for the given service then it is likely that their behavior will result in a normal (i.e. Gaussian) probability distribution function (as a result of the Central Limit Theorem from probability theory, according to which when independent random variables are added, their properly normalized sum tends toward a normal distribution even if the original variables themselves are not normally distributed) with a given mean value, which will the best choice to maximize the number of subscribers for the offered QoS (e.g. bitrates in the case of IAS).

Overall, one may note that QoS and pricing are directly related, i.e. higher QoS imposes higher price of services, and vice versa.

9.8   QoS Enforcement

Broadband and ultra-broadband networks are used for delivery of multiple heterogeneous services, so QoS is becoming more important. Also, more people on the global scale use telecom/ICT services and applications because more people have broadband access, either fixed or mobile, or both. With the saturation of horizontal markets in the telecom world (referring to the number of human subscribers), the expansion of the telecom/ICT world is targeted at connection of different "things" to the Internet (they are much more in number than humans), which when connected to the network are referred to as being "smart." And many smart services (for control of smart things) require QoS support also.

From the twentieth century the QoS provisioning in the telecom/ICT world has been based on standards (from the ITU, 3GPP, the IETF, the IEEE, etc.) regarding technical matters and licenses (based on telecommunication legislation, such as laws and rules), which refer to policy matters or interrelated policy-technical matters. One such example is spectrum allocation for mobile operators. Because spectrum is scarce it is given with licenses which impose certain obligations regarding QoS parameters. But for spectrum management there is a need for regional and global harmonization, with the aim of the same mobile devices (e.g. smartphones) being used in different networks in different countries (e.g. for roaming). However, QoS will not exist just by requesting telecom operators and generally service providers to include mandatory standards or licenses. In order to have the desired QoS in practice, there is a need to define a set of KPIs that are important and understandable to customers and then to make customers aware of them.

Overall, if legitimate concerns are addressed, QoS obligations are likely to be fulfilled. For the purpose of successful QoS monitoring, the regulators (i.e. NRAs) need to consider the opinions of both telecom operators and their customers. Typically, that can be achieved through a consultation process before applying obligations in regard to QoS. The consultation process starts (and ends) with circulation of documents on QoS policy options and proposals given by the NRA for discussion.

Figure 9.6 shows the possible techniques for ensuring that telecom operators will fulfill QoS obligations. Developed telecom/ICT markets prefer the encouragement option, while in less developed telecom/ICT markets, when QoS obligations seriously and persistently fail to be fulfilled, it is possible to use more radical enforcement techniques, which may include approaches not exclusively related to QoS, such as withdrawing licenses issued by the NRA. However, drastic enforcement techniques may be not available through the laws in a given country. Also, enforcement should be directed to provide benefits to the end customers and to the public in general, and it should not intentionally influence the business operations of telecom operators because the main idea is to improve QoS (while a non-operational network will certainly degrade further QoS, especially when QoS degradation is present in all telecom operators in a given country). Enforcement should be the last resort; timely indication of QoS degradation may trigger an appropriate reaction by the telecom operator (e.g. deploying new capacity, network optimization, repairing or replacing non-working equipment, and adopting a pricing strategy that reflects the WTP of customers and deployed capacity in the networks). Additionally, enforcement techniques may not be

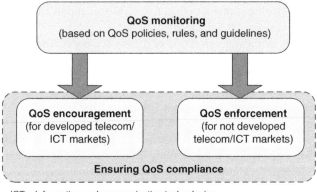

ICT – Information and communication technologies
QoS – Quality of service

Figure 9.6 Regulation approaches for ensuring QoS compliance.

possible always under the law in a given country, and typically enforcement procedures with financial intensities may involve longer court processes (dependent upon the legislation that regulates the QoS). Also, there is the possibility that telecom operators or service providers will consider financial penalties as operational costs, so when they are applied the penalties should be proportional to that provider's turnover (e.g. percentage of turnover). However, the best QoS enforcement is through market forces, when operators compete among themselves. Higher completion ensures higher QoS and lowest possible prices which are acceptable for most customers and still assure profitable businesses for the telecom operators.

Generally, the main aspects regarding QoS enforcement mechanisms [10] are given as follows:

- *Publishing all measurements.* It helps to show to all parties that the regulator is fair and open. By publishing measurement results (on the NRA's website, in the official gazette, etc.) the NRA helps customers to have information when choosing a provider. Also, the measurements are usually the main technique for encouraging compliance with QoS obligations.
- *Publicizing deficiencies to customers.* This is accomplished in various ways, such as notes on QoS levels in users' bills, or advertisements because QoS regulation is targeted to provide comparisons between the operators or against the given QoS targets in the country.
- *Demanding extra measurements.* This can be useful and appropriate when the actions needed to improve QoS can have a very fast effect (e.g. with network optimization).
- *Monitoring the implementation of QoS corrective plans.* This may require third party agencies with human capacity in telecom networks design and operation to assist NRAs. However, such an approach will be effective only with the approval of the telecom operators; otherwise it may lead to confusion about responsibilities and duties between the NRA and operators.
- *Requiring compensation to customers.* This approach is not always possible (e.g. a missed voice call cannot be compensated). It can be effective when customers can request compensation directly from operators without intervention by the NRA (or arbitrators). In that manner, compensation for QoS degradation is more likely to be paid for issues such as long service unavailability (e.g. due to longer repair times for the interruption) than in other cases of QoS degradation.
- *Imposing fines.* This approach is not always effective as it may seem because it can involve long legal processes (court processes), and is dependent upon the concrete legislation in the given country and the legal procedures for such cases. Also, fines may degrade the telecom operators' business (if they are too big) or have no effect at all (if they are too low). So, when fines are applied they should be proportional to the QoS degradations and should be beneficial to customers in obtaining better QoS.
- *Price changing (such as the introduction of quality factors into price controls, with certain rewards for good QoS or penalties for bad QoS).* Similar to imposing fines, this approach needs to be designed carefully to be helpful in improving QoS. This approach refers to the relation between QoS and pricing, which many times is not explicit or clear due to bundling of different services in the offers from telecom operators to their customers (e.g. triple-play services based on bundling of voice, TV, and IAS).

- *Excluding access to licenses needed for telecom/ICT services.* This is a drastic approach which is difficult to make it proportionate to QoS degradations by operators (because it intends to stop the given operators' operations). This measure can be effective when there is visible enough competition among the other telecom operators (with better QoS provisioning) to take over the market share (for the given services) from the operator that is sanctioned. Such measures may not be applicable to many operators at the same time because that can degrade the whole telecom/ICT market in the country or limit competition (which is essential for healthy telecom/ICT market development).

QoS enforcement mechanisms require expertise in various aspects (i.e. technical, regulatory, and business aspects) in different processes whenever telecom operators report QoS measurements. In developed telecom/ICT markets, the main regulatory approach is a "light-touch" one, which refers to monitoring the QoS and imposing certain recommendations, and when QoS degradations are detected and operators are not willing to comply with recommendations, the NRA may move toward stricter obligations when they are important and practical for improving the QoS.

As already stated before, the NRA can start from "naming and shaming" strategies and move on to stricter regulation. So, when all other measures do not give the desired effect in improving QoS, the regulator may proceed toward financial penalties, and finally use more drastic legal enforcements (e.g. license withdrawal). In general, in QoS regulation, encouragements and enforcements should be graduated and proportional [1], and targeted at the overall benefit of customers and society with better QoS.

9.9 Discussion

Common QoS standards are constantly being developed and updated. Typically, such standards can be regional (e.g. the QoS standard from ETSI) [11, 12]) or global (e.g. the QoS standards, guidelines [2], and manuals [1], from the ITU).

The main services which are subject to regulation from the QoS point of view are IAS (i.e. data) and carrier-grade VoIP services (i.e. VoIP as a PSTN/ISDN replacement), although SMS (in mobile networks) and TV/IPTV services (in fixed broadband networks) are also possible target services for QoS monitoring and regulation.

At the end of the second decade of the twentieth century there was a noticeable difference in the QoS regulation approaches between countries with developed (i.e. mature) telecom markets and those with developing (i.e. less mature) telecom/ICT markets. In competitive and developed telecom/ICT markets, QoS enforcement is targeted mainly at notices and publication of QoS measurements by using a given set of KPIs appropriate for targeted services, with the aim of informing customers (i.e. end-users for the given services) so they have enough necessary information when selecting or switching telecom operator. However, some customers with low WTP may choose lower QoS, and vice versa. Therefore, for critical services, which are important to society and to the public (not only to the operators and customers), the NRA may enforce measures to improve QoS if the QoS is essential for the service functioning, for public safety (e.g. intelligent transportation systems), or for the society as a whole (e.g. development of the digital society by moving public functions into the digital world, such as their provisioning over the public Internet). Also, emerging technologies such as SDN and NFV together with cloud computing (which is used for most of the data-based services) lead to network slicing and then different KPIs are required for different network slices that will be used for different types of services (e.g. massive machine type communication, ultra-reliable low-latency communication, ultra-broadband) [13].

In less mature telecom/ICT markets, the NRAs may and should consider applying financial penalties as a possible tool for QoS, after all previous options for improving the QoS of the given telecom operator have been implemented and are not showing results (e.g. notices sent to telecom operators to improve QoS in a given time period). However, one should note that there are no global rules for applying penalties for the purposes of QoS enforcement, so that should be done by considering all relevant factors that could influence QoS degradation in the given country. Also, QoS enforcement should bring benefits for all parties in the telecom/ICT chain, including end-users, the telecom/ICT market, and society in general.

References

1 ITU (2017). *Quality of Service Regulation Manual*. ITU.
2 ITU-T Supplement 9 to ITU-T E.800-series Recommendations, Guidelines on Regulatory Aspects of QoS, December 2013.
3 ITU Telecommunications/ICT Regulatory Database, 2017.
4 ITU-T Recommendation E.860, Framework of a Service Level Agreement , 2002.
5 ITU-T Recommendation E.800, Definition of Terms Related to Quality of Service, September 2008.
6 BEREC, Preliminary Report in View of a Common Position on Monitoring Mobile Coverage, October 2017.
7 Janevski, T. (2015). *Internet Technologies for Fixed and Mobile Networks*. Norwood, MA: Artech House.
8 European Commission, Study on Broadband and Infrastructure Mapping, Final Report, 2017.
9 ITU Report, IP-based Networks: Pricing of Telecommunication Services, 2003.
10 ITU, ICT Quality of Service Regulation: Practices and Proposals, September 2006.
11 ETSI ES 202 057-1 (2013). *Speech Processing, Transmission and Quality Aspects (STQ); User Related QoS Parameter Definitions and Measurements; Part 1: General.* ETSI.
12 ETSI EG 202 057-4 (2008). *Speech Processing, Transmission and Quality Aspects (STQ); User Related QoS Parameter Definitions and Measurements; Part 4: Internet Access.* ETSI.
13 Combo, *Monitoring Parameters Relation to QoS/QoE and KPIs*, http://www.ict-combo.eu/, 2014.

10

Conclusions

This book has focused on the quality of service for fixed and mobile ultra-broadband, including technologies, regulation, and business aspects. The QoS has been mandatory for traditional telecommunication services such as telephony (voice) and television since the first half of the last century; however, with the convergence of telecommunication networks and services onto Internet technologies, QoS provision remains a significant challenge for all ICT services, not only for traditional ones. Therefore the ITU and other regional organizations (e.g. organizations of regulators) are involved in activities for QoS regulation in all regions of the world, besides the standardized approaches for QoS provisioning in Internet-based environments which are done by different SDOs, including the ITU for global standardization and harmonization of telecommunications/ICTs, the IETF for Internet technologies from the network layer up to the application layer, the IEEE for Ethernet, WiFi, etc., 3GPP for broadband mobile networks (2G–5G and beyond), and others.

The introduction to this book, given with Chapter 1, aimed to provide information regarding the convergence of the telecom and Internet worlds which is being realized in the first two decades of the twenty-first century, and which is expected to continue toward an all-IP world. However, the traditional telecom world is based on strict end-to-end QoS provisioning, such as QoS in telephone networks for voice services (including fixed PSTN as well as PLMN such as GSM), broadcast networks for TV, and leased lines for business users. Meanwhile, the Internet was developed from the beginning for the best-effort services (i.e. each IP network does its best effort to transfer the IP packets through it) and it is network neutral by nature (e.g. no differentiation of IP packets by ISPs regarding the source or destination address, application, or type of content carried by the packets). The Internet and Internet technologies won the packet-switching technology battle in the 1990s (e.g. with ATM, as Internet major competitor from the European side – the European Telecommunications Standards Institute). So, from the end of the 1990s, and especially from the beginning of this century, it was recognized by all SDOs that the telecommunications (i.e. ICT) world and the native Internet world would converge toward the Internet technologies in both main aspects, networking and services. However, the approaches from both sides have merged, and this included QoS support as the main contribution and requirement from the telecom side (e.g. for real-time services, such as voice and TV) to influence the initially best-effort and network neutral Internet. Meanwhile, the convergence of all telecommunication services onto the Internet (including video, the most demanding service regarding

QoS for Fixed and Mobile Ultra-Broadband, First Edition. Toni Janevski.
© 2019 John Wiley & Sons Ltd. Published 2019 by John Wiley & Sons Ltd.

bitrates, i.e. bandwidth) started to happen with the development of broadband access to the IP-based networks and provision of IP-based services (via such broadband access).

Since the ICT world is transiting toward the Internet technologies, Chapter 2 was dedicated to Internet QoS. With the aim of having a standalone book, an overview of the main Internet technologies was given, including the main standardized protocols from the network layer up to the application layer (e.g. IP, IPv6, TCP, UDP, DHCP, DNS, HTTP, and email protocols), as well as fundamental Internet network architectures and networking approaches (client-server, and peer-to-peer). Further, to consider QoS issues for the Internet, the chapter discussed Internet traffic characterization, including voice, video, and data traffic. To provide QoS to different traffic types (where it is required), different QoS solutions on different protocol layers were given, from the data-link layer up to the application layer (going from the bottom to the top of the protocol stack). Further, Chapter 2 continued with different traffic management techniques nowadays, including packet prioritization and bandwidth allocation, as well as a description of relations between traffic management and network capacity, including positive and negative effects. The main SDO on the global scale is the ITU as the largest agency for ICTs; on the other side the main Internet technologies from the network layer (i.e. OSI layer 3) up to the application layer are being standardized by the IETF. Therefore, Chapter 2 gave a comparison of the IETF QoS framework (standardized solutions) and the ITU QoS framework. Further, it covered IETF standardized QoS solutions from the 1990s, such as DiffServ and IntServ, which are not implemented as such nowadays but influenced the different QoS approaches used today. The chapter further covered MPLS, which is the well proven QoS approach for core and transport networks, as well as DPI techniques related to QoS. Finally, the chapter showed a basic inter-provider QoS model (e.g. the ITU viewpoint) as well as IP network architectures for end-to-end QoS provisioning.

The telecommunication world's transition to an all-IP world for traditional services is well defined with the ITU's next generation networks umbrella of recommendations, covered in Chapter 3. The NGN in fact has copied the major Internet design philosophy, with separation of applications/services on the top from the underlying transport technologies (including access, core, and transport networks) at the bottom, by defining two main stratums: the service stratum and the transport stratum. The NGN was initially primarily targeted to transition of traditional telecommunication services (PSTN voice to VoIP and TV to IPTV) over IP networks end-to-end with QoS guarantees (similar to traditional telecommunication networks). That requires standardized signaling and control technologies. For that purpose the NGN has not re-invented the wheel but has used well standardized protocols for signaling, such as SIP and Diameter, put into a well defined IP multimedia subsystem (standardized by 3GPP). QoS end-to-end provisioning is directly related to standardized signaling, as well as to standardized functions between the two NGN stratums, where the main role is given to the resource and admission control function, which can be used in different transport environments in access, core, and transport networks (e.g. Ethernet access, MPLS for transport networks). However, the NGN is also targeted to standardized deployment of the Internet of Things, and has defined application interfaces for all services that require the service stratum functions of the NGN. Besides QoS functions in the NGN, Chapter 3 included standardized management for performance measurement in the NGN, as well as recently defined DPI performance models and metrics. Further, it gave a view into the future networks, as an evolution of the NGN, toward network virtualization. Finally, the chapter discussed business and regulatory aspects of the NGN and future networks.

The convergence onto the Internet as a single networking platform for all services is possible only with development of broadband access to the Internet, including fixed access and mobile access networks. Although broadband is a relative terminology with respect to the bitrates which provide seamless access to all existing applications and services at the present time, the term ultra-broadband is used (e.g. in many national broadband strategies for 2020 and beyond) to point to higher bitrates than the existing broadband worldwide. While the existing broadband access (on average) is in the range of Mbit/s/user to tens of Mbit/s per user, ultra-broadband access is targeted to bitrates in the range of 100 Mbit/s to Gbit/s per user. In that manner, Chapter 4 focused on fixed ultra-broadband, while Chapter 5 targeted mobile ultra-broadband technologies. Chapter 4 covered the very high speed (i.e. ultra-broadband) technologies over copper (e.g. VDSL, cable access with DOCSIS 3.x) and fiber (e.g. next generation passive optical network standards, and active WDM optical access), including QoS approaches in the given access networks. Also, the chapter 4 included QoS defined for Ethernet and Carrier Ethernet. Finally, different access networks are connected to core networks, core networks are connected to transport networks, and transport networks are interconnected, so the "holy grail" of QoS is end-to-end QoS network provisioning, which is also covered in that chapter. Chapter 4 ended with strategic aspects for ultra-broadband deployment, which refer to both developed and developing countries around the world.

Chapter 5 was focused on QoS technologies in the mobile ultra-broadband access network, including the most recent mobile standards, such as LTE-Advanced-Pro, as well as near future mobile technologies, such as 5G. Mobile broadband in the 2010s is typically represented by 3.5G (mainly UMTS/HSPA) and 4G (mainly LTE/LTE-Advanced) technologies, which in many developing countries around the world are in fact the only way to access the Internet with broadband or ultra-broadband speeds due to lack of physical network access infrastructures. Mobile communications are more personal than fixed ones (e.g. a fixed access is typically shared by many users at home, office, or public place, typically by using NAT translation between individual private IP addresses and the public IP address of the given connection toward the global Internet). So, mobile ultra-broadband is even more important than fixed access (although fixed access can always provide more bit/s because there is always the possibility to add another cable where it is needed, something that is not feasible in mobile networks due to limited and scarce radio spectrum that is usable for non-line-of-sight communication, e.g. the spectrum below 6 GHz). The IEEE 802.11 standards are targeted to gigabit rates (although these are aggregate bitrates for high-end WiFi equipment), such as IEEE 802.11ac, besides the well deployed IEEE 802.11n (regarding the WiFi physical layer). Chapter 5 also covered the QoS in giga speed WiFi, as well as considerations regarding competition from LTE/LTE-Advanced in unlicensed bands on 2.4 GHz. Further, the chapter covered the ITU umbrella for 5G, named IMT-2020, as well as QoS for 5G mobile standards by 3GPP. The future mobile networks are expected to have many small cells (e.g. femto cells) for increase in capacity and stability of bitrates (and hence higher QoS and QoE), which however have to be coupled with QoS mechanisms in the mobile core networks. Finally, Chapter 5 provided business and regulation aspects for mobile ultra-broadband.

Both fixed and mobile networks are deployed to carry services and user data to and from end-users (human users and machines or things). Chapter 6 offered the QoS vision for services provided over ultra-broadband fixed and mobile networks. Such services include real-time ones, with QoS guarantees end-to-end, such as QoS-enabled VoIP services, including VoIP over LTE networks (and over 5G mobile networks) as well as QoS-enabled IPTV (e.g. multicast of linear TV with E-MBMS (enhanced multimedia broadcast multicast services) and beyond) and unicast video streaming services. The chapter 6 considered QoS Internet access services, which directly impact OTT services, also called data services (by telecom operators and regulators), which include standardized services/applications such as the Web, email, etc., as well as proprietary OTT services/applications such as YouTube, Skype, Viber, Facebook, Twitter, BitTorrent, cloud services, and many others in different application ecosystems. This chapter included QoS for IoT services, including massive IoT (provided over public Internet access) and critical IoT (provided over managed IP networks with ultra-reliable low-latency communications), as well as cloud computing services (OTT and telco cloud services) with their QoS requirements and metrics. Finally, Chapter 6 covered business and regulatory challenges for services over ultra-broadband.

The next chapter outlined QoS parameters, key performance indicators and measurement approaches for different services. It first discussed QoS, QoE, and application needs, and also defined generic and service specific QoS parameters. End-to-end QoS provisioning is directly dependent upon the interconnection and mapping the QoS parameters from one telecom operator to another one. However, only a subset of many defined QoS parameters is important to end-users, and hence to operators and regulators, and such selected QoS parameters are called KPIs. In that regard, Chapter 7 covered KPIs for real-time services (e.g. voice, IPTV), KPIs for data services (i.e. for IAS), and VPNs, as well as KPIs for smart sustainable city services. The chapter covered broadband QoS measurement frameworks, as well as assessment methodologies (including objective and subjective assessment). Further, it gave a view into available quality measurement tools and platforms such as M-Lab, RIPE Atlas, and SamKnows.

Chapter 8 focused on network neutrality, which refers to open Internet access, also known as IAS. Network neutrality itself was invented with the Internet; however, with the convergence of all telecommunication services onto the Internet, the whole "story" has become more complex. So, telecom operators nowadays offer different services with QoS guarantees (e.g. VoIP as PSTN/ISDN replacement, IPTV as linear TV) over managed IP networks which are only virtually separated from the network-neutral Internet traffic (also referred to today as OTT or data traffic). Chapter 8 covered possible degradations to network neutrality by telecom operators, the main regulatory goals and the business aspects of network neutrality, the role of regulators (i.e. NRAs), as well as network neutrality approaches in Europe and the U.S. as leading regions in this regard. Further, this chapter discussed challenges and enforcement approaches for network neutrality.

Chapter 9 covered the QoS regulation framework. It included the scope of QoS regulation and its fundamentals, as well as the ITU's QoS regulation guidelines. The end-user QoS is established via a service level agreement, which may include different QoS parameters. Further, the chapter looked at approaches and guidelines for specifying parameters, levels, and measurement methods within a given QoS regulatory framework, as well as approaches for definition of KPIs for fixed and mobile services and their measurements on a periodical basis. This chapter further discussed relations between QoS and the pricing of services as well as customers' willingness to pay for services because QoS is also related to pricing (e.g. higher QoS may lead to higher price for services, or the same price for services but with a lower profit margin), which directly impacts telecom operators' business. Finally, there are many cases where legislation regarding QoS exists but it is not implemented in practice on a satisfactory level by the national telecom operators. For such cases, this chapter included QoS enforcement approaches, which can go from naming and shaming to more drastic financial penalties in the case of continuous degradation of certain KPIs for certain services. In that manner, Chapter 9 included various QoS enforcement approaches, applicable to different regions around the globe.

The last chapter of the book, this one, gives the main conclusions regarding QoS for fixed and mobile ultra-broadband.

Overall, this book provides all-round coverage for QoS for fixed and mobile ultra-broadband, including technologies, business, and regulation aspects, and also covers the basis for QoS deployments such as Internet technologies (as defined by the IETF), NGN (as defined by the ITU), fixed ultra-broadband (as defined by the ITU and the IEEE), and mobile ultra-broadband (as defined by the 3GPP and the ITU), for various services (standardized and OTT), including network neutrality aspects for the public Internet (as defined in Europe and the U.S.) and the QoS regulation framework and guidelines (as defined by the ITU).

Finally, the book is targeted at a widely diverse readership, ranging from students and professors/tutors at universities and training centers, via engineers, managers, and employees in the industry, telecom operators, service providers, toward ICT business readers, telecommunications regulators, ICT administrations, and all interested parties from companies and institutions which need capacity building in the area of QoS for fixed and mobile ultra-broadband, including technology, business, and regulation aspects.

Index